Materials and Energy Recovery from the Final Disposal of Organic Waste

Materials and Energy Recovery from the Final Disposal of Organic Waste

Editor

Gabriele Di Giacomo

MDPI • Basel • Beijing • Wuhan • Barcelona • Belgrade • Manchester • Tokyo • Cluj • Tianjin

Editor
Gabriele Di Giacomo
University of L'Aquila
Italy

Editorial Office
MDPI
St. Alban-Anlage 66
4052 Basel, Switzerland

This is a reprint of articles from the Special Issue published online in the open access journal *Energies* (ISSN 1996-1073) (available at: https://www.mdpi.com/journal/energies/special_issues/Final_Disposal_Organic_Waste).

For citation purposes, cite each article independently as indicated on the article page online and as indicated below:

LastName, A.A.; LastName, B.B.; LastName, C.C. Article Title. *Journal Name* **Year**, *Volume Number*, Page Range.

ISBN 978-3-0365-2852-6 (Hbk)
ISBN 978-3-0365-2853-3 (PDF)

Contents

About the Editor

Gabriele Di Giacomo has been a Full Professor of Chemical Engineering at the University of L'Aquila, Italy, since the 1990s. His main research interests include seawater desalination, the theoretical and experimental study of water and carbon dioxide applications under supercritical conditions, and the exploitation of production surpluses, by-products, and waste from agro-industrial activities. He was a member of the editorial board of Desalination for nearly two decades and currently sits on the editorial boards of *Food and Bioprocess Technology (FABT)* and Section C of *Energies*.

Editorial

Material and Energy Recovery from the Final Disposal of Organic Waste

Gabriele Di Giacomo

Department of Industrial and Information Engineering and of Economics, Campus of Roio, University of L'Aquila, 67100 L'Aquila, Italy; gabriele.digiacomo@univaq.it

Citation: Di Giacomo, G. Material and Energy Recovery from the Final Disposal of Organic Waste. *Energies* **2021**, *14*, 8459. https://doi.org/10.3390/en14248459

Received: 8 December 2021
Accepted: 13 December 2021
Published: 15 December 2021

While receiving nearly 10,000 times the energy that we presently need from the Sun, almost 600 EJ/a, developed and developing countries continue to mostly use fossil fuels even though the technologies available and the adaptation of individual and collective behaviours could make it possible to use only solar energy.

For example: in developed countries, the drinking water consumption has exceeded 0.2 m^3 per capita per day, even in arid areas where it is produced using a large amount of fossil-based energy; the consumption of fossil-derived fuels for automotive industries reached a level that was unimaginable until a few decades ago; food is produced and marketed, throwing half of the raw material. Additionally, it is absurdly packaged and consumed for a share that does not exceed 50% after keeping in the refrigerator, without any concern for the energy consumption.

All this and much more are part of a long list of irrational and unnatural behaviours and habits.

Each year, the solar energy accumulated as biomass is eight times greater than the current global energy consumption. Still, few appreciate this gift that nature continues to give us. Most of this precious and energy-intensive material is just discharged into the environment or landfilled as waste, thus losing its GHG emission neutral energetic content and the possibility of its mining for producing a variety of precious and renewable compounds.

Waste-to-energy can potentially contribute about 5% to energetic global consumption, and this is not such an enormous contribution; it would be better to avoid or reduce the production of waste drastically. However, some constraints, such as urbanization, make this impossible. Furthermore, the irreversible habits typical of modern living does not allow this. However, we have a moral duty to use the best technologies and strategies to value biomass waste.

We suggested that this Special Issue should meet this moral duty and received ten exciting contributions.

1. The Effective Management of Organic Waste Policy in Albania [1].
2. Anaerobic Degradation of Environmentally Hazardous Aquatic Plant Pistia stratiotes and Soluble Cu(II) Detoxification by Methanogenic Granular Microbial Preparation [2].
3. Influence of Valorization of Sewage Sludge on Energy Consumption in the Drying Process [3].
4. Challenges in Sustainable Degradability of Bio-Based and Oxo-Degradable Packaging Materials during Anaerobic Thermophilic Treatment [4].
5. Environmental and Economic Aspects of Biomethane Production from Organic Waste in Russia [5].
6. Adsorptive Recovery of Cd(II) Ions with the Use of Post-Production Waste Generated in the Brewing Industry [6].
7. Modern Use of Water Produced by Purification of Municipal Wastewater: A Case Study [7].

8. Odour Nuisance at Municipal Waste Biogas Plants and the Effect of Feedstock Modification on the Circular Economy—A Review [8].
9. Operational Parameters of Biogas Plants: A Review and Evaluation Study [9].
10. Pilot-Scale Experiences with Aerobic Treatment and Chemical Processes of Industrial Wastewaters from Electronics and Semiconductor Industry [10].

As can be seen from the titles of the papers, many problems have been addressed, and different technologies were critically evaluated to successfully afford the final disposal and valorization of a variety of waste biomass.

We have worked through the pandemic crisis with incredible difficulty: inaccessible research laboratories, complicated relationships, and very low morale. Nevertheless, we are happy with this work and this example. We extend thanks to all authors from different countries (Romania, Russia, Ukraine, Poland, Germany, Jordan, and Italy) with different degrees of development and problems and aims.

The involvement of many universities, national academies of sciences, and research centers give this Special Issue the unique value that comes from combining different experiences and ways of affording success and solving complex problems.

We hope that everyone will share the desire to name this Special Issue of *Energies* after Ida, a brilliant young colleague. She contributed to this work but failed to overcome a dark evil and gracefully walked away.

Funding: This research received no external funding.

Acknowledgments: I would like to thank MDPI for allowing this work to be published, and the reviewers.

Conflicts of Interest: The author declares no conflict of interest.

References

1. Oncioiu, I.; Căpuşneanu, S.; Topor, D.; Petrescu, M.; Petrescu, A.; Toader, M. The Effective Management of Organic Waste Policy in Albania. *Energies* **2020**, *13*, 4217. [CrossRef]
2. Havryliuk, O.; Hovorukha, V.; Savitsky, O.; Trilis, V.; Kalinichenko, A.; Dołhańczuk-Śródka, A.; Janecki, D.; Tashyrev, O. Anaerobic Degradation of Environmentally Hazardous Aquatic Plant Pistia stratiotes and Soluble Cu(II) Detoxification by Methanogenic Granular Microbial Preparation. *Energies* **2021**, *14*, 3849. [CrossRef]
3. Siedlecka, E.; Siedlecki, J. Influence of Valorization of Sewage Sludge on Energy Consumption in the Drying Process. *Energies* **2021**, *14*, 4511. [CrossRef]
4. Zaborowska, M.; Bernat, K.; Pszczółkowski, B.; Wojnowska-Baryła, I.; Kulikowska, D. Challenges in Sustainable Degradability of Bio-Based and Oxo-Degradable Packaging Materials during Anaerobic Thermophilic Treatment. *Energies* **2021**, *14*, 4775. [CrossRef]
5. Zueva, S.; Kovalev, A.; Litti, Y.; Ippolito, N.; Innocenzi, V.; De Michelis, I. Environmental and Economic Aspects of Biomethane Production from Organic Waste in Russia. *Energies* **2021**, *14*, 5244. [CrossRef]
6. Kalak, T.; Walczak, J.; Ulewicz, M. Adsorptive Recovery of Cd(II) Ions with the Use of Post-Production Waste Generated in the Brewing Industry. *Energies* **2021**, *14*, 5543. [CrossRef]
7. Tomassi, G.; Romano, P.; Di Giacomo, G. Modern Use of Water Produced by Purification of Municipal Wastewater: A Case Study. *Energies* **2021**, *14*, 7610. [CrossRef]
8. Wiśniewska, M.; Kulig, A.; Lelicińska-Serafin, K. Odour Nuisance at Municipal Waste Biogas Plants and the Effect of Feedstock Modification on the Circular Economy—A Review. *Energies* **2021**, *14*, 6470. [CrossRef]
9. Nsair, A.; Onen Cinar, S.; Alassali, A.; Abu Qdais, H.; Kuchta, K. Operational Parameters of Biogas Plants: A Review and Evaluation Study. *Energies* **2020**, *13*, 3761. [CrossRef]
10. Innocenzi, V.; Zueva, S.; Vegliò, F.; De Michelis, I. Pilot-Scale Experiences with Aerobic Treatment and Chemical Processes of Industrial Wastewaters from Electronics and Semiconductor Industry. *Energies* **2021**, *14*, 5340. [CrossRef]

Review

Operational Parameters of Biogas Plants: A Review and Evaluation Study

Abdullah Nsair [1], Senem Onen Cinar [1,*], Ayah Alassali [1], Hani Abu Qdais [2] and Kerstin Kuchta [1]

[1] Sustainable Resource and Waste Management, Hamburg University of Technology, Blohmstr. 15, 21079 Hamburg, Germany; abdullah.nsair@tuhh.de (A.N.); ayah.alassali@tuhh.de (A.A.); kuchta@tuhh.de (K.K.)

[2] Civil Engineering Department, Jordan University of Science and Technology, Irbid 22110, Jordan; hqdais@just.edu.jo

* Correspondence: senem.oenen@tuhh.de; Tel.: +49-40-42878-3527

Received: 28 May 2020; Accepted: 16 July 2020; Published: 22 July 2020

Abstract: The biogas production technology has improved over the last years for the aim of reducing the costs of the process, increasing the biogas yields, and minimizing the greenhouse gas emissions. To obtain a stable and efficient biogas production, there are several design considerations and operational parameters to be taken into account. Besides, adapting the process to unanticipated conditions can be achieved by adequate monitoring of various operational parameters. This paper reviews the research that has been conducted over the last years. This review paper summarizes the developments in biogas design and operation, while highlighting the main factors that affect the efficiency of the anaerobic digestion process. The study's outcomes revealed that the optimum operational values of the main parameters may vary from one biogas plant to another. Additionally, the negative conditions that should be avoided while operating a biogas plant were identified.

Keywords: biogas plants; anaerobic digestion; plant monitoring; bioenergy; process optimization

1. Introduction

To meet the increased demand for energy needs and to reduce greenhouse gas emissions, the capacity of worldwide installed renewable energy systems has been doubled over the last decade [1–5]. This also applies to biogas as a source of renewable energy, where the number of biogas plants installed in Europe has been increased from 6227 in 2009 to reach 18,202 by the end of 2018 [6]. The total produced electricity from biogas reached 88 TWh in 2017, 40% of which was generated in Germany [4]. Hence, Germany is a leading country in this field [6]. Biogas can be utilized—after treatment—in numerous applications, like electricity and heat generation, connection to the natural gas grid, or as biofuel in vehicles [7].

Anaerobic digestion is a biological process, in which the microorganisms degrade the complex organic matter to simpler components under anaerobic conditions to produce biogas and fertilizer [6,8]. This process has many environmental benefits, such as green energy production, organic waste treatment, environmental protection, and greenhouse gas emissions reduction [2,9–13]. The biodegradation of the complex organic matter undergoes four main steps. Namely, hydrolysis, acidogenesis, acetogenesis, and methanogenesis [3].

Biogas consists mainly of CH_4 and CO_2—the share of CH_4 is determined by the type of the feedstock fed into the biogas plant [2]. The different operational conditions also have significant effects on the biogas production potentials. In order to obtain the optimum biogas production with the lowest costs, the biogas plant design has to be optimized as per the needs and potentials. The design criteria of biogas plants (explained in the following section) should be considered for their construction [3]. Additionally, several parameters have to be controlled to prevent problems causing inhibition in biogas plants (Figure 1). Temperature, pH value, retention time, and organic loading rate have a direct effect on the microbial activity. Moreover, the physical features of the feedstock can vary, and it may contain toxic substances which can influence the microbial activities [4,14].

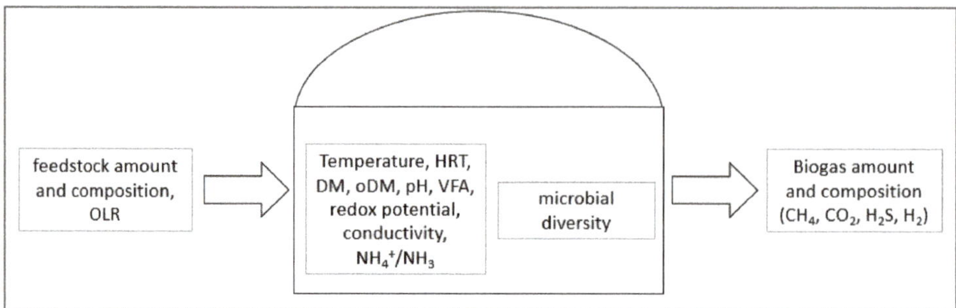

Figure 1. Operational parameters of the biogas plant. Adapted from Theurel and coauthors [9]. OLR: Organic Loading Rate, HRT: Hydraulic Retention Time, DM: Dry Matter, oDM: organic Dry Matter, VFA (Volatile Fatty Acids).

In this paper, a description of the anaerobic process that takes place in biogas plants is presented. Also, the operational conditions and the available technologies—including their advantages and disadvantages—are thoroughly discussed. Specifically, parameters defining the reactor's design, the operational conditions, and the monitoring procedures are summarized and discussed. The aim is to assess the different parameters affecting the biogas production for an optimized biogas plant design and operational conditions. The study was conducted by means of a literature review on articles, books, regulation agencies, and internet documents.

2. Reactor Design Considerations and Operational Conditions

To choose the optimum biogas reactor's design, following criteria should be considered.

- Dry matter (DM) content of the substrate: Wet digestion (DM < 12%) and dry digestion (DM > 12%)
- Mode of material feeding: Intermittent (no substrate addition during the dwell time), semi-continuous (at least once per working day) and continuous (flow)
- Number of process phases: Single-phase (all steps take place in the same reactor) and two-phase (hydrolysis and methanogenesis take place in separate reactors)
- Process temperature: Psychrophilic (<25 °C), mesophilic (37 to 42 °C) and thermophilic (50 to 60 °C) [3].

2.1. Substrates

In Europe, the biogas is mainly produced by the anaerobic digestion of agricultural residuals, manure, and energy crops. Additionally, the sludge of wastewater treatment plants, organic fraction of the municipal solid waste, or solid waste buried in landfills are possible feedstock sources [10,13,15]. The anaerobic digestion is an efficient method to treat the organic fraction of the municipal solid waste for energy production while mitigating the greenhouse emissions [16–18]. The anaerobic digestion of food waste is more sensitive than that of agricultural waste, where the volatile fatty acids (VFA) are rapidly produced in the initial stage, negatively affecting the anaerobic digestion process [19–21]. Generally, the fermentation technology depends on the utilized substrate [11]. When the biodegradable substrate has a dry matter content of ≤12% (in some research, 15%), then the wet fermentation process is conducted. Otherwise, the dry fermentation process is used. Around 90% of the biogas plants in Germany operate under wet fermentation conditions [3,12]. Figure 2 summarizes the mass fraction of substrates that are utilized in the biogas plants in Europe according to a study done by Scarlat and coauthors [22]. As shown in Figure 2, the highest share of the utilized feedstock in Europe is represented by energy crops, agricultural waste, and organic waste. In some European countries (e.g., France, Norway, Portugal, Spain, and the UK), the landfill gas signifies a tangible share of the utilized feedstock [22,23].

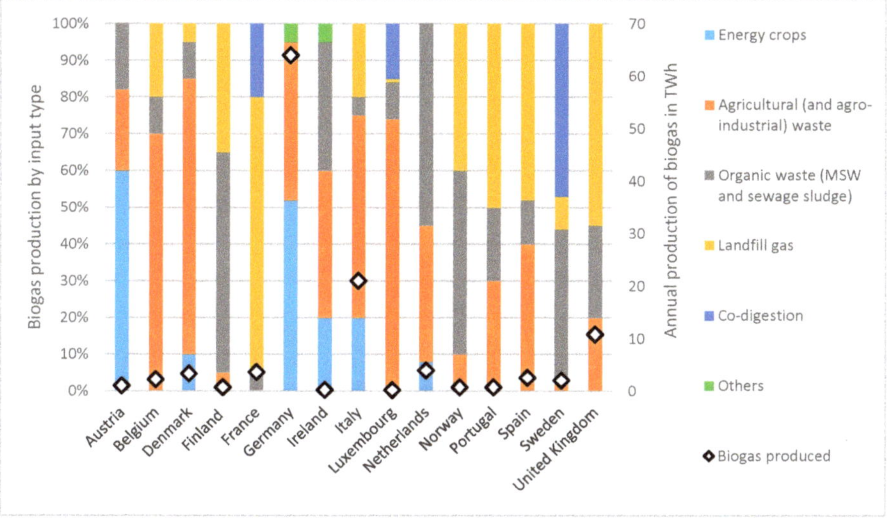

Figure 2. Mass fraction (wt.%) of utilized substrates in Europe. The white boxes represent the total annual biogas production in TWh [22,23].

The feedstock quantity and composition can affect the anaerobic digestion process as follows:

- The stirring technology is dependent on the dry matter content and the viscosity of the biodegradable feedstock [24,25].
- The feedstock's composition determines the content of the volatile solids as well as the ammonia concentration inside the reactor [26].
- Based on the feedstock quality and quantity, sedimentation and floating layers could be obtained [27,28]. The high amount of impurities in the substrates leads to sedimentation. The size of the substrate and its biodegradability are factors which determine sedimentation potentials. The floating layers can be created by the surfactants.
- The anaerobic digestion stability is dependent on the feedstock, because of their different chemical and physical properties. Therefore, suitable feeding is required to ensure the anaerobic digestion process [29].
- The technology used for the anaerobic digestion, as well as the digester's size and shape, are defined by the feedstock's properties, e.g., DM, oDM, biogas formation potentials, as well as carbon to nitrogen

ration (C:N ratio) [30]. Table 1 summarizes the main agricultural substrates that are globally utilized and their main properties.

Table 1. Main types of agricultural substrates utilized by the biogas plants worldwide. FM: Fresh Matter. NL: normal liter, FM: fresh matter, t: ton.

Substrate	DM %	oDM % (In DM)	Biogas Yield NL kg^{-1} FM	Methane Content NL kg^{-1} FM	Electricity Produced kWh t^{-1} FM	References
Pig slurry	4–19	73–86	20–35	10–21	40–71	[3,31,32]
Cattle slurry	6–11	75–82	20–30	11–19	40–61	[3]
Cattle manure	20–25	68–76	60–120	33–36	112–257	[3,26]
Poultry manure	34–50	60–75	130–270	70–140	257–551	[3,33]
Maize silage	28–39	85–98	170–230	68–120	347–469	[3,26,31,33,34]
Grass silage	15–50	70–95	102–200	46–109	208–408	[3,26,31,33]
Sugar beets	13–23	84–90	120–140	65–113	245–286	[3,26,34]
Olive pomace	57–90	55–86	92–147	65–104	188–300	[35]
Wheat straw	91–94	87–92		135–237	146–266	[36,37]
Corn (corn stover)	66–89	83–99		261–402	293–451	[33,34,38]
Rye	62–93	84–87	130	70	265	[33,34]

Using a mixture of different substrates enhances the content of nutrients, supplements, and phosphorus, while providing a balanced C:N ratio [35,39]. The increase in the C:N ratio results in a rapid consumption of the nitrogen before carbon digestion. Hence, methane potential drops [35,40]. However, the decrease in the C:N ratio leads to microorganisms' inhibition, as a result of ammonium accumulation [35]. The literature showed varying optimum C:N ratios for different substrates. The majority reported that the optimum C:N ratio is within the range of 20–30:1 for different substrates and different temperature conditions [41–47]. However, Guarino and coauthors [48] suggested a broader range of 9–50:1. Substrate biodegradability is another factor that affects the kinetics of the biodegradation process and consequently, the size of the digester [49].

2.2. Process Phases

The biogas plants can be operated in a single-stage mode or two-stage (multi-stage) anaerobic systems [50,51]. The decision of operating the biogas plant under single- or multi-stage systems can be made after evaluating the advantages and disadvantages of each choice. The decision criteria on the operational mode of the biogas plant are summarized below (Figure 3).

- Cost: the installation and maintenance of the multi-stage system are more expensive than that for the one-stage system [51].
- Operational conditions: the optimal operation conditions (e.g., temperature and pH) for the microorganisms in the multi-stage systems are more demanding than that for single-stage systems, since the operational parameters in the different stages are diverse. On the other hand, due to the separation of the phases, there is a better process control.
- The stability of the anaerobic digestion process is improved through using the multi-stage systems. The methanogenesis step is very sensitive to the changes in the organic load rate, the heterogeneity of the biodegradable feedstock, and the changes in the environmental conditions. Hence, the multi-stage is more advantageous than the single-stage system, where the control of these conditions is more efficient, and the flow of the biodegradable feedstock from the first digester to the others is more homogeneous in quantity and quality [51–57].
- The multi-stage systems have higher performance than single-stage systems in terms of the removal efficiency of the volatile solids and improving the biogas quality (methane content) [51,52,54,56,58–61].

- The single-stage systems are still the most used due to their simplicity [10].

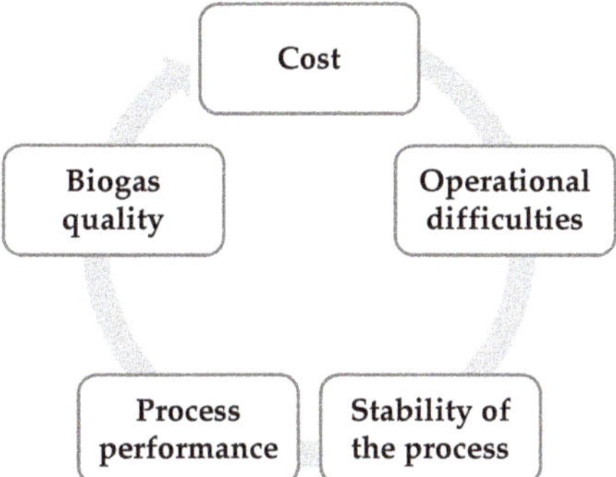

Figure 3. Decision-making criteria for selecting the mode of biogas plant operation under single- or multi-stage systems.

2.3. Process Temperature

Temperature is a significant parameter inside the reactor, which has a direct effect on the microbial performance. Biogas plants can be operated under psychrophilic (<25 °C), mesophilic (32 to 42 °C), or thermophilic (50 to 57 °C) conditions, defined based on the microorganism group used in the process [62]. Among all types, methanogens are defined as the most sensitive microorganisms to environmental conditions [63]. There is not a clear discrimination between the species living at different temperature ranges and the highest number of species within their optimum is observed under mesophilic conditions at 37 °C [64].

Not only the microorganisms but also reaction kinetics are affected by the temperature inside the reactor [65]. In the range of optimum temperatures, increasing the temperature leads to an increase in the enzymatic activities. Nevertheless, surpassing these defined optimum temperatures may lead to inhibition of enzymatic reactions. 37 °C is the optimum temperature for most of the mesophilic enzymes [64,66,67]. According to Streitwieser [68], the thermophilic range is a better option for easily degradable substrates, which leads to an increase in biogas production as well as an increase in the reaction rates. Besides, shorter start-up time is needed for the new operating biogas plants at thermophilic conditions [69–71]. To supply stable biogas production, the temperature should be maintained stable [72].

Several studies examined the effect of temperature changes, including both stepwise and abrupt changes. Wu and coauthors [73] studied the effects of sudden temperature decreases in lab-scale reactors to simulate the possible heating failures. Decreasing the temperature from 55 to 20 °C for 1, 5, 12, and 24 h in different reactors almost stopped the biogas production, which could be recovered after adjusting the temperature back to 55 °C. Other studies were conducted to examine the effect of abrupt temperature changes in thermophilic temperature conditions. The results showed that the recovery with daily temperature fluctuations under thermophilic conditions (below 60 °C) can be attained, however, temperatures above 60 °C have an adverse impact on hydrolysis and acidogenesis stages due to the high ammonia concentrations generated in the process [74–76]. The process microbiology was examined by Pap and coauthors [77], who observed the replacement of acetolastic methanogens by hydrogenotrophic archaea after changing the temperature conditions from mesophilic to thermophilic. Similar studies, conducted under mesophilic conditions, examined temperature fluctuation in the mesophilic conditions (35 to 30 °C, after 170 h to 32 °C) [78]. Recovery of the process was possible after 40 h. While a lower number of methanogenic archaea is observed at high temperatures, acidogenic bacteria are less affected by temperature changes [74,79].

On the other hand, stepwise temperature changes were examined to obtain a stable process in spite of the temperature fluctuations. The temperature change from 37 to 55 °C in 41 days was studied by Bousková and coauthors [80]. Results showed that 70 days was sufficient for process recovery. Thermophilic bacteria, which already exists in the mesophilic inoculum, can become a dominant group in the thermophilic conditions [81].

According to several studies reviewed, it was concluded that temperature fluctuations (i.e., more than ±3 °C d $(day)^{-1}$ under mesophilic conditions and more than ±1 °C d^{-1} under thermophilic conditions) should be avoided [82–86].

2.4. Mixing

The stirring (mixing) system inside the biogas plants has significant importance on the anaerobic process [87]. The core responsibilities of the stirring system are:

- Ensure the homogeneity of biodegradable substrates, temperature, and pH value inside the digester by mixing the fresh substrates with the existing one [26,27,88–90].
- Enhance the metabolism of the microorganisms and enhance the anaerobic process stability. Moreover, the mixing assists the gas-bubbles to flow upwards from the biodegradable feedstock at high total solids values [26,91,92].
- Reduce the creation of sediments on the bottom of the digesters to ensure the highest available volume for the anaerobic digestion process and reduce the need to clean the digester (on average, it is once every 4 to 7 years) [93].
- Break the foam layer on the top of the biodegradable substrate. The creation of this layer can inhibit 20% to 50% of the biogas production [94]. "Foam is generally a dispersion of a gas in a liquid consisting of a large proportion (approximately 95%) of gas. The liquid phase is located in a thin film which is present between the gas bubbles" [28,94]. Two groups of surface-active compounds are considered to be in charge of the foam formation, which are the surfactants and bio-surfactants. The surfactants are compounds inflowing the digester with the feeding, while the bio-surfactants are considered to be the outcomes of the activities of the microorganisms [27,28]. These foam layers have to be destroyed because their formation leads to a drop in the biogas yield, high-cost losses, and equipment damage, as well as operational disturbances [27,28,95–97].

Choosing the most suitable mixing technology relies on the conditions and requirements such as the value of the total solids of the biodegradable feedstock, the type of the used feedstock, the hydraulic retention time, and the size of the digester. Biogas plants operate with or without mixing, defined by the digester's structural concept: complete mix, plug flow, or batch concept [29]. The complete mix digesters are mainly used for biogas plants with biodegradable substrates with total solid values in the range of 2% to 10%, while plug-flow digesters are suitable for the range of 11% to 13% [98]. The completely mixed concept is a widely applied type in Germany. The complete mix concept contains a gas tight cover to collect the formed biogas inside a vertical cylindrical digester, which has walls from concrete, steel, or reinforced concrete [3]. The necessity of using a mixing system inside the digesters increases with increasing the total solid values of the biodegradable substrate [88–90].

The major mixing technologies used in the large-scale biogas plants are mechanical, pneumatic, and hydraulic mixing technologies [99]. The pneumatic (gas-lift) mixing is dependent on the formed biogas itself to move the biodegradable substrate [100]. The three types of pneumatic mixing are: The free gas-lift, the limited gas release from the bottom of the digester, and the creation of piston pumping by the use of big bubbles [101]. The main advantage of this mixing technology is the absence of any moving parts inside the digesters. This leads to a maximum utilization of the digester's volume for the biodegradable substrates. Additionally, it decreases the complexity and cost of maintenance [86,100]. The main disadvantage of this mixing technology is the inability to mix in the entire digester, especially near the bottom and the top. This results in sediments formation, and disability to break the floating layers. The poorly mixed zone might cover up to one-third of the digester [86,100–102].

The hydraulic mixing needs pumps (mainly airlift pumps) to mix the biodegradable substrates inside the digesters [26,100]. This technology has the advantage of installing the mixing components outside of the digesters. The main disadvantage of this technology that it is only suitable for small-scale digesters. Otherwise, sediments, floating layers, and poorly mixed zones will be obtained [3].

The dominant technology used in Germany is the mechanical mixing [91]. Different types of stirrers can be installed: submersible, long-axis, axial, and paddle stirrers [3]. The digester's form and size as well the used substrates are the main parameters used to choose the most suitable technology [3,103,104]. The previously mentioned mechanical stirrers vary, mainly according to their impeller velocities and sizes as well as the power consumption [91]. In the case of high viscosity of the biodegradable substrate (the total dry matter content is high as well), slow and large impellers (stirrers) are used. Yet, for low viscosities, the most suitable stirrers are the small and fast ones [33].

The researchers have been trying to discover the optimum stirring system and conditions in the anaerobic digesters for a long time. In 1982, the research of Hashimoto [105] focused on evaluating the stirring period inside the digesters. He found that stirring for only 2 h per day is more efficient than continuously stirring under laboratory thermophilic conditions. The results, however, showed similarities in the amount of biogas formed for the two different stirring conditions in pilot-scale anaerobic digesters. Table 2 summarizes the findings of different studies in the field of mixing inside the digesters.

On the basis of the information presented in Table 2, the optimal stirring inside the digester depends on several factors, as follows:

- The size of the digester,
- The operating temperature inside the digesters,
- The used mixing technology,
- The used feedstock and the DM value of the biodegradable feedstock.

Table 2. Findings of various studies conducted on the mixing characteristics of various biogas plants. 0: reference value (or no changes in the anaerobic process), +: improvement in the anaerobic process, −: worsening the anaerobic process, X: dependent on the stirring intensity, COD: chemical oxygen demand, MSW: municipal solid waste.

Scale of The Plant	Substrate	Stirring Period Daily (h)				Stirring Intensity		Temperature (°C)	Remarks	References
		0	1–8	9–23	24	High	Low			
A	Beef cattle waste		0		−			55		[105]
B	Beef cattle waste		0		0			55		[105]
A	Castor cake	0	+		+			30	Loading rates 4 and 8 g L^{-1} d^{-1}	[106]
A	Castor cake	0	−		−			37	Loading rates 4 and 12 g L^{-1} d^{-1}	[106]
A	Castor cake	0	+		+			37	Loading rate 8 g L^{-1} d^{-1}	[106]
B	Refuse-derived fuel and primary sludge	0			−			35		[107]
A	Water hyacinth and cattle dung	0	+		−			37		[108]
A	Unmodified olive mill wastewater	0		−		−	0	35		[109]
A	Fermented olive mill wastewater	0		0		0	0	35		[109]
A	Sewage sludge	0			−			28	Loading rates 2.4, 4.8, and 7.2 g COD L^{-1} d^{-1}	[110]
B	Animal manure	0			+			35		[111]
A	Dog food	0			−			35 & 55	Batch fed	[112]
A	Manure slurry	0			+			35		[88]
A	Animal waste	0			−	−	0	35	Biogas recirculation, low DM values	[89]
A	Cow manure	+	0		−			55		[113]
B	Cow manure		X		0	0	+	54		[113]
A	Lipid-rich waste	0			+			37	Batch reactor	[114]
A	Corn silage	0			+			37	Continuous reactor	[114]
A	Rice straw					+	0	35		[115]
A	Manure and MSW	0			+			55		[116]
B	Natural water, cow dung, rice straw and water hyacinth	0			+			31		[117]
A	MSW	0			X	−	+	33		[118]
C	Cow dung		X			+	−	27	Optimum mixing was for 3 h daily with 60 rpm	[119]
B	Cow manure and maize straw	0			X	−	+		Different DM and C:N ratios	[120]
A	Cow manure and vegetable waste	0	+							[121]
A	Cattle manure, tea waste	0			+			37		[122]
A	Municipal solid waste	−	+				0	34		[123]

2.5. The Energetic Potential of the Biogas Plants

The energetic potential of the biogas plants depends mainly on the operational parameters mentioned in the previous sections. In Table 1, the biogas formation potential for the primary agricultural substrates was presented. These values indicate the amount of biomehtane/biogas produced per one gram of fresh substrate. It is worth mentioning that the biogas has an average low calorific value (LCV) of 10 MJ kg^{-1} [124], and an average high calorific value (NCV) of 20–21 MJ kg^{-1} at methane content of 55% [75,125,126], where the methane has an NCV of 50 MJ kg^{-1} [127]. Table 1 adapts the calculation method used by Achinas and coauthors [126] in their research to estimate the electricity produced from fresh material (estimating 35% electrical efficiency combined heat power, heating value of 21 MJ m^{-3}, 55% methane content, 3.6 MJ $(kWh)^{-1}$). To evaluate the energy efficiency of the biogas plants, the researchers used different approaches, including or excluding the energy used for the cultivation, transport, and treatment of the substrates in the calculations. The internal electricity consumption of the biogas plants varies on average in the range of 4.9–9.3% [33,128]. The main sources of the electricity consumption are the stirring system, the feeding system, the combined heat and power unit, and the heating system. To optimize the electricity efficiency, researchers improved the stirring system (Table 2), others considered the temperature inside the digesters (Table 2), and others optimized the energy production unit. Further researchers were able to optimize the biogas yield through optimizing the mixing ratio of the substrates, the environmental conditions for the microorganisms, as well as the monitoring system.

3. The Conditions Inside the Reactor

3.1. Oxygen

Strictly anaerobic microorganism groups of acetogens and methanogens can be affected by an oxygen leak in the reactor, which can lead to inhibition [129]. On the other hand, micro-aeration can improve the efficiency of the hydrolysis step in the anaerobic digestion process [130–132]. Moreover, a micro oxygen injection to 50 L anaerobic digestion reactors provided a decrease in the H_2S concentration in the biogas (from 6000 to 30 ppm) [132]. Both experimental and simulation results of the study conducted by Botheju and coauthors [129] showed that an increase in the oxygen loads causes a decrease in the methane potential. More than 0.1 mg L^{-1} of oxygen concentration has caused inhibition of obligate anaerobic methanogenic archaea [3].

3.2. PH

There are various kinds of microbial groups taking part in the anaerobic digestion process, which have diverse optimum pH values for their optimal growth rates. For example, a pH range of 5.0 to 6.0 is suitable for acidogens, while pH from 6.5 to 8.0 is more convenient for the methanogens group [133]. The combined effect of pH and temperature were studied on the anaerobic digestion of grass silage by Sibiya and Muzenda [134]. Results showed that the highest efficiency was achieved at 45 °C and a pH value of 6.5. Usually, the biogas plants operate within a pH range of 6.5 to 8.4 [135,136]. Mpofu and coauthors [137] summarized the optimum temperatures and pH values for many acetogenic and methanogenic bacteria. The pH value is highly dependent on the VFAs, the ammonium content, and the alkalinity concentrations. The increase of the VFAs leads to a drop in the pH value. On the other hand, an increase in the alkalinity sources causes an increase in pH [33,62,138–143].

3.3. Dry Matter Content of the Biodegradable Feedstock

The total solids or dry matter content of the biodegradable substrate is highly connected to the feedstock. As mentioned earlier, these values play a crucial role in determining the fermentation technology (dry or wet fermentation), the digester or reactor design, as well as the stirring technology.

- The researchers consider that the wet fermentation process occurs at dry matter content of less than 15%, while the dry fermentation takes place at higher DM values.
- The mixing inside the digesters (technology as well as duration) is highly dependent on the dry matter content, as mentioned in Section 2.4. The dry matter content of the biodegradable feedstock is a critical factor in controlling the Bingham viscosity and the yield stress [144].
- The biogas formation potential might be dependent on the dry matter content of the biodegradable feedstock [145,146]

Most large-scale biogas plants operate in wet fermentation conditions, where the dry matter content is less than 12%. By the end of 2019, the worldwide cumulative installed biogas plants had the capacity of 19.5 TW_{el} [147,148]. The majority of the biogas plants are still adopting wet fermentation technology [149].

3.4. Organic Loading Rate (OLR)

Obtaining optimum biogas production with reasonable cost cannot be achieved without a well-planned organic loading rate (OLR). OLR represents the number of volatile solids fed per unit volume of digester per unit of time, as described in Equation (1) [3]:

$$OLR = \frac{m * c}{V_R * 100}\ \left(kg\ oDM\ m^{-3}\ d^{-1}\right) \quad (1)$$

where m is the amount of substrate fed in a unit of time (kg d^{-1}), c is the concentration of organic dry matter (% oDM), and V_R is the reactor's volume (m^3).

Keeping the OLR low can cause a decrease in the biogas production efficiency. On the other hand, a high organic loading rate can be a reason for process inhibition [150,151]. In order to obtain the optimum conditions for the specific biogas plant, OLR should be determined based on the feed substrate [152]. According to a study conducted by González-Fernández and coauthors [153], increasing the OLR of microalgae from 1.0 kg tCOD (total chemical oxygen demand) m^{-3} d^{-1} to 2.5 kg tCOD m^{-3} d^{-1} did not result in process inhibition, yet it provided higher methane production. Zuo and coauthors [154] studied two-stage anaerobic digestion of vegetable waste in lab-scale reactors. Increasing the OLR improved the methane content of the biogas. In order to prevent process inhibition at high OLR operation, implementing recirculation of biodegradable feedstock—which dilutes the biodegradable feedstock and supplies pH adjustment—can be a suitable solution. Overall, OLR is an essential parameter for designing the process, but increasing the OLR can cause an accumulation in the VFAs, resulting in process interruption [150,155]. Findings of the conducted studies that dealt with the effect of OLR on the anaerobic digestion process efficiency are summarized in Table 3.

Table 3. Studies performed to evaluate the effect of organic loading rate (OLR) on the anaerobic digestion process efficiency. CSTR: continuous stirred tank reactor, semi-CSTR: semi-continuous stirred tank reactor.

OLR ($kg\ m^{-3}\ d^{-1}$)	Stable/Optimum OLR	Reactor Type	Temperature	Substrate	Reference
1.0, 2.0, 3.0 and 4.0	3 kg m^{-3} d^{-1}	CSTR, Semi-CSTR, 8 L	35 °C	Maize, rye, fodder beets	[156]
1.0 and 2.5 (COD)	1 kg COD m^{-3} d^{-1}	CSTR, 1 L	35 °C	Thermally pretreated microalgae	[153]
1.8 to 4.0 (oDM)	2 kg oDM m^{-3} d^{-1}	CSTR, 4 L	35 °C	Animal by-products from the meat processing industry	[157]
Ranged from 1.2 to 8.0 (oDM)	8 kg oDM [76] m^{-3} d^{-1}	CSTR, 1.6 m^3	35 ± 2 °C	Municipal biomass waste and waste-activated sludge	[155]
3.0, 3.6, 4.2, 4.8, 6.0, 8.0, and 12.0 (oDM)	6–8 kg oDM m^{-3} d^{-1}	CSTR, 40 L	37 ± 2 °C	Rice straw and pig manure	[150]
Ranged from 4.2 to 12.8 (oDM)	1.12 g oDM L^{-1} d^{-1}	Semi-CSTR, 10 L	37 ± 0.5 °C	Dried pellets of exhausted sugar beet cossettes and pig manure	[145]
1.22, 1.46, 1.70, and 2.0 (oDM)	<2.00 kg oDM substrate.m^{-3} d^{-1}	Completely mixed bioreactor, 300 m^3	39 ± 1 °C	Rice straw	[158]
3.7 to 12.9 (oDM)	9.2 kg oDM m^{-3} d^{-1}	Semi-CSTR, 3000 mL	37 ± 1 °C	Food waste	[159]
Ranged from 0 to 10.0 (oDM)		CSTR, 5 L	37 ± 0.5 °C	Food waste	[160]

3.5. Hydraulic Retention Time (HRT)

Hydraulic retention time (HRT) is one of the determining parameters for the volume of the digester, which defines the remaining time of the feedstock until it is discharged (Equation (2)) [3].

$$HRT = \frac{V_R}{V}\ (d) \tag{2}$$

where V_R is the reactor volume (m^3), and V is the substrate volumetric feed rate in the reactor, daily ($m^3\ d^{-1}$).

Optimum biogas production can be obtained at different HRT's, depending on the used substrate [161]. Various HRTs were assessed by literature to find the optimum values for the different substrates. The adopted HRT varied from 0.75 to 60.00 days. The optimum HRT was suggested to be in the range of 16 to 60 days [162–167]. In order to prevent washouts of microorganisms required for the process, HRT should not be less than 10 to 25 days [166]. Kaosol and Sohgrathok [163] used seafood wastewater as anaerobic co-digestion material at different HRTs (10, 20, and 30). The maximum methane production was observed with the implementation of HRT of 20 days. HRT has fluctuated between 7.9 to 37.3 days during the study conducted by Krakat and coauthors [168], using 5.7 L lab-scale reactors. Decreasing the HRT leads to an increase in the number of species in the reactor and a slight increase in the CH_4 content. Schmidt and coauthors [166] studied the reduction of HRT from 6.0 to 1.5 days at three different reactor systems. The fastest inhibition was observed in the anaerobic sequencing batch reactor at an HRT of 3.0 days, and it was followed by a continuously stirred tank with HRT of 2.0 days. The experiment conducted using a fixed bed reactor was stable until the end of the experiment. Nevertheless, decreasing HRT resulted in a decrease in the specific biogas production.

3.6. Nutrients

The nutrients, which supply stability of microorganisms taking part in the anaerobic digestion process, can be classified under two categories: Micronutrients and macronutrients. Macronutrient ratio, carbon to nitrogen to phosphorus to sulfur ratio (C:N:P:S) = 600:15:5:1, is suitable to obtain a sustainable process [26]. The processes, where macronutrients are consumed, can be listed as follows:

- Carbon: For building cells' structure.
- Nitrogen: For protein biosynthesis.
- Sulfur: For the growth of methanogens and component of amino acids.
- Phosphate: To create energy carriers in the metabolism [169,170].

In addition to macronutrients, micronutrients such as iron, nickel, cobalt, selenium, molybdenum, and tungsten are needed in the process for microbial growth and should be added as supplemental nutrients to the process, in case the feedstock has a deficiency in such nutrients [26,171]. Conversely, the excess amount of trace elements can lead to inhibition of the process [171,172].

3.7. *Process Inhibitors*

According to Chen and coauthors [173], the main inhibitors of the anaerobic digestion process are the ammonia, sulfide, organics, and light and heavy metals.

3.7.1. Ammonia

The optimum concentration of the ammonia (as well the ammonium ion) inside the digesters has a critical role in the stability of the anaerobic digestion process. The optimal ammonia concentration increases the stability process through ensuring adequate buffer capacity of the methanogenic medium. It is worth mentioning that the ammonia is considered to be an end product of the biological degradation of the nitrogenous part of the substrates, such as the proteins and the urea [173,174]. The microorganisms need the ammonia as a nutrient for their metabolisms [175].

Nonetheless, high ammonia concentration is considered to be an inhibitor to the anaerobic digestion process and to the microbial activity inside the digesters [173–175]. Researchers have been trying to find the optimum total ammonia-nitrogen concentration (TAN) in the anaerobic digestion process, and their inhibitory levels. Chen and coauthors [173] summarized the results of other research mentioning that concentrations lower than 200 mg L^{-1} are beneficial for the anaerobic digestion. However, the TAN concentrations 1.7 to 14.0 mg L^{-1} can cause a drop in the methane formation. The decrease in efficiency can reach up to 50%, depending on the substrates, inoculum, and environmental conditions.

3.7.2. Sulfide

Sulfide is considered as one of the inorganic inhibitors of the anaerobic digestion process [176]. The sulfide is typically an output of the sulfate-producing bacterial (SRB) activities. These bacteria are responsible for: (I) sulfides generation, which may cause inhibition of SRB or methane-producing bacteria, (II) alkalinity sources, causing a change in the pH value, (III) accelerating the oxidation of the organics, (IV) reducing the efficiency of the methanogenesis process, and (V) decreasing the methane formation [177].

3.7.3. Light and Heavy Metals

The primary light metals influencing the anaerobic digestion process are sodium (Na), potassium (K), magnesium (Mg), calcium (Ca), and aluminum (Al) [173]. These metals are required in the process for the microbial growth, enhancement of the bacterial cell immobilization (Ca), and formation of adenosine phosphate (Na^+) [178–181]. However, the concentrations of these light metals should be controlled where: Mg^{2+} is responsible for limiting the production of double cells [178], K^+ is in charge of neutralizing the cell membrane potential [173], Na^+ can inhibit the acetoclastic methanogens [178], Ca^{2+} is accountable for the destabilization of the buffering system through precipitation of phosphates and carbonates [180], and finally, Al^{3+} is considered as an inhibitor of the anaerobic process—it can also compete with the adsorption of other metals [182].

The heavy metals affecting the anaerobic digestion are mainly iron (Fe), copper (Cu), zinc (Zn), nickel (Ni), cobalt (Co), molybdenum (Mo), chromium (Cr), cadmium (Cd), lead (Pb), and mercury (Hg) [173,178]. Metals with inhibitory effects on the anaerobic digestion process are summarized in Table 4 together with their inhibitory concentrations. Generally, the inhibitory and optimum metals concentrations can vary for different temperature conditions of the digesters.

Table 4. Metal concentration in the feedstock. Adapted from References [183–187].

Metal	Inhibitory Concentration in mg L^{-1}	Positive Concentration in mg L^{-1}	Reference
Aluminum	1000–2500		[188]
Cadmium	36–3400	0.1–0.3	[189–193]
Calcium	300–8000	100–1035	[173,188,194–196]
Chromium	27–2500	0.1–15	[191–193,197]
Cobalt	35–950	0.03–19	[197–202]
Copper	12.5–1000	0–10	[190,191,193,197,201,203]
Iron		0.3–4000	[189,202]
Lead	67.2–8000	0.2	[191,196,204]
Magnesium	750–4000	0–720	[183,188]
Mercury	125		
Molybdenum	1000	0–0.1	[202,205]
Nickel	35–1600	0.03–27	[191,193,198–201]
Potassium	400–28,934	0–400	[188,196,206]
Sodium	3000–16,000	100–350	[188,206,207]
Zinc	5–1500	0–5	[189,191,193,198,201,202]

3.7.4. Organics

The anaerobic digestion process can be inhibited by the poorly soluble organic compounds, or by the organic compounds that can be adsorbed to the surface of the biodegradable feedstock [173]. The following organic compounds are already listed to be toxic to the anaerobic digestion process [173,208]: alkylbenzenes, halogenated benzenes, nitro benzenes, phenol, alkylphenols, nitrophenols, halogenated phenols, alcohols, halogenated alcohols, alkanes, halogenated aliphatics, aldehydes, ethers, ketones, acrylates, carboxylic acids, amines, nitriles, amides, pyridine and its derivatives, as well as long-chain fatty acids [209–234].

3.7.5. Secondary Metabolites

Anaerobic digestion can be considered as a suitable treatment of food waste for biogas production, due to their high moisture and organic content. Plants contain components, such as flavor substances, which may affect the digestion process [235]. The plants' secondary metabolites are classified into carotenoids, terpenes, phenolics, alkaloids, and glucosinolates. Terpenes are one of the largest groups, with more than 30,000 compounds synthesized in a myriad of plants [236]. In a study done by Wikandari and coauthors [228], fruit flavor compounds including hexanal, α-Pinene, Car-3-ene, Nonanal, E-2-hexenal, myrcene, and octanol were found to be inhibiting the methane production. The results of this work stress that all the tested terpenoids are inhibiting the anaerobic bacteria. Generally, the type of inhibition is depending on the concentration of inhibitory substances entering the biogas digester [236].

4. Monitoring of the Operational Conditions in Biogas Plants

As mentioned before, the anaerobic digestion process includes four stages that come in sequence, where different kinds of microorganisms take part at each stage. In order to obtain a stable and efficient process, process monitoring is necessary [4,62,237]. Monitoring enables early detection of problems and disturbances and indicates the required adjustments to the operational parameters (to be within acceptable ranges).

In general, monitoring parameters of the biogas plants can be classified under three categories: parameters characterizing the process (feedstock type and quantity, biogas production amount and its quality, reactor temperature, dry matter concentration, ammonia concentration, and pH), parameters supplying early detection of instability (VFA, alkalinity, hydrogen concentration, redox potential, and other complex monitoring parameters), and variable process parameters defined by plant operators (OLR and HRT) [62].

The monitoring of the biogas plant's operational parameters can be achieved by on-line, at-line, and off-line analyzers. There is an increased interest in the on-line monitoring applications, due to the fast and automated process control. There are several parameters that can be monitored at the biogas plant on a real-time basis. The frequently on-line-monitored parameters in biogas plants are summarized in Table 5.

Table 5. Studies conducted on biogas plants' operational parameters and their monitoring methods.

Parameter	Measurement Method	Reference
Cobalt concentration in the high presence of iron concentrations	Total reflection X-ray fluorescence spectroscopy	[238]
VOCs (volatile organic compounds) emitted from different units of food waste anaerobic digestion plant	Portable GC-MS (gas chromatography–mass spectroscopy)	[239]
CH_4 emissions from pressure relief valves of an agricultural biogas plant	Flow velocity and temperature sensors	[240]
Ammonia in biogas	Impedance measurement of biogas condensate in the gas room above the digester	[241]
Dissolved active trace elements in biogas	Total reflection X-ray fluorescence spectroscopy in dried digester slurry	[238]
H_2S in biogas	Gas responsive nano-switch (copper oxide composite)	[242]
Microbial communities depend on the substrate combinations	Sequencing of the 16S rRNA, biodegradable feedstock samples from eight different biogas plants	[243]
Controlling gas pressure in the digester	Programmable logic controller (PCL)	[244]
Ammonia in biomethane	Luminescent ammonia sensor based on an imidazole-containing Ru(II) polypyridyl complex immobilized on silica microspheres	[183]
pH, temperature, oxidation-reduction potential (ORP)	via electrodes, on-line monitoring with PCL	[245]
CO_2, CH_4, H_2O	On-line monitoring with a Supercontinuum laser-based off-resonant broadband photoacoustic spectroscopy	[246]
Different volatile fatty acids	On-line monitoring with total-reflectance Fourier-transformed infrared spectroscopy (ATR-MIR-FTIR)	[2]

The currently available technologies do not enable the monitoring of all operational parameters of the biogas plant. Therefore, samples have to be collected from the biogas plant and analyzed in individual facilities (off-line monitoring) [142]. Off-line monitoring takes place in the laboratory, where samples should be taken for the defined test. Unlike the off-line monitoring, on-line monitoring can provide real-time data on the plant's operation without any time loss for sampling, transfer, and analysis. A study about the on-line monitoring system was done in 2013 in Germany, and it showed that the majority of the biogas plants are equipped with on-line systems to monitor the electricity generation. Additionally, on-line systems were used to determine produced heat, input solid feedstock, biogas temperature, parasitic electricity demand, biogas volume, biogas composition and input liquid feedstock [62].

To improve monitoring systems of the biogas reactors, near infrared spectroscopy (NIR) and mid-infrared spectroscopy (MIR) are seen as promising technologies [9]. In order to obtain operation flexibility at biogas plants (e.g., changing operational parameters and feedstock type and amount), improvements to the biogas plants' monitoring technologies and applications are necessary [2,9,247].

5. Conclusions

Anaerobic digestion is an established technology, used to treat a wide variety of organic wastes. It is one of several biological processes that deliver economic and environmental benefits (i.e., producing bioenergy and/or biochemical while treating the organic fraction of waste). The anaerobic digestion process is complex—it includes various physical and biochemical reactions. The stability of the anaerobic digestion process is affected by many factors (e.g., the conditions inside and surrounding the reactor, the reactor's design, the operational parameters, etc.). In order to maintain a stable, efficient, and sustainable biogas production, the operational parameters should be determined and controlled.

The aim of this paper was to review and evaluate recent studies in the field to determine the critical parameters and their impacts on the anaerobic digestion process, and consequently, on the biogas production. This paper presented a summary to the design parameters of the biogas plant, the significant environmental conditions in the reactor, and the available monitoring and controlling technologies of the anaerobic digestion process (Figure 4).

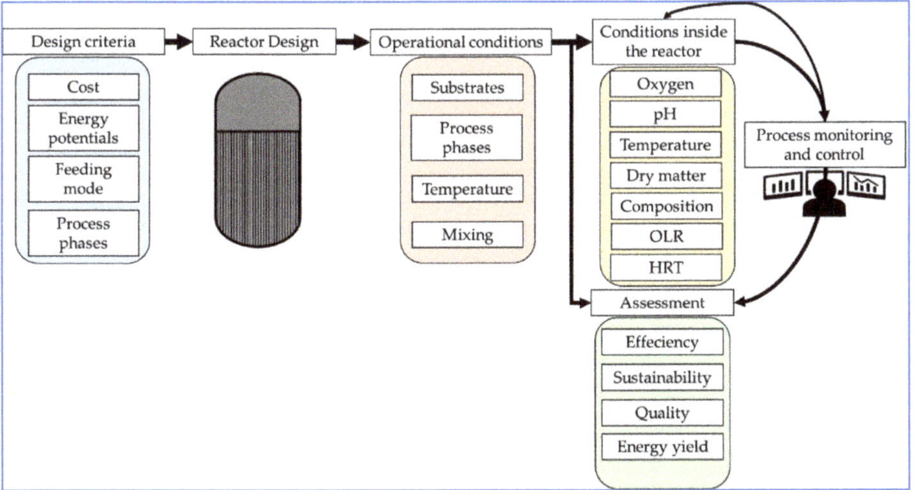

Figure 4. A summary of the paper's discussed aspect.

This review concludes that decisions regarding biogas plants' design, operation, and monitoring conditions depend on many factors (e.g., feedstock, temperature, pH, OLR, HRT, nutrients, inhibitors, biogas quality, etc.). However, the optimal range of the operational parameters varies from one biogas plant to another. Therefore, an inclusive monitoring system is required to enhance the performance of the anaerobic digestion process. Based on this review, it is recommended to improve and expand the available monitoring methods of the process in order to obtain an efficient, sustainable, and flexible operation of the biogas plants. To achieve that, further research needs to focus on the development of on-line, at-line, and off-line monitoring analyzers in the biogas plants.

Author Contributions: Conceptualization, A.N. and S.O.C.; methodology, A.N. and S.O.C.; validation, A.N., S.O.C., A.A., H.A.Q. and K.K.; writing—original draft preparation, A.N., S.O.C. and A.A.; writing—review and editing, A.N., S.O.C., A.A., H.A.Q. and K.K.; visualization, A.N., S.O.C. and A.A.; supervision, H.A.Q. and K.K.; funding acquisition, K.K. All authors have read and agreed to the published version of the manuscript.

Funding: This research received no external funding.

Acknowledgments: Publishing fees were supported by Funding Programme *Open Access Publishing* of Hamburg University of Technology. Thanks to the German Academic Exchange Service for their scholarship to the researcher (SOC).

Conflicts of Interest: The authors declare no conflict of interest.

References

1. Hren, R.; Petrovič, A.; Čuček, L.; Simonič, M. Determination of Various Parameters during Thermal and Biological Pretreatment of Waste Materials. *Energies* **2020**, *13*, 2262. [CrossRef]
2. Falk, H.M.; Benz, H.C. *Monitoring the Anaerobic Digestion Process*; IRC-Library; Information Resource Center der Jacobs University Bremen: Bremen, Germany, 2011.
3. Rohstoffe, F.N. *Guide to Biogas from Production to Use*; Federal Ministry of Food; Agriculture and Consumer Protection; Fachagentur Nachwachsende Rohstoffe E.V. (FNR): Gülzow, Germany, 2012.
4. Refai, S. Development of Efficient Tools for Monitoring and Improvement of Biogas Production. Ph.D. Thesis, Universitäts-und Landesbibliothek Bonn, Bonn, Germany, 2016.
5. Ashraf, M.T.; Fang, C.; Alassali, A.; Sowunmi, A.; Farzanah, R.; Brudecki, G.; Chaturvedi, T.; Haris, S.; Bochenski, T.; Cybulska, I.; et al. Estimation of Bioenergy Potential for Local Biomass in the United Arab Emirates. *Emir. J. Food Agric.* **2016**, *28*, 99. [CrossRef]
6. Piwowar, A. Agricultural Biogas—An Important Element in the Circular and Low-Carbon Development in Poland. *Energies* **2020**, *13*, 1733. [CrossRef]
7. Rosén, T.; Ödlund, L. System Perspective on Biogas Use for Transport and Electricity Production. *Energies* **2019**, *12*, 4159. [CrossRef]
8. Gómez, D.; Ramos-Suárez, J.L.; Fernández, B.; Muñoz, E.; Tey, L.; Romero-Güiza, M.; Hansen, F. Development of a Modified Plug-Flow Anaerobic Digester for Biogas Production from Animal Manures. *Energies* **2019**, *12*, 2628. [CrossRef]
9. Theuerl, S.; Herrmann, C.; Heiermann, M.; Grundmann, P.; Landwehr, N.; Kreidenweis, U.; Prochnow, A. The Future Agricultural Biogas Plant in Germany: A Vision. *Energies* **2019**, *12*, 396. [CrossRef]
10. Sarker, S.; Lamb, J.J.; Hjelme, D.R.; Lien, K.M. A Review of the Role of Critical Parameters in the Design and Operation of Biogas Production Plants. *Appl. Sci.* **2019**, *9*, 1915. [CrossRef]
11. Daniel-Gromke, J.; Rensberg, N.; Denysenko, V.; Stinner, W.; Schmalfuß, T.; Scheftelowitz, M.; Nelles, M.; Liebetrau, J. Current Developments in Production and Utilization of Biogas and Biomethane in Germany. *Chem. Ing. Tech.* **2017**, *90*, 17–35. [CrossRef]
12. Gemmeke, B.; Rieger, C.; Weiland, P.; Schröder, J. *Biogas-Messprogramm II, 61 Biogasanlagen im Vergleich*; Fachagentur Nachwachsende Rohstoffe E.V. (FNR): Gülzow, Germany, 2009.
13. Stolze, Y.; Bremges, A.; Maus, I.; Pühler, A.; Sczyrba, A.; Schlüter, A. Targeted in situ metatranscriptomics for selected taxa from mesophilic and thermophilic biogas plants. *Microb. Biotechnol.* **2017**, *11*, 667–679. [CrossRef]
14. Annibaldi, V.; Cucchiella, F.; Gastaldi, M.; Rotilio, M.; Stornelli, V. Sustainability of Biogas Based Projects: Technical and Economic Analysis. In Proceedings of the E3S Web of Conferences, Kitahiroshima, Japan, 27–29 August 2018; EDP Sciences: Les Ulis, France, 2019; Volume 93, p. 03001.
15. Heerenklage, J.; Rechtenbach, D.; Atamaniuk, I.; Alassali, A.; Raga, R.; Koch, K.; Kuchta, K. Development of a method to produce standardised and storable inocula for biomethane potential tests—Preliminary steps. *Renew. Energy* **2019**, *143*, 753–761. [CrossRef]

16. Al-Addous, M.; Saidan, M.N.; Bdour, M.; Alnaief, M. Evaluation of Biogas Production from the Co-Digestion of Municipal Food Waste and Wastewater Sludge at Refugee Camps Using an Automated Methane Potential Test System. *Energies* **2018**, *12*, 32. [CrossRef]
17. Pavi, S.; Kramer, L.E.; Gomes, L.P.; Miranda, L.A.S. Biogas production from co-digestion of organic fraction of municipal solid waste and fruit and vegetable waste. *Bioresour. Technol.* **2017**, *228*, 362–367. [CrossRef] [PubMed]
18. Campuzano, R.; González-Martínez, S. Characteristics of the organic fraction of municipal solid waste and methane production: A review. *Waste Manag.* **2016**, *54*, 3–12. [CrossRef]
19. Li, W.; Loh, K.-C.; Zhang, J.; Tong, Y.W.; Dai, Y. Two-stage anaerobic digestion of food waste and horticultural waste in high-solid system. *Appl. Energy* **2018**, *209*, 400–408. [CrossRef]
20. Chandra, R.; Takeuchi, H.; Hasegawa, T. Methane production from lignocellulosic agricultural crop wastes: A review in context to second generation of biofuel production. *Renew. Sustain. Energy Rev.* **2012**, *16*, 1462–1476. [CrossRef]
21. Yang, Y.-Q.; Shen, D.-S.; Li, N.; Xu, N.; Long, Y.; Lu, X.-Y. Co-digestion of kitchen waste and fruit–vegetable waste by two-phase anaerobic digestion. *Environ. Sci. Pollut. Res.* **2013**, *20*, 2162–2171. [CrossRef]
22. Scarlat, N.; Dallemand, J.-F.; Fahl, F. Biogas: Developments and perspectives in Europe. *Renew. Energy* **2018**, *129*, 457–472. [CrossRef]
23. Geerolf, L. The Biogas Sector Development: Current and Future Trends in Western and Northern Europe. In *Master of Science*; KTH School of Industrial Engineering and Management: Stockholm, Sweden, 2018.
24. Tixier, N.; Guibaud, G.; Baudu, M. Determination of some rheological parameters for the characterization of activated sludge. *Bioresour. Technol.* **2003**, *90*, 215–220. [CrossRef]
25. Björn, A.; de la Monja, P.S.; Karlsson, A.; Ejlertsson, J.; Svensson, B.H. Rheological characterization. In *Biogas*; IntechOpen: London, UK, 2012.
26. Weiland, P. Biogas production: Current state and perspectives. *Appl. Microbiol. Biotechnol.* **2009**, *85*, 849–860. [CrossRef]
27. Ganidi, N.; Tyrrel, S.; Cartmell, E. Anaerobic digestion foaming causes—A review. *Bioresour. Technol.* **2009**, *100*, 5546–5554. [CrossRef]
28. Moeller, L.; Goersch, K.; Neuhaus, J.; Zehnsdorf, A.; Mueller, R.A. Comparative review of foam formation in biogas plants and ruminant bloat. *Energy Sustain. Soc.* **2012**, *2*, 12. [CrossRef]
29. Li, Y.; Park, S.Y.; Zhu, J. Solid-state anaerobic digestion for methane production from organic waste. *Renew. Sustain. Energy Rev.* **2011**, *15*, 821–826. [CrossRef]
30. Wang, X.; Yang, G.; Feng, Y.; Ren, G.; Han, X. Optimizing feeding composition and carbon–nitrogen ratios for improved methane yield during anaerobic co-digestion of dairy, chicken manure and wheat straw. *Bioresour. Technol.* **2012**, *120*, 78–83. [CrossRef] [PubMed]
31. Asam, Z.-U.-Z.; Poulsen, T.G.; Nizami, A.-S.; Rafique, R.; Kiely, G.; Murphy, J.D. How can we improve biomethane production per unit of feedstock in biogas plants? *Appl. Energy* **2011**, *88*, 2013–2018. [CrossRef]
32. Vasmara, C.; Cianchetta, S.; Marchetti, R.; Galletti, S. Biogas production from wheat straw pre-treated with ligninolytic fungi and co-digestion with pig slurry. *Environ. Eng. Manag. J.* **2015**, *14*, 1751–1760. [CrossRef]
33. Nsair, A.; Cinar, S.Ö.; Abu-Qdais, H.; Kuchta, K. Optimizing the performance of a large-scale biogas plant by controlling stirring process: A case study. *Energy Convers. Manag.* **2019**, *198*, 111931. [CrossRef]
34. Nsair, A. Improving the performance of biogas systems. In *Case Study: Applying Enhanced Stirring Strategies*, 51st ed.; Abfall Aktuell: Hamburg, Germany, 2020.
35. Al-Addous, M.; Alnaief, M.; Class, C.; Nsair, A.; Kuchta, K.; Alkasrawi, M. Technical possibilities of biogas production from Olive and Date Waste in Jordan. *BioResources* **2017**, *12*, 9383–9395.
36. Kaparaju, P.; Serrano, M.; Thomsen, A.B.; Kongjan, P.; Angelidaki, I. Bioethanol, biohydrogen and biogas production from wheat straw in a biorefinery concept. *Bioresour. Technol.* **2009**, *100*, 2562–2568. [CrossRef]
37. Risberg, K.; Sun, L.; Levén, L.; Horn, S.J.; Schnürer, A. Biogas production from wheat straw and manure—Impact of pretreatment and process operating parameters. *Bioresour. Technol.* **2013**, *149*, 232–237. [CrossRef]

38. Li, Y.; Zhang, R.; Chen, C.; Liu, G.; He, Y.; Liu, X. Biogas production from co-digestion of corn stover and chicken manure under anaerobic wet, hemi-solid, and solid-state conditions. *Bioresour. Technol.* **2013**, *149*, 406–412. [CrossRef]
39. Fantozzi, F.; Buratti, C. Biogas production from different substrates in an experimental Continuously Stirred Tank Reactor anaerobic digester. *Bioresour. Technol.* **2009**, *100*, 5783–5789. [CrossRef] [PubMed]
40. Hills, D.J. Effects of carbon: Nitrogen ratio on anaerobic digestion of dairy manure. *Agric. Wastes* **1979**, *1*, 267–278. [CrossRef]
41. Yuan, Y.; Bian, A.; Zhang, L.; Chen, T.; Pan, M.; He, L.; Wang, A.; Ding, C. A combined process for efficient biomethane production from corn straw and cattle manure: Optimizing C/N Ratio of mixed hydrolysates. *BioResources* **2019**, *14*, 1347–1363.
42. Zhang, Z.; Zhang, G.; Li, W.; Li, C.; Xu, G. Enhanced biogas production from sorghum stem by co-digestion with cow manure. *Int. J. Hydrogen Energy* **2016**, *41*, 9153–9158. [CrossRef]
43. Zahan, Z.; Othman, M.Z.; Muster, T. Anaerobic digestion/co-digestion kinetic potentials of different agro-industrial wastes: A comparative batch study for C/N optimisation. *Waste Manag.* **2018**, *71*, 663–674. [CrossRef] [PubMed]
44. Yan, Z.; Song, Z.; Li, D.; Yuan, Y.; Liu, X.; Zheng, T. The effects of initial substrate concentration, C/N ratio, and temperature on solid-state anaerobic digestion from composting rice straw. *Bioresour. Technol.* **2015**, *177*, 266–273. [CrossRef]
45. Riya, S.; Suzuki, K.; Terada, A.; Hosomi, M.; Zhou, S. Influence of C/N Ratio on Performance and Microbial Community Structure of Dry-Thermophilic Anaerobic Co-Digestion of Swine Manure and Rice Straw. *J. Med. Bioeng.* **2016**, *5*, 11–14. [CrossRef]
46. Fernandez-Bayo, J.D.; Yazdani, R.; Simmons, C.W.; Vander-Gheynst, J.S. Comparison of thermophilic anaerobic and aerobic treatment processes for stabilization of green and food wastes and production of soil amendments. *Waste Manag.* **2018**, *77*, 555–564. [CrossRef]
47. Dębowski, M.; Kisielewska, M.; Kazimierowicz, J.; Rudnicka, A.; Dudek, M.; Romanowska-Duda, Z.; Zieliński, M. The effects of Microalgae Biomass Co-Substrate on Biogas Production from the Common Agricultural Biogas Plants Feedstock. *Energies* **2020**, *13*, 2186. [CrossRef]
48. Guarino, G.; Carotenuto, C.; di Cristofaro, F.; Papa, S.; Morrone, B.; Minale, M. Does the C/N ratio really affect the Bio-methane Yield? A three years investigation of Buffalo Manure Digestion. *Chem. Eng. Trans.* **2016**, *49*, 463–468.
49. Abu-Qdais, H.; Bani-Hani, K.A.; Shatnawi, N. Modeling and optimization of biogas production from a waste digester using artificial neural network and genetic algorithm. *Resour. Conserv. Recycl.* **2010**, *54*, 359–363. [CrossRef]
50. Ganesh, R.; Torrijos, M.; Sousbie, P.; Lugardon, A.; Steyer, J.P.; Delgenes, J.P. Single-phase and two-phase anaerobic digestion of fruit and vegetable waste: Comparison of start-up, reactor stability and process performance. *Waste Manag.* **2014**, *34*, 875–885. [CrossRef] [PubMed]
51. Ward, A.J.; Hobbs, P.J.; Holliman, P.J.; Jones, D.L. Optimisation of the anaerobic digestion of agricultural resources. *Bioresour. Technol.* **2008**, *99*, 7928–7940. [CrossRef] [PubMed]
52. Demirel, B.; Yenigun, O. Two-phase anaerobic digestion processes: A review. *J. Chem. Technol. Biotechnol.* **2002**, *77*, 743–755. [CrossRef]
53. Mata-Alvarez, J. *Biomethanization of the Organic Fraction of Municipal Solid Wastes*; IWA publishing: London, UK, 2002.
54. Voelklein, M.; Jacob, A.; Shea, R.O.; Murphy, J.D. Assessment of increasing loading rate on two-stage digestion of food waste. *Bioresour. Technol.* **2016**, *202*, 172–180. [CrossRef]
55. Wu, L.-J.; Kobayashi, T.; Li, Y.-Y.; Xu, K.-Q. Comparison of single-stage and temperature-phased two-stage anaerobic digestion of oily food waste. *Energy Convers. Manag.* **2015**, *106*, 1174–1182. [CrossRef]
56. Xu, F.; Li, Y.; Ge, X.; Yang, L.; Li, Y. Anaerobic digestion of food waste—Challenges and opportunities. *Bioresour. Technol.* **2018**, *247*, 1047–1058. [CrossRef]
57. Bouallagui, H.; Touhami, Y.; Ben-Cheikh, R.; Hamdi, M. Bioreactor performance in anaerobic digestion of fruit and vegetable wastes. *Process Biochem.* **2005**, *40*, 989–995. [CrossRef]

58. Liu, D.; Liu, D.; Zeng, R.J.; Angelidaki, I. Hydrogen and methane production from household solid waste in the two-stage fermentation process. *Water Res.* **2006**, *40*, 2230–2236. [CrossRef]
59. Nielsen, H.B.; Mladenovska, Z.; Westermann, P.; Ahring, B. Comparison of two-stage thermophilic (68 °C/55 °C) anaerobic digestion with one-stage thermophilic (55 °C) digestion of cattle manure. *Biotechnol. Bioeng.* **2004**, *86*, 291–300. [CrossRef]
60. Zhang, J.; Loh, K.-C.; Li, W.; Lim, J.W.; Dai, Y.; Tong, Y.W. Three-stage anaerobic digester for food waste. *Appl. Energy* **2017**, *194*, 287–295. [CrossRef]
61. de Gioannis, G.; Muntoni, A.; Polettini, A.; Pomi, R.; Spiga, D. Energy recovery from one- and two-stage anaerobic digestion of food waste. *Waste Manag.* **2017**, *68*, 595–602. [CrossRef] [PubMed]
62. Drosg, B. *Process Monitoring in Biogas Plants*; IEA Bioenergy Paris: Paris, France, 2013.
63. Adekunle, K.F.; Okolie, J.A. A Review of Biochemical Process of Anaerobic Digestion. *Adv. Biosci. Biotechnol.* **2015**, *6*, 205–212. [CrossRef]
64. Jabłoński, S.; Rodowicz, P.; Łukaszewicz, M.; Ski, S.J.J.O.; Lukaszewicz, M. Methanogenic archaea database containing physiological and biochemical characteristics. *Int. J. Syst. Evol. Microbiol.* **2015**, *65*, 1360–1368. [CrossRef] [PubMed]
65. Mondal, C.; Biswas, G.K. Effect of Temperature on Kinetic Constants in Anaerobic Bio-digestion. *Chitkara Chem. Rev.* **2013**, *1*, 19–28. [CrossRef]
66. Hans, B. *Enzyme Kinetics Principles and Methods*; Wiley Vch Valag: Weinheim, Germany, 2008.
67. Caballero-Arzápalo, N. *Untersuchungen zum Anaeroben Abbauprozess Ausgewählter Abfallsubstrate mit Hilfe Spezieller Mikroorganismen und Enzyme*; Technische Universität München: Munich, Germany, 2015.
68. Streitwieser, D.A. Comparison of the anaerobic digestion at the mesophilic and thermophilic temperature regime of organic wastes from the agribusiness. *Bioresour. Technol.* **2017**, *241*, 985–992. [CrossRef]
69. Pandey, P.K.; Soupir, M.L. Impacts of Temperatures on Biogas Production in Dairy Manure Anaerobic Digestion. *Int. J. Eng. Technol.* **2012**, *4*, 629–631. [CrossRef]
70. Zhang, J.-S.; Sun, K.-W.; Wu, M.-C.; Zhang, L. Influence of temperature on performance of anaerobic digestion of municipal solid waste. *J. Environ. Sci.* **2006**, *18*, 810–815.
71. Hamzah, M.A.F.; Jahim, J.M.; Abdul, P.M.; Asis, A.J. Investigation of Temperature Effect on Start-Up Operation from Anaerobic Digestion of Acidified Palm Oil Mill Effluent. *Energies* **2019**, *12*, 2473. [CrossRef]
72. Rohstoffe, F.N. *Leitfaden Biogas: Von der Gewinnung zur Nutzung*; Fachagentur Nachwachsende Rohstoffe E.V. (FNR): Gülzow-Prüzen, Germany, 2016; pp. 156–157.
73. Wu, M.-C.; Sun, K.-W.; Zhang, Y. Influence of temperature fluctuation on thermophilic anaerobic digestion of municipal organic solid waste. *J. Zhejiang Univ. Sci. B* **2006**, *7*, 180–185. [CrossRef]
74. el Mashad, H.M. Effect of temperature and temperature fluctuation on thermophilic anaerobic digestion of cattle manure. *Bioresour. Technol.* **2004**, *95*, 191–201. [CrossRef] [PubMed]
75. Ahring, B.K.; Sandberg, M.; Angelidaki, I. Volatile fatty acids as indicators of process imbalance in anaerobic digestors. *Appl. Microbiol. Biotechnol.* **1995**, *43*, 559–565. [CrossRef]
76. Ahring, B.K.; Ibrahim, A.A.; Mladenovska, Z. Effect of temperature increase from 55 to 65 °C on performance and microbial population dynamics of an anaerobic reactor treating cattle manure. *Water Res.* **2001**, *35*, 2446–2452. [CrossRef]
77. Pap, B.; Györkei, A.; Boboescu, I.Z.; Nagy, I.K.; Bíró, T.; Kondorosi, E.; Maróti, G. Temperature-dependent transformation of biogas-producing microbial communities' points to the increased importance of hydrogenotrophic methanogenesis under thermophilic operation. *Bioresour. Technol.* **2015**, *177*, 375–380. [CrossRef] [PubMed]
78. Chae, K.-J.; Jang, A.; Yim, S.; Kim, I.S. The effects of digestion temperature and temperature shock on the biogas yields from the mesophilic anaerobic digestion of swine manure. *Bioresour. Technol.* **2008**, *99*, 1–6. [CrossRef]
79. Kim, M.-S.; Kim, D.-H.; Yun, Y.-M. Effect of operation temperature on anaerobic digestion of food waste: Performance and microbial analysis. *Fuel* **2017**, *209*, 598–605. [CrossRef]

80. Boušková, A.; Dohányos, M.; Schmidt, J.E.; Angelidaki, I. Strategies for changing temperature from mesophilic to thermophilic conditions in anaerobic CSTR reactors treating sewage sludge. *Water Res.* **2005**, *39*, 1481–1488. [CrossRef]
81. Chachkhiani, M.; Dabert, P.; Abzianidze, T.; Partskhaladze, G.; Tsiklauri, L.; Dudauri, T.; Godon, J.-J. 16S rDNA characterisation of bacterial and archaeal communities during start-up of anaerobic thermophilic digestion of cattle manure. *Bioresour. Technol.* **2004**, *93*, 227–232. [CrossRef]
82. Gerber, M. *Ganzheitliche Stoffliche und Energetische Modellierung des Biogasbildungsprozesses*; Ruhr-Universität Bochum: Bochum, Germany, 2010.
83. Wang, B. *Factors that Influence the Biochemical Methane Potential (BMP) Test*; Lund University: Lund, Sweden, 2016.
84. Al-Seadi, T.; Rutz, D.; Prassl, H.; Köttner, M.; Finsterwalder, T.; Volk, S.; Janssen, R. *Biogas Handbook*; ICRISAT: Esbjerg, Denmark, 2008.
85. Besgen, S. *Energie-und Stoffumsetzung in Biogasanlagen-Ergebnisse Messtechnischer Untersuchungen an Landwirtschaftlichen Biogasanlagen im Rheinland*; Universitäts-und Landesbibliothek Bonn: Bonn, Germany, 2005.
86. Gerardi, M.H. *The Microbiology of Anaerobic Digesters*; Wiley: Hoboken, NJ, USA, 2003.
87. Nsair, A.; Bade, O.; Kuchta, K. Development of Velocity Sensor to Optimize the Energy Yield in a Biogas Plant. *Environ. Sci. Technol.* **2018**, 51–56.
88. Karim, K.; Hoffmann, R.; Klasson, T.; Al-Dahhan, M.; Klasson, K. Anaerobic digestion of animal waste: Waste strength versus impact of mixing. *Bioresour. Technol.* **2005**, *96*, 1771–1781. [CrossRef]
89. Karim, K.; Klasson, K.; Hoffmann, R.; Drescher, S.; de Paoli, D.; Al-Dahhan, M. Anaerobic digestion of animal waste: Effect of mixing. *Bioresour. Technol.* **2005**, *96*, 1607–1612. [CrossRef] [PubMed]
90. Karim, K.; Varma, R.; Vesvikar, M.; Al-Dahhan, M. Flow pattern visualization of a simulated digester. *Water Res.* **2004**, *38*, 3659–3670. [CrossRef] [PubMed]
91. Lemmer, A.; Naegele, H.-J.; Sondermann, J. How Efficient are Agitators in Biogas Digesters? Determination of the Efficiency of Submersible Motor Mixers and Incline Agitators by Measuring Nutrient Distribution in Full-Scale Agricultural Biogas Digesters. *Energies* **2013**, *6*, 6255–6273. [CrossRef]
92. Wiedemann, L.; Conti, F.; Janus, T.; Sonnleitner, M.; Zörner, W.; Goldbrunner, M. Mixing in Biogas Digesters and Development of an Artificial Substrate for Laboratory-Scale Mixing Optimization. *Chem. Eng. Technol.* **2016**, *40*, 238–247. [CrossRef]
93. Last, S. The Anaerobic Digestion Biofuels Blog. Available online: https://blog.anaerobic-digestion.com/digester-cleaning-services/ (accessed on 28 January 2019).
94. Nandi, R.; Saha, C.K.; Huda, M.S.; Alam, M.M. Effect of mixing on biogas production from cowdung. *Eco-Friendly Agril J.* **2017**, *10*, 7–13.
95. Kopplow, O. *Maßnahmen zur Minderung des Schäumens im Faulbehälter Unter Besonderer Berücksichtigung der Klärschlammdesintegration*; Inst. für Umweltingenieurwesen: Rostock, Germany, 2006.
96. Westlund, A.D.; Hagland, E.; Rothman, M. Foaming in anaerobic digesters caused by Microthrix parvicella. *Water Sci. Technol.* **1998**, *37*, 51–55. [CrossRef]
97. Barjenbruch, M.; Hoffmann, H.; Kopplow, O.; Tränckner, J. Minimizing of foaming in digesters by pre-treatment of the surplus-sludge. *Water Sci. Technol.* **2000**, *42*, 235–241. [CrossRef]
98. Mir, M.A.; Hussain, A.; Verma, C. Design considerations and operational performance of anaerobic digester: A review. *Cogent Eng.* **2016**, *3*, 795. [CrossRef]
99. Hopfner-Sixt, K.; Amon, T. Monitoring of agricultural biogas plants in Austria—Mixing technology and specific values of essential process parameters. In Proceedings of the 15th European Biomass Conference and Exhibition, Berlin, Germany, 7–11 May 2007; Springer: Berlin/Heidelberg, Germany, 2007; Volume 711, p. 17181728.
100. Thorin, E.; Nordlander, E.; Lindmark, J.; Dahlquist, E.; Yan, J.; Bel-Fdhila, R. Modeling of the Biogas Production process—A Review. In Proceedings of the International Conference on Applied Energy ICAE, Suzhou, China, 5–8 July 2012.
101. Black, C.; United States Environmental Protection Agency, Office of Technology Transfer. *Process Design Manual for Sludge Treatment and Disposal*; US Environmental Protection Agency, Technology Transfer: Washington, DC, USA, 1979.

102. Karim, K.; Thoma, G.J.; Al-Dahhan, M. Gas-lift digester configuration effects on mixing effectiveness. *Water Res.* **2007**, *41*, 3051–3060. [CrossRef]
103. Weiland, P. Biomass Digestion in Agriculture: A Successful Pathway for the Energy Production and Waste Treatment in Germany. *Eng. Life Sci.* **2006**, *6*, 302–309. [CrossRef]
104. Bártfai, Z.; Oldal, I.; Tóth, L.; Szabó, I.; Beke, J. Conditions of using propeller stirring in biogas reactors. *Hung. Agric. Eng.* **2015**, 5–10. [CrossRef]
105. Hashimoto, A.G. Effect of mixing duration and vacuum on methane production rate from beef cattle waste. *Biotechnol. Bioeng.* **1982**, *24*, 9–23. [CrossRef] [PubMed]
106. Gollakota, K.; Meher, K. Effect of particle size, temperature, loading rate and stirring on biogas production from castor cake (oil expelled). *Boil. Wastes* **1988**, *24*, 243–249. [CrossRef]
107. Chen, T.H.; Chynoweth, P.; Biljetina, R. Anaerobic digestion of municipal solid waste in a nonmixed solids concentrating digestor. *Appl. Biochem. Biotechnol.* **1990**, *24*, 533–544. [CrossRef]
108. Madamwar, D.; Patel, A.; Patel, V. Effect of temperature and retention time on methane recovery from water hyacinth-cattle dung. *J. Ferment. Bioeng.* **1990**, *70*, 340–342. [CrossRef]
109. Hamdi, M. Effects of agitation and pretreatment on the batch anaerobic digestion of olive mil. *Bioresour. Technol.* **1991**, *36*, 173–178. [CrossRef]
110. Nasr, F.A. Treatment, and reuse of sewage sludge. *Environmentalist* **1997**, *17*, 109–113. [CrossRef]
111. Rodriguez-Andara, A.; Esteban, J.L. Kinetic study of the anaerobic digestion of the solid fraction of piggery slurries. *Biomass Bioenergy* **1999**, *17*, 435–443. [CrossRef]
112. Kim, M.; Ahn, Y.-H.; Speece, R.E. Comparative process stability and efficiency of anaerobic digestion; mesophilic vs. thermophilic. *Water Res.* **2002**, *36*, 4369–4385. [CrossRef]
113. Kaparaju, P.; Buendia, I.; Ellegaard, L.; Angelidakia, I. Effects of mixing on methane production during thermophilic anaerobic digestion of manure: Lab-scale and pilot-scale studies. *Bioresour. Technol.* **2008**, *99*, 4919–4928. [CrossRef]
114. Rojas, C.; Fang, S.; Uhlenhut, F.; Borchert, A.; Stein, I.; Schlaak, M. Stirring and biomass starter influences the anaerobic digestion of different substrates for biogas production. *Eng. Life Sci.* **2010**, *10*, 339–347. [CrossRef]
115. Chen, J.; Li, X.; Liu, Y.; Zhu, B.; Yuan, H.; Pang, Y. Effect of mixing rates on anaerobic digestion performance of rice straw. *Transact. CSAE* **2011**, *27*, 144–148.
116. Ghanimeh, S.; el Fadel, M.; Saikaly, P.E. Mixing effect on thermophilic anaerobic digestion of source-sorted organic fraction of municipal solid waste. *Bioresour. Technol.* **2012**, *117*, 63–71. [CrossRef] [PubMed]
117. Keanoi, N.; Hussaro, K.; Teekasap, S. Effect of with/without agitation of agricultural waste on biogas production from anaerobic co-digestion-a small scale. *Am. J. Environ. Sci.* **2014**, *10*, 74–85. [CrossRef]
118. Lindmark, J.; Thorin, E.; Fdhila, R.B.; Dahlquist, E. Effects of mixing on the result of anaerobic digestion: Review. *Renew. Sustain. Energy Rev.* **2014**, *40*, 1030–1047. [CrossRef]
119. El-Bakhshwan, M.; El-Ghafar, S.A.; Zayed, M.; El-Shazly, A. Effect of mechanical stirring on biogas production efficiency in large scale digesters. *J. Soil Sci. Agric. Eng.* **2015**, *6*, 47–63. [CrossRef]
120. Zareei, S.; Khodaei, J. Modeling and optimization of biogas production from cow manure and maize straw using an adaptive neuro-fuzzy inference system. *Renew. Energy* **2017**, *114*, 423–427. [CrossRef]
121. Abdullah, N.O.; Pandebesie, E.S. The Influences of Stirring and Cow Manure Added on Biogas Production from Vegetable Waste Using Anaerobic Digester. *IOP Conf. Ser. Earth Environ. Sci.* **2018**, *135*, 012005. [CrossRef]
122. Aksay, M.V.; Ozkaymak, M.; Calhan, R. Co-digestion of cattle manure and tea waste for biogas production. *Int. J. Energ. Res.* **2018**, *8*, 1246–1353.
123. Babaei, A.; Shayegan, J. Effects of temperature and mixing modes on the performance of municipal solid waste anaerobic slurry digester. *J. Environ. Heal. Sci. Eng.* **2020**, *17*, 1077–1084. [CrossRef]
124. Ioelovich, M. Recent findings and the energetic potential of plant biomass as a renewable source of biofuels–a review. *Bio. Resour.* **2015**, *10*, 1879–1914.
125. Agrahari, R.P.; Tiwari, G.N. The Production of Biogas Using Kitchen Waste. *Int. J. Energy Sci.* **2013**, *3*, 408. [CrossRef]

126. Achinas, S.; Achinas, V.; Euverink, G.J.W. A Technological Overview of Biogas Production from Biowaste. *BioRixv* **2017**, *3*, 299–307. [CrossRef]
127. Jaber, J.; Probert, S.; Williams, P.T. Gaseous fuels (derived from oil shale) for heavy-duty gas turbines and combined-cycle power generators. *Appl. Energy* **1998**, *60*, 1–20. [CrossRef]
128. Frey, J.; Grüssing, F.; Nägele, H.-J.; Oechsner, H. Cutting the electric power consumption of biogas plants: The impact of new technologies. *Landtechnik. Agric. Eng.* **2013**, *68*, 58–63.
129. Botheju, D. Oxygen Effects in Anaerobic Digestion—A Review. *Open Waste Manag. J.* **2011**, *4*, 1–19. [CrossRef]
130. Jagadabhi, P.S.; Kaparaju, P.; Rintala, J. Effect of micro-aeration and leachate replacement on COD solubilization and VFA production during mono-digestion of grass-silage in one-stage leach-bed reactors. *Bioresour. Technol.* **2010**, *101*, 2818–2824. [CrossRef]
131. Jenicek, P.; Keclik, F.; Máca, J.; Bindzar, J. Use of microaerobic conditions for the improvement of anaerobic digestion of solid wastes. *Water Sci. Technol.* **2008**, *58*, 1491–1496. [CrossRef]
132. Nghiem, L.; Manassa, P.; Dawson, M.; Fitzgerald, S.K. Oxidation reduction potential as a parameter to regulate micro-oxygen injection into anaerobic digester for reducing hydrogen sulphide concentration in biogas. *Bioresour. Technol.* **2014**, *173*, 443–447. [CrossRef]
133. Tabatabaei, M.; Ghanavati, H. Biogas: Fundamentals, process, and operation. In *Prominent Parameters in Biogas Production Systems*; Springer: Berlin/Heidelberg, Germany, 2018.
134. Sibiya, N.T.; Muzenda, E.; Tesfagiorgis, H.B. Effect of temperature and pH on the anaerobic digestion of grass silage. In Proceedings of the 6th International Conference on Green Technology, Renewable Energy and Environmental Engineering, Cape Town, South Africa, 15–16 April 2014.
135. Zhang, C.; Su, H.; Baeyens, J.; Tan, T. Reviewing the anaerobic digestion of food waste for biogas production. *Renew. Sustain. Energy Rev.* **2014**, *38*, 383–392. [CrossRef]
136. Voß, E. Prozessanalyse und Optimierung von Landwirtschaftlichen Biogasanlagen. Ph.D. Thesis, Institut für Siedlungswasserwirtschaft und Abfalltechnik, Hanover, Germany, 2015.
137. Mpofu, A.B.; Welz, P.J.; Oyekola, O.O. Anaerobic Digestion of Secondary Tannery Sludge: Optimisation of Initial pH and Temperature and Evaluation of Kinetics. *Waste Biomass Valorizat.* **2019**, *11*, 873–885. [CrossRef]
138. Ren, Y.; Yu, M.; Wu, C.; Wang, Q.; Gao, M.; Huang, Q.; Liu, Y. A comprehensive review on food waste anaerobic digestion: Research updates and tendencies. *Bioresour. Technol.* **2018**, *247*, 1069–1076. [CrossRef] [PubMed]
139. Önen, S.; Nsair, A.; Kuchta, K. Innovative operational strategies for biogas plant including temperature and stirring management. *Waste Manag. Res.* **2018**, *37*, 237–246. [CrossRef] [PubMed]
140. Murto, M.; Björnsson, L.; Mattiasson, B. Impact of food industrial waste on anaerobic co-digestion of sewage sludge and pig manure. *J. Environ. Manag.* **2004**, *70*, 101–107. [CrossRef] [PubMed]
141. Wang, C.; Hong, F.; Lü, Y.; Li, X.; Liu, H. Improved biogas production and biodegradation of oilseed rape straw by using kitchen waste and duck droppings as co-substrates in two-phase anaerobic digestion. *PLoS ONE* **2017**, *12*, e0182361. [CrossRef] [PubMed]
142. Boe, K. Online Monitoring and Control of the Biogas Process. Ph.D. Thesis, Technical University of Denmark, Copenhagen, Denmark, 2006.
143. Cecchi, F.; Pavan, P.; Alvarez, J.M.; Bassetti, A.; Cozzolino, C. Anaerobic digestion of municipal solid waste: Thermophilic vs. mesophilic performance at high solids. *Waste Manag. Res.* **1991**, *9*, 305–315. [CrossRef]
144. Baudez, J.-C.; Markis, F.; Eshtiaghi, N.; Slatter, P. The rheological behaviour of anaerobic digested sludge. *Water Res.* **2011**, *45*, 5675–5680. [CrossRef]
145. Aboudi, K.; Álvarez-Gallego, C.J.; García, L.I.R. Semi-continuous anaerobic co-digestion of sugar beet byproduct and pig manure: Effect of the organic loading rate (OLR) on process performance. *Bioresour. Technol.* **2015**, *194*, 283–290. [CrossRef]
146. Dhar, H.; Kumar, P.; Kumar, S.; Mukherjee, S.; Vaidya, A.N. Effect of organic loading rate during anaerobic digestion of municipal solid waste. *Bioresour. Technol.* **2016**, *217*, 56–61. [CrossRef]
147. IRENA. Bioenergy. 2020. Available online: https://www.irena.org/bioenergy (accessed on 27 April 2020).

148. Statista GmbH. Installierte Elektrische Leistung der Biogasanlagen in Deutschland in den Jahren 1999 bis 2019. 2020. Available online: https://de.statista.com/statistik/daten/studie/167673/umfrage/installierte-elektrische-leistung-von-biogasanlagen-seit-1999/ (accessed on 27 April 2020).
149. Chiumenti, A.; da Borso, F.; Limina, S. Dry anaerobic digestion of cow manure and agricultural products in a full-scale plant: Efficiency and comparison with wet fermentation. *Waste Manag.* **2018**, *71*, 704–710. [CrossRef]
150. Li, N.; Liu, S.; Mi, L.; Li, Z.; Yuan, Y.; Yan, Z.; Liu, X. Effects of feedstock ratio and organic loading rate on the anaerobic mesophilic co-digestion of rice straw and pig manure. *Bioresour. Technol.* **2015**, *187*, 120–127. [CrossRef] [PubMed]
151. Sun, M.-T.; Fan, X.-L.; Zhao, X.-X.; Fu, S.; He, S.; Manasa, M.; Guo, R.-B. Effects of organic loading rate on biogas production from macroalgae: Performance and microbial community structure. *Bioresour. Technol.* **2017**, *235*, 292–300. [CrossRef] [PubMed]
152. Montingelli, M.; Tedesco, S.; Olabi, A.G. Biogas production from algal biomass: A review. *Renew. Sustain. Energy Rev.* **2015**, *43*, 961–972. [CrossRef]
153. González-Fernández, C.; Sialve, B.; Bernet, N.; Steyer, J.-P. Effect of organic loading rate on anaerobic digestion of thermally pretreated Scenedesmus sp. biomass. *Bioresour. Technol.* **2013**, *129*, 219–223. [CrossRef] [PubMed]
154. Zuo, Z.; Wu, S.; Zhang, W.; Dong, R. Effects of organic loading rate and effluent recirculation on the performance of two-stage anaerobic digestion of vegetable waste. *Bioresour. Technol.* **2013**, *146*, 556–561. [CrossRef]
155. Liu, X.; Wang, W.; Shi, Y.; Zheng, L.; Gao, X.; Qiao, W.; Zhou, Y. Pilot-scale anaerobic co-digestion of municipal biomass waste and waste activated sludge in China: Effect of organic loading rate. *Waste Manag.* **2012**, *32*, 2056–2060. [CrossRef]
156. Mähnert, P.; Linke, B. Kinetic study of biogas production from energy crops and animal waste slurry: Effect of organic loading rate and reactor size. *Environ. Technol.* **2009**, *30*, 93–99. [CrossRef]
157. Luste, S.; Luostarinen, S. Anaerobic co-digestion of meat-processing by-products and sewage sludge—Effect of hygienization and organic loading rate. *Bioresour. Technol.* **2010**, *101*, 2657–2664. [CrossRef]
158. Zhou, J.; Yang, J.; Yu, Q.; Yong, X.; Xie, X.; Zhang, L.; Wei, P.; Jia, H. Different organic loading rates on the biogas production during the anaerobic digestion of rice straw: A pilot study. *Bioresour. Technol.* **2017**, *244*, 865–871. [CrossRef]
159. Nagao, N.; Tajima, N.; Kawai, M.; Niwa, C.; Kurosawa, N.; Matsuyama, T.; Yusoff, F.M.; Toda, T. Maximum organic loading rate for the single-stage wet anaerobic digestion of food waste. *Bioresour. Technol.* **2012**, *118*, 210–218. [CrossRef]
160. Song, H.; Zhang, Y.; Kusch-Brandt, S.; Banks, C. Comparison of Variable and Constant Loading for Mesophilic Food Waste Digestion in a Long-Term Experiment. *Energies* **2020**, *13*, 1279. [CrossRef]
161. Ezekoye, V.A.; Ezekoye, B.A.; Offor, P.O. Effect of retention time on biogas production from poultry droppings and cassava peels. *Niger. J. Biotechnol.* **2011**, *22*, 53–59.
162. Li, C.; Champagne, P.; Anderson, B.C. Biogas production performance of mesophilic and thermophilic anaerobic co-digestion with fat, oil, and grease in semi-continuous flow digesters: Effects of temperature, hydraulic retention time, and organic loading rate. *Environ. Technol.* **2013**, *34*, 2125–2133. [CrossRef] [PubMed]
163. Kaosol, T.; Sohgrathok, N. Influence of Hydraulic Retention Time on Biogas Production from Frozen Seafood Wastewater Using Decanter Cake as Anaerobic Co-digestion Material. *Int. J. Environ. Eng.* **2012**, *20*.
164. Dareioti, M.A.; Kornaros, M. Anaerobic mesophilic co-digestion of ensiled sorghum, cheese whey and liquid cow manure in a two-stage CSTR system: Effect of hydraulic retention time. *Bioresour. Technol.* **2015**, *175*, 553–562. [CrossRef] [PubMed]
165. Dareioti, M.A.; Kornaros, M. Effect of hydraulic retention time (HRT) on the anaerobic co-digestion of agro-industrial wastes in a two-stage CSTR system. *Bioresour. Technol.* **2014**, *167*, 407–415. [CrossRef]
166. Schmidt, T.; Ziganshin, A.M.; Nikolausz, M.; Scholwin, F.; Nelles, M.; Kleinsteuber, S.; Pröter, J. Effects of the reduction of the hydraulic retention time to 1.5 days at constant organic loading in CSTR, ASBR, and fixed-bed reactors—Performance and methanogenic community composition. *Biomass Bioenergy* **2014**, *69*, 241–248. [CrossRef]

167. Shi, X.-S.; Dong, J.-J.; Yu, J.-H.; Yin, H.; Hu, S.-M.; Huang, S.-X.; Yuan, X.-Z. Effect of Hydraulic Retention Time on Anaerobic Digestion of Wheat Straw in the Semicontinuous Continuous Stirred-Tank Reactors. *BioMed Res. Int.* **2017**, 1–6. [CrossRef]
168. Krakat, N.; Schmidt, S.; Scherer, P. Mesophilic Fermentation of Renewable Biomass: Does Hydraulic Retention Time Regulate Methanogen Diversity? *Appl. Environ. Microbiol.* **2010**, *76*, 6322–6326. [CrossRef]
169. Vintiloiu, A.; Lemmer, A.; Oechsner, H.; Jungbluth, T. Mineral substances and macronutrients in the anaerobic conversion of biomass: An impact evaluation. *Eng. Life Sci.* **2012**, *12*, 287–294. [CrossRef]
170. Sibiya, N.T.; Tesfagiorgis, H.B.; Muzenda, E. Influence of nutrients addition for enhanced biogas production from energy crops: A review. *Magnesium* **2015**, *1*, 1–5.
171. Demirel, B.; Scherer, P. Trace element requirements of agricultural biogas digesters during biological conversion of renewable biomass to methane. *Biomass Bioenergy* **2011**, *35*, 992–998. [CrossRef]
172. Bougrier, C.; Dognin, D.; Laroche, C.; Gonzalez, V.; Benali-Raclot, D.; Rivero, J.A.C. Anaerobic digestion of Brewery Spent Grains: Trace elements addition requirement. *Bioresour. Technol.* **2018**, *247*, 1193–1196. [CrossRef] [PubMed]
173. Chen, Y.; Cheng, J.; Creamer, K.S. Inhibition of anaerobic digestion process: A review. *Bioresour. Technol.* **2008**, *99*, 4044–4064. [CrossRef] [PubMed]
174. Kayhanian, M. Ammonia Inhibition in High-Solids Biogasification: An Overview and Practical Solutions. *Environ. Technol.* **1999**, *20*, 355–365. [CrossRef]
175. Yenigun, O.; Demirel, B. Ammonia inhibition in anaerobic digestion: A review. *Process. Biochem.* **2013**, *48*, 901–911. [CrossRef]
176. Chen, J.L.; Ortiz, R.; Steele, T.W.; Stuckey, D.C. Toxicants inhibiting anaerobic digestion: A review. *Biotechnol. Adv.* **2014**, *32*, 1523–1534. [CrossRef]
177. McCartney, D.; Oleszkiewicz, J. Sulfide inhibition of anaerobic degradation of lactate and acetate. *Water Res.* **1991**, *25*, 203–209. [CrossRef]
178. Fagbohungbe, M.O.; Herbert, B.M.; Hurst, L.; Ibeto, C.N.; Li, H.; Usmani, S.Q.; Semple, K.T. The challenges of anaerobic digestion and the role of biochar in optimizing anaerobic digestion. *Waste Manag.* **2017**, *61*, 236–249. [CrossRef]
179. Thiele, J.H.; Wu, W.-M.; Jain, M.K.; Zeikus, J.G. Ecoengineering high rate anaerobic digestion systems: Analysis of improved syntrophic biomethanation catalysts. *Biotechnol. Bioeng.* **1990**, *35*, 990–999. [CrossRef]
180. van Langerak, E.; Gonzalez-Gil, G.; van Aelst, A.; van Lier, J.; Hamelers, H.; Lettinga, G. Effects of high calcium concentrations on the development of methanogenic sludge in upflow anaerobic sludge bed (UASB) reactors. *Water Res.* **1998**, *32*, 1255–1263. [CrossRef]
181. Dimroth, P.; Thomer, A. A primary respiratory Na^+ pump of an anaerobic bacterium: The Na^+-dependent NADH: Quinone oxidoreductase of Klebsiella pneumoniae. *Arch. Microbiol.* **1989**, *151*, 439–444. [CrossRef] [PubMed]
182. Cabirol, N.; Barragán, E.; Durán, A.; Noyola, A. Effect of aluminium and sulphate on anaerobic digestion of sludge from wastewater enhanced primary treatment. *Water Sci. Technol.* **2003**, *48*, 235–240. [CrossRef]
183. Urriza-Arsuaga, I.; Bedoya, M.; Orellana, G. Tailored luminescent sensing of NH3 in biomethane productions. *Sens. Actuators B Chem.* **2019**, *292*, 210–216. [CrossRef]
184. Romero-Güiza, M.; Vila, J.; Mata-Alvarez, J.; Simon, F.-G.; Astals, S. The role of additives on anaerobic digestion: A review. *Renew. Sustain. Energy Rev.* **2016**, *58*, 1486–1499. [CrossRef]
185. Romero-Güiza, M.; Astals, S.; Mata-Alvarez, J.; Simon, F.-G. Feasibility of coupling anaerobic digestion and struvite precipitation in the same reactor: Evaluation of different magnesium sources. *Chem. Eng. J.* **2015**, *270*, 542–548. [CrossRef]
186. Schattauer, A.; Abdoun, E.; Weiland, P.; Plöchl, M.; Heiermann, M. Abundance of trace elements in demonstration biogas plants. *Biosyst. Eng.* **2011**, *108*, 57–65. [CrossRef]
187. Lo, H.; Chiang, C.; Tsao, H.; Pai, T.; Liu, M.; Kurniawan, T.; Chao, K.; Liou, C.; Lin, K.; Chang, C.; et al. Effects of spiked metals on the MSW anaerobic digestion. *Waste Manag. Res.* **2012**, *30*, 32–48. [CrossRef]

188. Lo, H.-M.; Chiu, H.; Lo, S.; Lo, F. Effects of different SRT on anaerobic digestion of MSW dosed with various MSWI ashes. *Bioresour. Technol.* **2012**, *125*, 233–238. [CrossRef]
189. Guo, Q.; Majeed, S.; Xu, R.; Zhang, K.; Kakade, A.; Khan, A.; Hafeez, F.Y.; Mao, C.; Liu, P.; Li, X. Heavy metals interact with the microbial community and affect biogas production in anaerobic digestion: A review. *J. Environ. Manag.* **2019**, *240*, 266–272. [CrossRef]
190. Mueller, R.F.; Steiner, A. Inhibition of Anaerobic Digestion Caused by Heavy Metals. *Water Sci. Technol.* **1992**, *26*, 835–846. [CrossRef]
191. Altaş, L. Inhibitory effect of heavy metals on methane-producing anaerobic granular sludge. *J. Hazard. Mater.* **2009**, *162*, 1551–1556. [CrossRef] [PubMed]
192. Abdel-Shafy, H.I.; Mansour, M. Biogas production as affected by heavy metals in the anaerobic digestion of sludge. *Egypt. J. Pet.* **2014**, *23*, 409–417. [CrossRef]
193. Li, C.; Fang, H.H. Inhibition of heavy metals on fermentative hydrogen production by granular sludge. *Chemosphere* **2007**, *67*, 668–673. [CrossRef] [PubMed]
194. Gagliano, M.C.; Sudmalis, D.; Temmink, H.; Plugge, C.M. Calcium effect on microbial activity and biomass aggregation during anaerobic digestion at high salinity. *New Biotechnol.* **2020**, *56*, 114–122. [CrossRef]
195. Yuan, Z.; Yang, H.; Zhi, X.; Shena, J. Increased performance of continuous stirred tank reactor with calcium supplementation. *Int. J. Hydrogen Energy* **2010**, *35*, 2622–2626. [CrossRef]
196. Tan, L.; Qu, Y.; Zhou, J.; Ma, F.; Li, A. Dynamics of microbial community for X-3B wastewater decolorization coping with high-salt and metal ions conditions. *Bioresour. Technol.* **2009**, *100*, 3003–3009. [CrossRef]
197. Lin, C. Heavy metal effects on fermentative hydrogen production using natural mixed microflora. *Int. J. Hydrogen Energy* **2008**, *33*, 587–593. [CrossRef]
198. Fermoso, F.G.; Bartacek, J.; Jansen, S.; Lens, P.N. Metal supplementation to UASB bioreactors: From cell-metal interactions to full-scale application. *Sci. Total. Environ.* **2009**, *407*, 3652–3667. [CrossRef]
199. Gikas, P. Kinetic responses of activated sludge to individual and joint nickel (Ni (II)) and cobalt (Co (II)): An isobolographic approach. *J. Hazard. Mater.* **2007**, *143*, 246–256. [CrossRef]
200. Kida, K.; Shigematsu, T.; Kijima, J.; Numaguchi, M.; Mochinaga, Y.; Abe, N.; Morimura, S. Influence of Ni2+ and Co2+ on Methanogenic Activity and the Amounts of Coenzymes Involved in Methanogenesis. *J. Biosci. Bioeng.* **2001**, *91*, 590–595. [CrossRef]
201. Ma, J.; Mungoni, L.J.; Verstraete, W.; Carballa, M. Maximum removal rate of propionic acid as a sole carbon source in UASB reactors and the importance of the macro- and micro-nutrients stimulation. *Bioresour. Technol.* **2009**, *100*, 3477–3482. [CrossRef] [PubMed]
202. Worm, P.; Fermoso, F.G.; Lens, P.N.; Plugge, C.M. Decreased activity of a propionate degrading community in a UASB reactor fed with synthetic medium without molybdenum, tungsten, and selenium. *Enzym. Microb. Technol.* **2009**, *45*, 139–145. [CrossRef]
203. Chan, P.C.; Lu, Q.; Toledo, R.A.; Gu, J.-D.; Shim, H. Improved anaerobic co-digestion of food waste and domestic wastewater by copper supplementation—Microbial community change and enhanced effluent quality. *Sci. Total Environ.* **2019**, *670*, 337–344. [CrossRef] [PubMed]
204. Yuan, Q.; Sparling, R.; Oleszkiewicz, J.A. VFA generation from waste activated sludge: Effect of temperature and mixing. *Chemosphere* **2011**, *82*, 603–607. [CrossRef] [PubMed]
205. Cai, Y.; Zheng, Z.; Zhao, Y.; Zhang, Y.; Guo, S.; Cui, Z.; Wang, X. Effects of molybdenum, selenium and manganese supplementation on the performance of anaerobic digestion and the characteristics of bacterial community in acidogenic stage. *Bioresour. Technol.* **2018**, *266*, 166–175. [CrossRef]
206. Fang, C.; Boe, K.; Angelidaki, I. Anaerobic co-digestion of desugared molasses with cow manure; focusing on sodium and potassium inhibition. *Bioresour. Technol.* **2011**, *102*, 1005–1011. [CrossRef]
207. Feijoo, G.; Soto, M.; Mendez, R.; Lema, J.M. Sodium inhibition in the anaerobic digestion process: Antagonism and adaptation phenomena. *Enzym. Microb. Technol.* **1995**, *17*, 180–188. [CrossRef]
208. Stamatelatou, K.; Antonopoulou, G.; Lyberatos, G. *Production of Biogas via Anaerobic Digestion*; Elsevier BV: Amsterdam, The Netherlands, 2011; pp. 266–304.

209. Yang, J.; Speece, R. The effects of chloroform toxicity on methane fermentation. *Water Res.* **1986**, *20*, 1273–1279. [CrossRef]
210. Renard, P.; Bouillon, C.; Naveau, H.; Nyns, E.-J. Toxicity of a mixture of polychlorinated organic compounds towards an unacclimated methanogenic consortium. *Biotechnol. Lett.* **1993**, *15*, 195–200. [CrossRef]
211. van Beelen, P.; van Vlaardingen, P. Toxic effects of pollutants on the mineralization of 4-chlorophenol and benzoate in methanogenic river sediment. *Environ. Toxicol. Chem.* **1994**, *13*, 1051–1060. [CrossRef]
212. Sierra-Alvarez, R.; Lettinga, G. The effect of aromatic structure on the inhibition of acetoclastic methanogenesis in granular sludge. *Appl. Microbiol. Biotechnol.* **1991**, *34*, 544–550. [CrossRef]
213. Soto, M.; Mendez, R.; Lema, J. Biodegradability and toxicity in the anaerobic treatment of fish canning wastewaters. *Environ. Technol.* **1991**, *12*, 669–677. [CrossRef]
214. Fang, H.H.P.; Chen, T.; Chan, O.C. Toxic effects of phenolic pollutants on anaerobic benzoate-degrading granules. *Biotechnol. Lett.* **1995**, *17*, 117–120. [CrossRef]
215. Shin, H.-S.; Kwon, J.-C. Degredation and interaction between organic concentrations and toxicity of 2,4,6-trichlorophenol in anaerobic system. *Biotechnol. Tech.* **1998**, *12*, 39–43. [CrossRef]
216. Uberoi, V.; Bhattacharya, S.K. Toxicity, and degradability of nitrophenols in anaerobic systems. *Water Environ. Res.* **1997**, *69*, 146–156. [CrossRef]
217. McCue, T.; Hoxworth, S.; Randall, A.A. Degradation of halogenated aliphatic compounds utilizing sequential anaerobic/aerobic treatments. *Water Sci. Technol.* **2003**, *47*, 79–84. [CrossRef]
218. Mormile, M.R.; Suflita, J.M. The Toxicity of Selected Gasoline Components to Glucose Methanogenesis by Aquifer Microorganisms. *Anaerobe* **1996**, *2*, 299–303. [CrossRef]
219. Stuckey, D.C.; Owen, W.F.; McCarty, P.L.; Parkin, G.F. Anaerobic toxicity evaluation by batch and semi-continuous assays. *J. Water Pollut. Control Fed.* **1980**, 720–729.
220. Demirer, G.; Speece, R. Anaerobic biotransformation of four3-carbon compounds (acrolein, acrylic acid, allyl alcohol and n-propanol) in UASB reactors. *Water Res.* **1998**, *32*, 747–759. [CrossRef]
221. Gonzalez-Gil, G.; Kleerebezem, R.; Lettinga, G. Conversion and toxicity characteristics of formaldehyde in acetoclastic methanogenic sludge. *Biotechnol. Bioeng.* **2002**, *79*, 314–322. [CrossRef]
222. Playne, M.J.; Smith, B.R. Toxicity of organic extraction reagents to anaerobic bacteria. *Biotechnol. Bioeng.* **1983**, *25*, 1251–1265. [CrossRef]
223. Hayward, G.; Lau, I. Toxicity of organic solvents to fatty acid forming bacteria. *Can. J. Chem. Eng.* **1989**, *67*, 157–161. [CrossRef]
224. Stergar, V.; Zagorc-Končan, J.; Zgajnar-Gotvanj, A. Laboratory scale and pilot plant study on treatment of toxic wastewater from the petrochemical industry by UASB reactors. *Water Sci. Technol.* **2003**, *48*, 97–102. [CrossRef]
225. Liu, S.-M.; Wu, C.-H.; Huang, H.-J. Toxicity and anaerobic biodegradability of pyridine and its derivatives under sulfidogenic conditions. *Chemosphere* **1998**, *36*, 2345–2357. [CrossRef]
226. Surerus, V.; Giordano, G.; Teixeira, L.A. Activated sludge inhibition capacity index. *Braz. J. Chem. Eng.* **2014**, *31*, 385–392. [CrossRef]
227. Hwu, C.-S.; Lettinga, G. Acute toxicity of oleate to acetate-utilizing methanogens in mesophilic and thermophilic anaerobic sludges. *Enzym. Microb. Technol.* **1997**, *21*, 297–301. [CrossRef]
228. Wikandari, R.; Gudipudi, S.; Pandiyan, I.; Millati, R.; Taherzadeh, M. Inhibitory effects of fruit flavors on methane production during anaerobic digestion. *Bioresour. Technol.* **2013**, *145*, 188–192. [CrossRef]
229. Nie, Y.; Tian, X.; Zhou, Z.; Cheng, J. Impact of food to microorganism ratio and alcohol ethoxylate dosage on methane production in treatment of low-strength wastewater by a submerged anaerobic membrane bioreactor. *Front. Environ. Sci. Eng.* **2017**, *11*, 6. [CrossRef]
230. Garcia-Bernet, D.; Loisel, D.; Guizard, G.; Buffiere, P.; Steyer, J.-P.; Escudie, R. Rapid measurement of the yield stress of anaerobically digested solid waste using slump tests. *Waste Manag.* **2011**, *31*, 631–635. [CrossRef]
231. Bhattacharya, S.K.; Qu, M.; Madura, R.L. Effects of nitrobenzene and zinc on acetate utilizing methanogens. *Water Res.* **1996**, *30*, 3099–3105. [CrossRef]
232. Blum, D.J.W.; Speece, R.E. A Database of Chemical Toxicity to Environmental Bacteria and Its Use in Interspecies Comparisons and Correlations. *J. Water Pollut. Control Fed.* **1991**, *63*, 198–207.

233. Borja, R.; Alba, J.; Banks, C. Impact of the main phenolic compounds of olive mill wastewater (OMW) on the kinetics of acetoclastic methanogenesis. *Process. Biochem.* **1997**, *32*, 121–133. [CrossRef]
234. Boucquey, J.-B.; Renard, P.; Amerlynck, P.; Filho, P.M.; Agathos, S.N.; Naveau, H.; Nyns, E.-J. High-rate continuous biodegradation of concentrated chlorinated aliphatics by a durable enrichment of methanogenic origin under carrier-dependent conditions. *Biotechnol. Bioeng.* **1995**, *47*, 298–307. [CrossRef] [PubMed]
235. Mahanty, B.; Zafar, M.; Han, M.J.; Park, H.-S. Optimization of co-digestion of various industrial sludges for biogas production and sludge treatment: Methane production potential experiments and modeling. *Waste Manag.* **2014**, *34*, 1018–1024. [CrossRef]
236. Mohamed, S.D.Y. *Influence of Oregano (Origanum vulgare L.), Fennel (Foeniculum vulgare L.) and Hop Cones (Humulus lupulus L.) on Biogas and Methane Production*; Universitat Giessen: Giessen, Germany, 2014.
237. Hess, J.; Bernard, O. Advanced dynamical risk analysis for monitoring anaerobic digestion process. *Biotechnol. Prog.* **2009**, *25*, 643–653. [CrossRef]
238. Arthur, R.; Scherer, P. Application of total reflection X-Ray fluorescence spectrometry to quantify cobalt concentration in the presence of high iron concentration in biogas plants. *Spectrosc. Lett.* **2019**, *53*, 100–113. [CrossRef]
239. Zheng, G.; Liu, J.; Shao, Z.; Chen, T. Emission characteristics and health risk assessment of VOC's from a food waste anaerobic digestion plant: A case study of Suzhou, China. *Environ. Pollut.* **2019**, *257*, 113546. [CrossRef]
240. Reinelt, T.; Liebetrau, J. Monitoring and Mitigation of Methane Emissions from Pressure Relief Valves of a Biogas Plant. *Chem. Eng. Technol.* **2019**, *43*, 7–18. [CrossRef]
241. Wünscher, H.; Frank, T.; Cyriax, A.; Tobehn-Steinhäuser, I.; Ortlepp, T.; Kirner, T. Monitoring of Ammonia in Biogas. *Chem. Eng. Technol.* **2019**, *43*, 99–103. [CrossRef]
242. Paul, A.; Schwind, B.; Weinberger, C.; Tiemann, M.; Wagner, T. Gas Responsive Nanoswitch: Copper Oxide Composite for Highly Selective H 2 S Detection. *Adv. Funct. Mater.* **2019**, *29*. [CrossRef]
243. Kushkevych, I.; Kobzová, E.; Vítězová, M.; Vítěz, T.; Dordević, D.; Bartoš, M. Acetogenic microorganisms in operating biogas plants depending on substrate combinations. *Boilogia* **2019**, *74*, 1229–1236. [CrossRef]
244. Mudaheranwa, E.; Rwigema, A.; Ntagwirumugara, E.; Masengo, G.; Singh, R.; Biziyaremye, J. Development of PLC based monitoring and control of pressure in Biogas Power Plant Digester. In Proceedings of the 2019 International Conference on Advances in Big Data, Computing and Data Communication Systems (icABCD), Winterton, South Africa, 5–6 August 2019; Institute of Electrical and Electronics Engineers (IEEE): Piscataway, NJ, USA, 2019; pp. 1–7.
245. Logan, M.; Safi, M.; Lens, P.; Visvanathan, C. Investigating the performance of internet of things based anaerobic digestion of food waste. *Process. Saf. Environ. Prot.* **2019**, *127*, 277–287. [CrossRef]
246. Selvaraj, R.; Vasa, N.J.; Nagendra, S.M.S. Off-Resonant Broadband Photoacoustic Spectroscopy for Online Monitoring of Biogas Concentration with a Wide Dynamic Range. In Proceedings of the Conference on Lasers and Electro-Optics, San Jose, CA, USA, 5–10 May 2019; The Optical Society: Washington, DC, USA, 2019; p. JW2A.20.
247. Flexibierung von Biogasanlagen. *Federal Ministry of Food, Agriculture and Consumer Protection*; Fachagentur Nachwachsende Rohstoffe E.V. (FNR): Gülzow, Germany, 2018.

Article

The Effective Management of Organic Waste Policy in Albania

Ionica Oncioiu [1,*], Sorinel Căpuşneanu [1], Dan Ioan Topor [2], Marius Petrescu [3], Anca-Gabriela Petrescu [3] and Monica Ioana Toader [2]

1 Faculty of Finance-Banking, Accountancy and Business Administration, Titu Maiorescu University, 040051 Bucharest, Romania; capusneanu.sorin@gmail.com
2 Faculty of Economic Sciences, 1 Decembrie 1918 University, 510009 Alba-Iulia, Romania; topor.dan.ioan@gmail.com (D.I.T.); monica.ioana.toader.edu@gmail.com (M.I.T.)
3 Faculty of Economic Sciences, Valahia University, 130024 Targoviste, Romania; petrescu.marius_m@yahoo.com (M.P.); anki.p_2007@yahoo.com (A.-G.P.)
* Correspondence: nelly_oncioiu@yahoo.com; Tel.: +40-744-322-911

Received: 5 July 2020; Accepted: 13 August 2020; Published: 14 August 2020

Abstract: Following a recycling or continuous recycling process, there is always waste with no material or market value that can be converted into energy or other fossil fuel substitutes. The present study aimed to evaluate the management of organic waste policy and to predict the trend of organic waste generation in Albania. The research used an appropriate Box–Jenkins Auto Regressive Integrated Moving Average (ARIMA) to determine the quantification of organic waste to be generated. The main results obtained can support the decision-making process in the planning, change and short-term implementation of organic waste management, and the information provided is very useful in collecting, transporting, storing and managing waste in Albanian cities (Tirana, Durrës, Kukës, Berat, Shkodra, Dibër, Gjirokastër and Elbasan). Furthermore, the high percentage of the organic waste generation until 2025 constitutes good premises to raising public awareness related to their energy recovery.

Keywords: environment technologies; waste management; organic waste; organic waste projection

1. Introduction

In the current context of the transition to the bio economy, the biggest environmental issue for all countries is organic waste management [1–5]. One of the main driving forces for this trend in all countries is the general increase in consumption [6–8]. The level of organic waste production in cities seems to be correlated with the level of income as well as with economic growth [9–11].

In this context, the European Union's new waste management guidelines include measures aimed at greater recycling and reuse during the life cycle of products to benefit both the environment and the economy [12–14]. Withal, most countries understand that the recycling strategy—that is, the 4Rs (recovery, reuse, regeneration and recycle)—can lead any country to transition to a circular economy [15–18].

Specialized studies have highlighted the importance of the social aspect in the efficiency of organic waste management systems [19–22]. Regarding recycling, some authors showed that social influences and economic factors are only some of the reasons some communities are developing strong recycling habits [15,23]. Additionally, in most cases, as the distance to recycling bins decreases, the number of citizens who collect waste at home increases while the government should support markets for recycled materials to increase recycling rates [17,24].

Other studies have indicated various advantages offered by Waste To Energy (WTE) technologies: (1) reduces the amount of waste disposal with an impact on climate change [25,26]; (2) helps to improve recycling rates; (3) reduces dependence on fossil fuels to generate electricity; and (4) prevents contamination of air/water content [27–30].

The aim of our research was to develop a prognosis model for organic waste generation in Albania to help evaluate organic waste management options from the point of view of environmental implications. Furthermore, this study should arouse the consciousness of the municipal decision makers to implement ecologically sustainable measures (e.g., increasing recycling quotas). Our research was based on numerous studies and reports that have examined waste recycling and organic waste management to reduce negative environmental impact. This study provides the possibilities of developing alternate solutions for the issue of urban organic waste management in Albania.

The structure of the paper is as follows. Section 2 presents the literature review. Section 3 presents the potential of organic waste recycling in Albania. Section 4 provides the proposed methodology and data. Section 5 analyzes the empirical results and discusses implications. Section 6 draws the conclusions.

2. Literature Review

In the circular economy, waste must be properly treated and converted into natural resources in order to protect the environment and the public health using different WTE technologies. These technologies are responsible for converting non-recyclable waste (semi-solid, liquid and gaseous) into valuable products such as different fuels, heat and electricity, which are part of the waste management hierarchy [31]. Considered unnecessary long ago, these WTE technologies have become the main source of saving landfills, reducing costs by disposing and managing organic waste and transforming it into valuable fuels, fertilizers and electricity [32]. Specialists have identified two categories of WTE technologies: thermochemical technologies and biochemical technologies [33].

Thermochemical technologies are those technologies by which a sufficient amount of thermal energy is applied to organic waste components in a closed vessel and the bonds of molecular structures are broken and broken down into smaller molecules. Following this process occurs the recombination by which the carbon and hydrogen atoms released from the decomposed molecules combine with oxygen, releasing more energy than that consumed to decompose the molecular structure of waste components [34]. The most well-known thermochemical technologies are incineration, gasification, pyrolysis, plasma arc gasification, thermal depolymerization and hydrothermal carbonization [35].

2.1. Incineration

Incineration is the technique by which organic substances present in waste are burned and as a result of this process heat is recovered to reduce the volume of Municipal Solid Waste (MSW) [36,37] and reduce the infectious properties and potential toxicity of hazardous medical waste [38]. A very large quantity of oxygen is needed to ensure the complete oxidation of the waste incineration, and the combustion temperature in the incineration plant is about 8500 °C, forming CO_2 and H_2O. The incineration process consists of the fuel system, the combustion chamber, the exhaust system and the waste disposal system. Before incineration, it is necessary to separate the non-combustible substances (glass, metal, etc.) from MSW because after incineration the bottom ash of the incinerator or solid slag results [37]. According to the studies of specialists [38–45], several thermochemical incineration technologies have been identified: (1) the principle of the process involving the complete oxidation of waste; (2) the type of exothermic reaction; (3) requirements for raw materials taking into account dry waste of biological and synthetic origin; (4) the method of pre-processing the raw material by drying and pelletizing; (5) the permitted moisture content of the raw material of 25–30%; (6) temperature (700–14,000 °C); (7) endurance time (minutes/seconds); (8) final products (heat and ash); (9) environmental issues (ash leakage and toxic gases); (10) cost of capital (medium-high); (11) degree of efficiency (50–60%); (12) product applications (heat and power applications, aggregate and filler); and (13) future (moderate) potential. The advantages of thermochemical incineration technology include: (1) the resulting ash can be used as a low-cost aggregate or filler for construction works on bridges, roads and highways; (2) it has relatively low capital requirements, requires less skilled labor compared to other WTE technologies and is more suitable for rural and urban areas; (3) it significantly reduces waste storage space; and (4) it has high energy generation efficiency and low emission rates due to the installation of pollution content control devices in incinerators that help maintain the emission limits required by law [44,46–51].

As disadvantages, the specialists identified the following: (1) the modifications made to the combustion and pollution control equipment to reduce emissions contribute to the increase of construction and operating costs; (2) the impossibility of avoiding the emission of toxic gases (dioxins and furans) even if some incineration plants have been modified and developed; (3) some risks of emission of flue gases due to incorrect handling of heavy metals produced by incineration; and (4) lack of professional staff and variety of waste can cause problems over time in the waste incineration process [44,47,51–54].

2.2. Gasification

Gasification is a way of transferring organic material or converting (liquid and solid materials) into clean and useful syntheses [32,55,56] or other forms of energy through a cleaning reaction. Before entering the raw material processing, it is necessary to extract all inorganic materials (glass, metal, contaminants and inert materials) that cannot be transformed into gaseous products [32,57] and then grind them into very small particles. Types of waste that act as raw material for gasification include: MSW, RDF, coal, sludge, black liquor, organic waste streams, tires, PVC, biomass, refinery waste and shredded car waste (ASR) [58]. According to specialist studies, gasification as a thermochemical technology offers several advantages: (1) reduces the formation of dangerous products (furans and dioxins) because the chemical reaction occurs in an environment with low oxygen consumption; (2) an energy efficient method using only a small part of the stoichiometric amount of oxygen; (3) requires low equipment cleaning costs; (4) the gas obtained can be used in combined cycle turbines or fuel cells, which have a tendency to convert combustible energy into electricity compared to conventional steam boilers; and (5) contributes to the significant reduction of waste storage space (by about 90%) and slag content (about 10%) [47,59].

Among the disadvantages of gasification, the specialists mentioned: (1) the release of polluting compounds (alkaline, tar, halogens and heavy metals) in the syngas produced can cause operational and environmental problems; (2) causing acid rain (if corrosive halogens are released into the atmosphere) or the accumulation of heavy metals (if emitted into the environment); (3) gas turbines may be destroyed during combustion due to alkalis; and (4) thick (tar) and high molecular weight organic gases spoil ceramic filters, sulfur removal systems, reform catalysts and increase the existence of shakes in refractory surfaces, metals and boilers [59].

2.3. Pyrolysis

By using oxygen-free thermochemical decomposition or other reactive materials, pyrolysis consists of the transformation of carbonaceous materials into syngas (combination of CO, H_2, CO_2 and CH_4) at higher temperatures, the association of solids (char) and at lower temperatures of liquids (oxygenated oils) [58,60–62]. The pyrolysis process requires constant raw materials, where the feed particles for a longer or shorter period of time have the same moisture content, size and composition [41]. The types of waste that act as raw materials for pyrolysis include: MSW, RDF, coal, sludge, tires, wastewater, biomass, ASR, chloride and polyvinyl [58,60,63]. The final products of pyrolysis depend on the heating contribution, the pyrolysis temperature, the dimension of the waste particles and the vapor resistance time [60,64].

According to the studies of specialists, the advantages of pyrolysis as a thermochemical technology include: (1) reduced pollution and good environmental values compared to other thermochemical technologies (except for gasification with plasma arc); (2) in the absence of oxygen, it contributes to the formation of dangerous products (ransom and dioxin); (3) contributes to the reduction of the volume of MSW type waste (70–90%); (4) the use in oil or powerplants of the oils and syngas produced by MSW pyrolysis for the production of electricity; (5) the use of pyrolytic products as fuels for boilers and other products (adhesives, chemicals, motor fuels and other products after refining); and (6) the low temperatures allow the recovery of metals and reduction of flue gas [60].

Among the disadvantages of pyrolysis specialists are the following: (1) causes pollution, destruction of nature, natural resources and climate change by releasing toxic residues into the air, toxic ash and tarring; (2) risks of explosive reactions due to accidental air intrusions or installation failures due to tarsal deposits; (3) pyrolysis products contain low energy and commercial scale pyrolysis plants that accept MSW are very limited in the world; and (4) requires high operational and capital costs [37,47,50,60].

2.4. Plasma arc Gasification

Plasma arc gasification is the process that takes place in an oxygen-free atmosphere by decomposing waste by partial oxidation into molecules of H_2O, H_2 and CO [65], using a plasma torch reactor for processing MSW carbon materials [66]. As a heat source, it uses a plasma arc flame and an electric arc provides power to the plasma torch for the ionization of gases and organic matter that decompose into syngas and solid waste. The plasma torch converts electricity into intense thermal energy [65]. Almost all types of waste act as raw materials for plasma arc gasification, such as MSW, dangerous waste, RDF, coal ash, tires, biomass, ship waste and ASR [59,65]. No initial processing of the raw materials is required for plasma arc gasification because almost all types of waste (except nuclear) can be treated and processed directly [66].

The advantages offered by this thermochemical technology include: (1) it offers the highest energy efficiency in terms of technology and the most advantages in terms of storage surface requirements (approximately 99% by transforming the residual waste into vitrified slag); (2) converts into syngas-type products any type of waste that is difficult to treat (toxic ash from incinerators, electronic components, hazardous medical waste, etc.); (3) the use of syngas-type products for the production of electricity by means of gas turbines or the production of transport fuels; (4) production of 0.1 tons of waste per 1 tons of waste introduced (depending on the waste structure); (5) the absence of methane and low environmental (carbon) emissions compared to other facilities; and (6) converts inorganic components into vitrified slag used in varicose applications (roofing materials and road construction) [50,60,64–67].

Specialists have also identified a series of disadvantages of plasma gasification: (1) high costs with high electricity consumption through the use of plasma; and (2) high capital costs due to the use of plasma torch being one of the most expensive WTE technologies compared to other facilities [50,60,66].

2.5. Thermal Depolymerization

Thermal depolymerization consists in the depolymerization of various organic materials using water at high temperature and pressure to form petroleum products. In this process, the long polymer chain is broken down into a shorter monomer chain, simulating natural geographic processes that produce fossil fuels [68]. In other words, this process involves chopping the fodder into very small pieces which are then combined with water and heated to a temperature of 2500 °C and subjected to high pressure [36].

The products obtained (solid carbon, gaseous components, water, light oils and heavy oils) following this process are separated using a fractional distillation technique from an oil refinery [68]. The water resulting from obtaining the products is returned to the front, where it is mixed with the next unit of waste, forming a gas that is used to heat it. Therefore, 15% of the energy produced from thermal depolymerization is used to run the process, which makes thermal depolymerization a process with an efficiency of 85% [69].

The raw materials used in thermal depolymerization are varied and include carbon waste (except nuclear waste), plastic bottles, agricultural biomass, tires, electronic waste, medical waste and municipal liquid waste [69–71].

Thermal depolymerization involves the following steps [71]: (1) After introducing the raw materials into the chamber, they are combined with water and heated under high pressure to a temperature of 200–3000 °C. (2) The pressure of the combination is rapidly reduced, which allows the oils to be separated from the water and the volatile gases are removed and used for heating boilers or rotating turbines. (3) The remaining oil is heated to 5000 °C to obtain light hydrocarbons.

According to specialists, thermal depolymerization brings a number of advantages: (1) recovery of polyamides, polyurethanes and PETs; (2) recycles the energy of organic components by easily separating liquid fuel from water without pre-drying; (3) removal of heavy metals by transforming the ionized form of metals into stable oxides; (4) aids in the processing of shales, tar sands and heavy metals (considered profitable) and a modified version of thermal depolymerization could extract a variety of minerals from coal treatment; (5) the generated gases have low temperature, avoiding energy loss for gas cleaning; and (6) the fuel produced is not harmful to gas turbines and does not contain alkali metals [68,69,72].

The disadvantages of thermal depolymerization include: (1) high processing costs, which influences the production of liquid crude oil; (2) generates risks of the release of dioxins, furans, CO_2 and CH_4 at temperatures above 4000 °C; and (3) requires additional refining steps because monomers cannot be transformed by this process into oil [68].

2.6. Hydrothermal Carbonization

Hydrothermal carbonization is the process that transforms organic substances into hydrocarbons (with mesoporous texture, aromatic structure and moderate caloric value) [73] under moderate pressure (2–10 MPa) and temperature (180–3500 °C) in the presence of water.

The raw materials used in the hydrothermal carbonization process are: wood, food and poultry waste with high carbon content and humidity over 20%; cardboard and paper; biomass; MSW; and sewage sludge [74]. The raw materials go through a succession of stages: hydrolysis, dehydration, decarboxylation, flavoring and condensation. The result consists in the formation of products in liquid, solid and gaseous state [40].

Pre-processing of raw materials consists in crushing the raw material, mixing it with water (75–90%) and subjecting it to a saturation pressure that allows carbonization. The presence of adequate water in the hydrothermal carbonization process is essential because, as the temperature increases, their physical and chemical properties change, favoring the stimulation of organic solvents [40].

The products obtained from the hydrothermal carbonization process consist of higher solid yields (hydro char with 75–90% carbon content), products in the aquatic phase (residues, organic acids and sugars with 15–20% carbon) and gases (CO_2 and 5% carbon-rich hydrocarbons) [75]. Most of the carbon content of the raw material is produced in hydrocarbons, which improves the energy density of hydro char [40] that provides various alternative energy sources, soil growth and environmental remediation. HTC's gaseous and liquid products also provide energy [76].

According to specialists, hydrothermal carbonization offers a number of advantages: (1) modifies hybrid nanostructures and designer carbons along with practical applications replacing petrochemical processes with a scalable green process; (2) completely separates the physical structure of the waste compared to other WTE techniques and does not require an intensive drying process; (3) hydrocarbons improve water retention capacity, increase hydraulic conductivity and reduce the tensile strength of hard fixing soils; (4) increased reactivity compared to natural coal due to the chemical structure of hydrocarbons including olefin and aliphatic construction units; (5) significant reduction in GHG emissions and odor due to the preservation in the solid material of a carbon content of 75–90%; and (6) energy efficient (energy inputs) compared to other WTE technologies [40,72–77].

The disadvantages of hydrothermal carbonization include: (1) high costs due to the supply of pressure in a continuous regime to which are added the safety and material aspects of the reactor; (2) requires subsequent treatments for the separation of solid water products but also treatments for water processing; (3) energy inefficiency when processing raw materials with a moisture content of less than 40%; (4) risks of dust formation and fungal degradation of hydrocarbons in the absence of a water content of at least 10–15%; (5) high hydrocarbon acidity due to high ash content; and (6) negative impact caused by water emissions from processing [74,76].

3. The Waste Management in Albania

3.1. The Status and Waste Morphology in Albania

In Albania, renewable energy is composed of geothermal energy, solar energy, wind energy, hydroenergy and biomass. Relying mainly on hydroenergy resources (geographically advantageous), Albania faces problems due to prolonged droughts and declining river water flows [78]. Albania also benefits from year-round solar energy (about 2100–2700 h of sunshine) [79], with a United Nations Development Program (UNDP) program installation of about 50,000 m^2 of solar panels at the end of 2018 [80].

According to the studies of specialists, from the point of view of waste management, Albania has a deficient and unsatisfactory infrastructure, not being able to keep up with rapid economic growth and urban expansion [81]. Although considerable efforts have been made, the state of waste management in Albania remains sub-standard and partial as there is no separate or segregated waste collection [82,83]. In 2016, Albania made legislative progress through assistance from European Commission regulatory agencies by adopting legislation, standards and compliance on waste management aligned with that of the EU [84]. Waste to Fuel (WTF) Transformation is a direct form of energy recovery, being practically the process by which fuel is obtained from the primary treatment of waste. At the local level, the recycling of biomass is considered by transforming organic products (food, wood, paper and textiles) into biofuels (bioethanol and biodiesel), and, at the industrial level, it is obtaining them from other organic products such as corn, sugar cane and cereals.

According to an Albanian study, municipal and urban waste is composed of 66% biodegradable waste (of which 47% is organic: cardboard, paper, wood and certain textile waste), 7% glass and metal and 31% energy-convertible waste (pyrolysis oil and industrial synthetic diesel) and other general components (rubber, paper and plastics) [81]. Based on the above, Albania has viable options for recycling waste and converting it into fuel. Several specialists have dedicated their studies to exploring the use of Biomass to Energy (BTE) and the potential of renewable energy in Albania [85,86] and indicating the methods of evaluation, implementation and benefits of following the use of recycling their own organic waste [87]. In addition, inorganic waste such as plastic waste and rubber waste can be reprocessed according to WTF producing pyrolysis oil, industrial diesel not taking into account the reprocessing of oil-based waste or the import of plastic materials or packaging [83]. Using waste management practices, Albania can rely on renewable resources for biofuel and pyrolysis fuel. This can lead to a reduction in energy imports by increasing domestic production of fossil fuels with a negative impact on the average consumer [81].

3.2. The Potential of Organic Waste Recycling in Albania

The concept of green growth gained importance in Albania with the search for a new growth model that started on a global scale in order to achieve sustainable development goals. By using this concept, the environment is protected and the competitiveness is increased by obtaining in the production sectors a clean production and eco-efficiency.

In the Second Environmental Assessment Report-Albania-Second Review [88], the UN Economic Commission for Europe states that the state of waste management in Albania is at a low level, with waste collection systems being implemented only in cities. The disparity is based on the inability of national, regional and local administration to achieve a sustainable waste collection and on limited human resources, even if the legal framework in accordance with European Union Acquis Communautaire exists [89].

The responsibility of applying the policies and legislation on waste management is on the Ministry of Environment, Forests and Water Administration [90]. Decentralization predominates in organic waste management: the financing of waste collection and transport to disposal facilities is done by the municipality and undertaken by private companies. The municipality selects on the basis of public auction private companies (about 2/3) with which it concludes contracts for a period of 3–5 years [91]. Albania provides waste collection services using its own companies to 1/3 of municipalities [92].

On the other hand, this strategy involves expertise at national level to prepare and implement the waste management plan, focusing on source waste decomposition, increasing producer responsibility and the improvement the waste information system. However, the lack of collection education exists because collectors and individual companies have difficulty identifying clean and decomposed waste. Recyclable waste comes from most urban areas and only partially from the industrial sector.

The evolution of the organic waste in Albania is the consequence of rapid development (Figure 1).

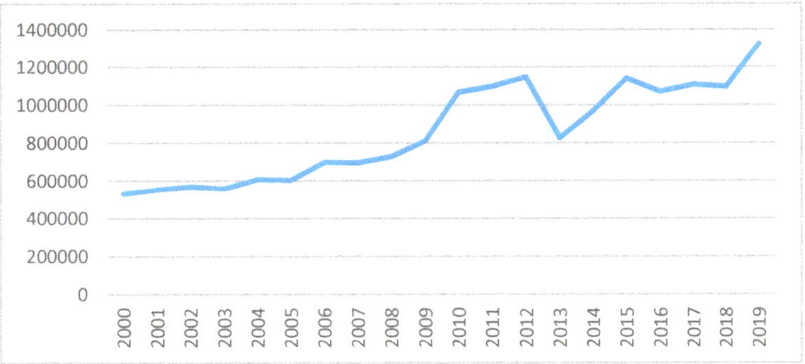

Figure 1. Total generation of organic waste (thousand tones) [91,92].

As a consequence of these statistics, Albania developed the National Waste Management Strategy for 2010–2025, which sets the direction of the Albanian government's policy for sustainable waste management by 2025. This strategy is divided into three operational phases of five years each, as follows: by 2020, it aims to stop the growth of municipal waste produced (recycling/composting 55% of organic waste); by 2025, it aims to recover energy from 15% of organic waste [93]. Nevertheless, some progress has been made in Albania [94], notably:

- Organic waste: Dumping in uncontrolled sites is the main method of organic waste disposal. The municipality has led considerable work to improve the separate collection through awareness campaigns of citizens.
- Hotspots: The waste deposits are a priority for Albania. The United Nations Development Programme Project Identification and Prioritization of Environmental Hotspots in Albania (January 2008–August 2011), funded by the Government of the Netherlands, identified 35 hotspots, while nine priority hotspots were selected for assessment and the preparation of remediation action plans.

On the economic side, the opportunity cost of depositing the wastes is considerably high when considering the fact that Albania is producing more than 10 MW installed capacity from 1500 tons of household waste per day and that 300 tons of packaging waste can be recycled daily [95].

As a result, recycling of packaging wastes and recycling of organic wastes through legal regulations are encouraged; it is understood that only wastes which are not recoverable and pre-treatment residue should be stored regularly.

4. Data and Methodology

Based on the direction of the Albanian government's policy for sustainable waste management and the dates collection, we formulated the objective of the study by which we developed an appropriate Autoregressive Integrated Moving Average (ARIMA) model for time models for organic waste collection in Albanian cities (Tirana, Durrës, Kukës, Berat, Shkodra, Dibër, Gjirokastër and Elbasan). We marked five-year forecasts with an appropriate prediction interval. Through this model, we aimed to identify the stochastic process of the time series and accurately forecast future values.

For our study, we selected the period between 2000 and 2019 for organic waste collection in Albanian cities (Tirana, Durrës, Kukës, Berat, Shkodra, Dibër, Gjirokastër and Elbasan) from datasets on the Eurostat website [96], the Albania Institute of Statistics [97], the European Environment Agency data portal [93] and the data provided by the Center for Regional and Local Development Studies Albania (CRLDS).

The ARIMA methodology was developed in this research because it uses a combination of autoregressive (AR), integration (I) (referring to the reverse process of differencing to produce the forecast) and moving average (MA) operations, which is usually stated as ARIMA (p, d, q), where p is the number of autoregressive terms, d is the number of nonseasonal differences needed for stationarity and q is the number of lagged forecast errors in the prediction equation).

To predict the trend of organic waste generation in Albanian cities (Tirana, Durrës, Kukës, Berat, Shkodra, Dibër, Gjirokastër and Elbasan), an ARIMA is expressed in the form:

$$\text{if } \omega_t = (1 - X)^d r_t, \tag{1}$$

Then,

$$\omega_t = \beta_0 + \beta_1 \omega_{t-1} + \beta_2 \omega_{t-2} + \ldots + \beta_p \omega_{p-1} + \epsilon_t - \mu_1 \epsilon_{t-1} - \mu_2 \epsilon_{t-2} - \ldots - \mu_p \epsilon_{t-p}, \tag{2}$$

where ω_t is regressors, $\beta_0, \beta_1, \beta_2, \ldots, \beta_p$ are model parameters, ϵ_t is an error term and $\mu_1, \mu_2, \ldots, \mu_p$ are the moving average parameters to be estimated.

The first thing to note is that the ARIMA model refers only to a stationary time series, thus the first stage of this model is reducing non-stationary series to a stationary series by taking first-order differences.

The stationary verification of the time series data was followed by the identification of the candidate ARIMA models by finding the initial values for the orders of the parameters "p" and "q". Then, the accurate estimation of the model parameter was obtained by at least squares, low Akaike information criteria (AIC) were selected and Schwartz Bayesian Criterion (SBC) was used and estimated by SBC = logσ2 + (m log n)/n. Therefore, AIC = (−2 log L + 2 m), where m = p + q and L is the likelihood function.

To examine the trend over time, the Autocorrelation Function (ACF) and Partial Autocorrelation Function (PACF) were used. The accuracy was checked using two measures, namely RMSE and MAPE, and the generation of organic waste model parameters were estimated using SPSS package.

5. Results and Discussion

Following the proposed model and data presented above, the first step in the analysis was to plot the given data. Figure 2 shows generation of organic waste centralized data, Auto Correlation Function (ACF) and Partial Auto Correlation Function (PACF). After checking the AIC and BIC of the tentative ARIMA models, we selected the suitable model ARIMA (0, 1, 1) for generation of organic waste.

The model verification was concerned with investigating thoroughly the auto correlations and partial auto correlations of the residuals to see if they contain any systematic pattern which could still be removed to improve on the chosen ARIMA (Figure 3).

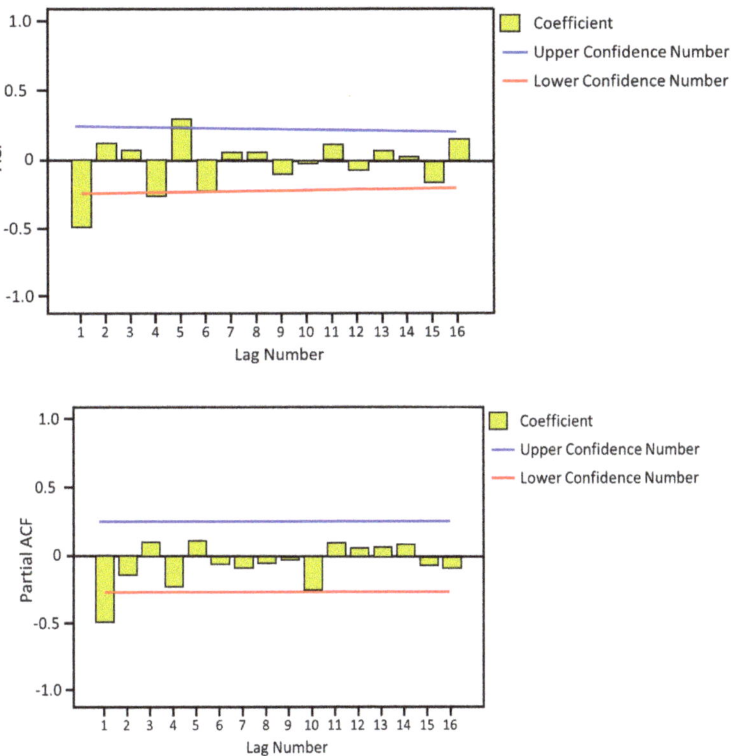

Figure 2. ACF and PACF plots for generation of organic waste.

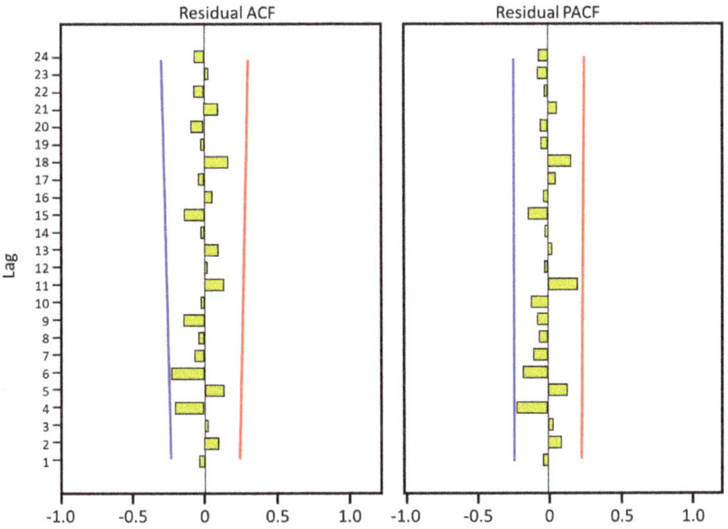

Figure 3. Residuals ACF and PACF plots for generation of organic waste.

On the other hand, forecasting of generation of organic waste in Albania was done for the six years using the ARIMA (0, 1, 1) model with 95% confidence level (Table 1).

Table 1. Forecasted values of generation of organic waste in Albania with 95% confidence level (CL).

Year	Forecasted Value	LCL	UCL
2020	1,325,020	1,317,595	1,325,265
2021	1,279,380	1,272,186	1,287,038
2022	1,296,788	1,288,770	1,305,358
2023	1,296,400	1,287,225	1,305,794
2024	1,297,969	1,287,818	1,308,120
2025	1,299,323	1,288,468	1,310,178

LCL, Lower Confidence Level; UCL, Upper Confidence Level.

Obviously, the present prognosis model for organic waste generation is often used to provide justification for the adoption of waste policy measures and in the planning of recycling facilities and collection service. Accurate predictions of organic waste quantities can determine successful planning and operation of an organic waste management system in the countries included in the study. The prediction findings of future organic waste generation rates can facilitate significant changes in regulations regarding waste minimization and recycling. Thus, the projections on domestic waste disposal rates stated in this study will be valid only if there is no significant change in the Albanian environmental authorities (Tirana, Durrës, Kukës, Berat, Shkodra, Dibër, Gjirokastër and Elbasan) to promote waste recycling and reduction and that the assumed maximum number of housing estates participating in waste separation program is attained.

Taking this prediction into account requires progress in increasing recycling and composting by the widespread provision and relevant authorities of the segregated organic waste collection services and mechanized post collection separation across the country. However, in all Albanian cities (Tirana, Durrës, Kukës, Berat, Shkodra, Dibër, Gjirokastër and Elbasan), the responsibility for the provision of waste services has been delegated to local governments suffering from a lack of financial resources to support and improve public services, including the waste service.

In accordance with the data presented in Table 1, to reduce the gaps in the reuse and recycling of organic waste, Albanian authorities must provide the necessary resources to develop and implement effective waste management policies, establish the necessary infrastructure for the collection and recycling of waste and set up business partnerships that support and improve the recycling process.

Additionally, organic waste recycling involves costs that cannot be immediately recovered. Therefore, the way in which financial resources are managed, as well as the good management of infrastructure and appropriate maintenance of equipment, can ensure a sustainable modern recycling system that will contribute to the development of the economy and society.

Based on the results of this forecast, a strategic reflection should be made to build an efficient recyclable waste sector because even the privatization of this sector has led to social and urban problems.

Despite all these inherent limitations, the benefits observed in the organic waste management policy in Albania have not been verified across all the regions, because the interregional diversity observed in the management policy indicates that several regions are favored over others.

Moreover, this study contains elements that have the potential to open new avenues for future research. To analyse how to recycle organic waste after sorting, a study should be conducted that considers its composition, which depends on several factors (e.g., the degree of culture, the geographical area, the climatic zone and the total weight of residuals).

6. Conclusions

The circular economy creates added value through its benefits: (1) the potential to create new jobs, new products and new services; and (2) improved competitiveness of the economy [98–100]. To create added value, the circular economy has to be operational on a local or regional basis. Moreover, circular economic initiatives: (1) reduce the level of dependence on fossil fuels of a region or a country; (2) reduce to a minimum waste production; (3) decrease carbon emissions; and (4) contribute to combating climate change [101].

The results of this model can be implemented as a useful decision support tool for organic waste of Tirana, Durrës, Kukës, Berat, Shkodra, Dibër, Gjirokastër and Elbasan. The public and the decision-makers must be aware that recycling brings benefits to both the environment and the economy by providing raw materials to create new products and fostering innovation and job creation. In this respect, Albania should pay more attention to forecast models of organic waste.

As this study demonstrated, WTE plays a key role in the circular economy by contributing to synergies in three EU policies, namely waste management, environment (climate change) and Member States' energy union policies, which encourages meeting the targets related to these policies in the context of energy and resource efficiency.

In conclusion, Albanian authorities can learn several lessons from this study regarding organic waste disposal rates. First, it is recommended that Albanian authorities devote more resources to collecting statistics on waste recycling because accurate waste projections cannot be made without a source of data.

Second, the results generated by the study's models merely show that the waste reduction is not impossible for Tirana, Durrës, Kukës, Berat, Shkodra, Dibër, Gjirokastër and Elbasan (and similar Albanian cities as well) because predictive models do not always determine the future. Moreover, the main objective of this model is to identify the stochastic process of the time series and predict the future values accurately.

Third, even when valid solutions are identified, empowering Albanian cities in the fight against organic waste also means disposing of charismatic politicians without a vision and with concerns related for their pockets. Local politicians are much more in the picture and more vulnerable to criticism. However, their capacity to change things quickly is much higher.

Based on the results of this forecast, a strategic reflection should be made to build an efficient recyclable waste sector because even the privatization of this sector has led to social and urban problems.

Author Contributions: Conceptualization, I.O. and S.C.; Methodology, I.O. and S.C.; Validation, I.O., M.P. and D.I.T.; Formal Analysis, A.-G.P., M.P. and D.I.T.; Writing—Original Draft Preparation, A.-G.P. and M.I.T.; and Writing—Review and Editing, I.O. All authors have read and agreed to the published version of the manuscript.

Funding: This research received no external funding.

Acknowledgments: We thank Luiza Hoxhaj, executive director of Center for Regional and Local Development Studies Albania-CRLDS, for her valuable help for providing data presented in this article.

Conflicts of Interest: The authors declare no conflict of interest.

References

1. Lakatos, E.S.; Cioca, L.-I.; Dan, V.; Ciomos, A.O.; Crisan, O.A.; Barsan, G. Studies and Investigation about the Attitude towards Sustainable Production, Consumption and Waste Generation in Line with Circular Economy in Romania. *Sustainability* **2018**, *10*, 865. [CrossRef]
2. Rada, E.C.; Cioca, L.I. Optimizing the Methodology of Characterization of Municipal Solid Waste in EU under a Circular Economy Perspective. *Energy Procedia* **2017**, *119*, 72–85. [CrossRef]
3. Eriksson, O.; Bisaillon, M.; Haraldsson, M.; Sundberg, J. Integrated waste management as a mean to promote renewable energy. *Renew. Energy* **2014**, *61*, 38–42. [CrossRef]
4. Oduro-Appiah, K.; Afful, A.; Kotey, V.N.; De Vries, N. Working with the Informal Service Chain as a Locally Appropriate Strategy for Sustainable Modernization of Municipal Solid Waste Management Systems in Lower-Middle Income Cities: Lessons from Accra, Ghana. *Resources* **2019**, *8*, 12. [CrossRef]
5. Rada, E.C.; Zatelli, C.; Cioca, L.I.; Torretta, V. Selective Collection Quality Index for Municipal Solid Waste Management. *Sustainability* **2018**, *10*, 257. Available online: http://www.mdpi.com/2071-1050/10/1/257 (accessed on 2 May 2020). [CrossRef]

6. Laurent, A.; Bakas, I.; Clavreul, J.; Bernstad, A.; Niero, M.; Gentil, E.; Hauschild, M.Z.; Christensen, T.H. Review of lca studies of solid waste management systems–Part i: Lessons learned and perspectives. *Waste Manag.* **2014**, *34*, 573–588. [CrossRef]
7. Navarro-Esbrı, J.; Diamadopoulos, E.; Ginestar, D. Time series analysis and forecasting techniques for municipal solid waste management. *Resour. Conserv. Recycl.* **2002**, *35*, 201–214. [CrossRef]
8. Hoornweg, D.; Bhada, P. What a waste. A Global review of solid waste management. *Urban Dev. Ser. Knowl. Pap.* **2012**, *281*, 44. [CrossRef]
9. Li, Z.; Fu, H.; Qu, X. Estimating municipal solid waste generation by different activities and various resident groups: A case study of Beijing. *Sci. Total Environ.* **2011**, *409*, 4406–4414. [CrossRef]
10. Abd Kadir, S.A.S.; Yin, C.Y.; Rosli Sulaiman, M.; Chen, X.; El-Harbawi, M. Incineration of municipal solid waste in Malaysia: Salient issues, policies and waste to energy initiatives. *Renew. Sustain. Energy Rev.* **2013**, *24*, 181e6. [CrossRef]
11. Denafas, G.; Ruzgas, T.; Martuzeviˇcius, D.; Shmarin, S.; Hoffmann, M.; Mykhaylenko, V.; Ogorodnik, S.; Romanov, M.; Neguliaeva, E.; Chusov, A.; et al. Seasonal variation of municipal solid waste generation and composition in four east European cities. *Resour. Conserv. Recycl.* **2014**, *89*, 22–30. [CrossRef]
12. Lazarevic, D.; Buclet, N.; Brandt, N. The application of life cycle thinking in the context of European waste policy. *J. Clean. Prod.* **2012**, *29*, 199–207. [CrossRef]
13. Huang, J.; Zhao, R.; Huang, T.; Wang, X.; Tseng, M.-L. Sustainable Municipal Solid Waste Disposal in the Belt and Road Initiative: A Preliminary Proposal for Chengdu City. *Sustainability* **2018**, *10*, 1147. [CrossRef]
14. den Boer, E.; Jędrczak, A.; Kowalski, Z.; Kulczycka, J.; Szpadt, R. A review of municipal solid waste composition and quantities in Poland. *Waste Manag.* **2010**, *30*, 369–377. [CrossRef]
15. Troschinetz, A.M.; Mihelcic, J.R. Sustainable recycling of municipal solid waste in developing countries. *Waste Manag.* **2009**, *29*, 915–923. [CrossRef]
16. Challcharoenwattana, A.; Pharino, C. Co-Benefits of Household Waste Recycling for Local Community's Sustainable Waste Management in Thailand. *Sustainability* **2015**, *7*, 7417–7437. [CrossRef]
17. Ávila, C.; Cedano, K.; Martínez, M. Sustainability analysis of waste to energy strategies for municipal solid waste treatment. *Int. J. Environ. Sustain.* **2017**, *13*, 1–14. [CrossRef]
18. Ghesla, P.L.; Gomes, L.P.; Caetano, M.O.; Miranda, L.A.S.; Dai-Prá, L.B. Municipal Solid Waste Management from the Experience of São Leopoldo/Brazil and Zurich/Switzerland. *Sustainability* **2018**, *10*, 3716. [CrossRef]
19. Montevecchi, F. Policy mixes to achieve absolute decoupling: A case study of municipal waste management. *Sustainability* **2016**, *8*, 442. [CrossRef]
20. Johnstone, N.; Labonne, J. Generation of Household Solid Waste in OECD Countries: An Empirical Analysis Using Macroeconomic Data. *Land Econ.* **2004**, *80*, 529–538. [CrossRef]
21. Zaman, A.U.; Lehmann, S. Challenges and Opportunities in Transforming a City into a "Zero Waste City". *Challenges* **2011**, *2*, 73–93. [CrossRef]
22. Aranda Usón, A.; Ferreira, G.; Zambrana Vásquez, D.; Zabalza Bribián, I.; Llera Sastresa, E. Environmental-benefit analysis of two urban waste collection systems. *Sci. Total Environ.* **2013**, *463–464*, 72–77. [CrossRef] [PubMed]
23. Kawai, K.; Tasaki, T. Revisiting estimates of municipal solid waste generation per capita and their reliability. *J. Mater. Cycles Waste Manag.* **2016**, *18*, 1–13. Available online: https://link.springer.com/article/10.1007%2Fs10163-015-0355-1 (accessed on 22 May 2020). [CrossRef]
24. Xu, L.; Gao, P.; Cui, S.; Liu, C. A hybrid procedure for MSW generation forecasting at multiple time scales in Xiamen City, China. *Waste Manag.* **2013**, *33*, 1324–1331. [CrossRef] [PubMed]
25. Zaman, A. Life Cycle Environmental Assessment of Municipal Solid Waste to Energy Technologies. *Glob. J. Environ. Res.* **2009**, *3*, 155–163.
26. Malinauskaite, J.; Jouhara, H.; Czajczynska, D.; Stanchev, P.; Katsou, E.; Rostkowski, P.; Thorne, R.J.; Colón, J.; Ponsá, S.; Al-Mansour, F.; et al. Municipal solid waste management and waste-to-energy in the context of a circular economy and energy recycling in Europe. *Energy* **2017**, *141*, 2013–2044. [CrossRef]
27. Klein, A. Gasification: An Alternative Process for Energy Recovery and Disposal of Municipal Solid Wastes. Master's Thesis, Columbia University, New York, NY, USA, 2002. Available online: http://www.seas.columbia.edu/earth/wtert/sofos/klein_thesis.pdf (accessed on 2 March 2020).
28. Fruergaard, T.; Astrup, T. Optimal Utilization of Waste-To-Energy in an LCA Perspective. *Waste Manag.* **2011**, *31*, 572–582. [CrossRef]

29. Turconi, R.; Butera, S.; Boldrin, A.; Grosso, M.; Rigamonti, L.; Astrup, T.F. Life Cycle Assessment of Waste Incineration in Denmark and Italy Using Two LCA Models. *Waste Manag. Res.* **2011**, *29*, S78–S90. [CrossRef]
30. Tabata, T. Waste-To-Energy Incineration Plants as Greenhouse Gas Reducers: A Case Study of Seven Japanese Metropolises. *Waste Manag. Res.* **2013**, *31*, 1110–1117. [CrossRef]
31. Shareefdeen, Z.; Elkamel, A.; Tse, S. Review of current technologies used in municipal solid waste-to-energy facilities in Canada. *Clean Technol. Environ. Policy* **2015**, *17*, 1837–1846. [CrossRef]
32. Gasification Technologies Council (GTC). *Gasification: The Waste and Energy Solution*; Gasification Technologies Council (GTC): Penwell, TX, USA, 2011.
33. World Energy Council (WEC). *World Energy Resources: Waste to Energy*; World Energy Council: London, UK, 2013. Available online: https://www.worldenergy.org/wp-content/uploads/2013/10/WER_2013_7b_Waste_to_Energy.pdf (accessed on 2 December 2019).
34. Theroux, M. Gasification vs. Incineration. 2014. Available online: http://www.houstontx.gov/onebinforall/Gasification_vs_Incineration.pdf (accessed on 22 February 2020).
35. Qazi, W.A.; Abushammala, F.M.; Azam, M.-H. Multi-criteria decision analysis of waste-to-energy technologies for municipal solid waste management in Sultanate of Oman. *Waste Manag. Res.* **2018**, *36*, 594–605. [CrossRef] [PubMed]
36. European Commission (EC). *Integrated Pollution Pre-vent and Control: Reference Document on the Best Available Techniques for Waste Incineration*; European Commission: Brussels, Belgium, 2006. Available online: http://eippcb.jrc.ec.euro-pa.eu/reference/BREF/wi_bref_0806.pdf (accessed on 10 December 2019).
37. DEFRA. Advanced Thermal Treatment of Municipal Solid Waste. Department for Environment Food & Rural Affairs, 2013. Available online: https://www.gov.uk/gov-ernment/uploads/system/uploads/attachment_data/file/221035/pb13888-thermal-treatment-waste.pdf (accessed on 11 October 2019).
38. National Research Council (NRC). Waste Incineration Overview. In *Waste Incineration and Public Health*; National Academy Press: Washington, DC, USA, 2000; p. 364.
39. Atwadkar, R.R.; Jafhav, L.D.; Wagh, M.M.; Shinde, N.N. A Multi Criteria Ranking of Different Technologies for the Waste to Energy of Municipal Solid Waste in the City of Kolhapur. *Int. J. Emerg. Technol. Adv. Eng.* **2014**, *4*, 937–942.
40. Lu, X. Understanding Hydrothermal Carbonization of Mixed Feedstock for Waste Conversion. Ph.D. Thesis, University of South Carolina, Columbia, SC, USA, 2014.
41. Wong, D.; Tam, S. *Thermal Waste Treatment in the European Union*; The Legislative Council: Hong Kong, China, 2016.
42. Chiu, H.Y.; Pai, T.Y.; Liu, M.H.; Chang, C.A.; Lo, F.C.; Chang, T.C.; Lo, H.M.; Chiang, C.F.; Chao, K.P.; Lo, W.Y.; et al. Electricity Production from Municipal Solid Waste using Microbial Fuel Cells. *Waste Manag. Res.* **2016**, *34*, 619–629. [CrossRef] [PubMed]
43. *Ricardo Energy & Environment Thermal Treatment*; Ricardo-AEA Ltd.: Harwell, UK, 2016. Available online: https://d2oc0ihd6a5bt.cloudfront.net/wp-content/up-loads/sites/837/2016/03/B4_0_RICARDO_THER-MAL.pdf (accessed on 12 December 2019).
44. Waste Management Resources (WMR). Incineration. 2009. Available online: http://www.wrfound.org.uk/articles/in-cineration.html (accessed on 20 October 2019).
45. Yuan, J.; Emura, K.; Farnham, C. Effects of Recent Climate Change on Hourly Weather Data for HVAC Design: A Case Study of Osaka. *Sustainability* **2018**, *10*, 861. [CrossRef]
46. Salvato, J.; Nemerow, N.; Agardy, F. *Environmental Engineering and Sanitation*, 5th ed.; Wiley: New York, NY, USA, 2003.
47. Tatarniuk, C. The Feasibility of Waste-To-Energy in Saskatchewan based on Waste Composition and Quantity. Master's Thesis, University of Saskatchewan, Saskatoon, SK, Canada, 2007. Available online: http://citeseerx.ist.psu.edu/viewdoc/download?pdf (accessed on 15 October 2019).
48. Tolis, A.I.; Rentizelas, A.; Aravossis, K.G.; Tatsiopoulos, I.P. Electricity and Combined Heat and Power from Municipal Solid Waste; Theoretically Optimal Investment Decision Time and Emissions Trading Implications. *Waste Manag. Res.* **2010**, *28*, 985–995. [CrossRef]
49. Pirotta, F.; Ferreira, E.; Bernardo, C. Energy Recovery and Impact on Land use of Maltese Municipal Solid Waste Incineration. *Energy* **2013**, *49*, 1–11. [CrossRef]

50. Ouda, O.K.M.; Raza, S.A. Waste-to-Energy: Solution for Municipal Solid Waste Challenges-Global Perspective. In Proceedings of the IEEE, International Symposium on Technology Management and Emerging Technologies (ISTMET), Bandung, Indonesia, 27–29 May 2014.
51. Souza, S.D.; Horttanainen, M.; Antonelli, J.; Klaus, O.; Lindino, C.A.; Nogueira, C.E.C. Technical Potential of Electricity Production from Municipal Solid Waste Disposed in The Biggest Cities in Brazil: Landfill Gas, Biogas and Thermal Treatment. *Waste Manag. Res.* **2014**, *32*, 1015–1023. [CrossRef]
52. Morselli, L.; De Robertis, C.; Luzi, J.; Passarini, F.; Vassura, I. Environmental Impacts of Waste In-cineration in A Regional System (Emilia Romagna, Italy) Evaluated from A Life Cycle Perspective. *J. Hazard. Mater.* **2008**, *159*, 505–511. [CrossRef]
53. Margallo, M.; Aldaco, R.; Irabien, A.; Carrillo, V.; Fischer, M.; Bala, A.; Palmar, P.F. Life Cycle Assessment Modelling of Waste-To-Energy Incineration in Spain and Portugal. *Waste Manag. Res.* **2014**, *32*, 492–499. [CrossRef]
54. Kairytė, A.; Kremensas, A.; Vaitkus, S.; Członka, S.; Strąkowska, A. Fire Suppression and Thermal Behavior of Biobased Rigid Polyurethane Foam Filled with Biomass Incineration Waste Ash. *Polymers* **2020**, *12*, 683. [CrossRef]
55. Ouadi, M.; Brammer, J.G.; Kay, M.; Hornung, A. Fixed Bed Downdraft Gasification of Paper Industry Wastes. *Appl. Energy* **2013**, *103*, 692–699. [CrossRef]
56. Narnaware, S.; Srivastava, N.; Vahora, S. Gasification: An alternative solution for energy recovery and utilization of vegetable market waste. *Waste Manag. Res.* **2017**, *35*, 276–284. [CrossRef] [PubMed]
57. Warren, K.; Gandy, S.; Davis, G.; Read, A.; Fitzgerald, J.; Emelia Holdaway, E. *Waste to Energy Background Paper*; Ricardo-AEA: Oxfordshire, UK, 2013. Available online: https://www.greenindustries.sa.gov.au/_literature_165466/Waste_to_Energy_Background_Paper_ (accessed on 22 October 2019).
58. Ni, F.; Chen, M. Studies on Pyrolysis and Gasification of Automobile Shredder Residue in China. *Waste Manag. Res.* **2014**, *32*, 980–987. [CrossRef] [PubMed]
59. Zafar, S. Waste to Energy (WTE) Conversion. 2008. Available online: http://www.altenergymag.com/content.php?issue_number=08.08.01&article=wastetoenergy (accessed on 2 November 2019).
60. Stantec. *Waste to Energy: A Technical Review of Municipal Solid Waste Thermal Treatment Practices*; Stantec Consulting Ltd.: Burnaby, BC, Canada, 2011. Available online: http://www.incinerateur.qc.ca/documents/bcmoe-wte-emmissions-rev-mar2011.pdf (accessed on 12 December 2019).
61. Velghe, I.; Carleer, R.; Yperman, J.; Schreurs, S. Study of the Pyrolysis of Municipal Solid Waste for the Production of Valuable Products. *J. Anal. Appl. Pyrolysis* **2011**, *92*, 366–375. [CrossRef]
62. Raclavska, H.; Corsaro, A.; Hlavsová, A.; Juchelková, D.; Zajonc, O. The Effect of Moisture on the Release and Enrichment of Heavy Metals during Pyrolysis of Municipal Solid Waste. *Waste Manag. Res.* **2015**, *33*, 267–274. [CrossRef] [PubMed]
63. Gao, F. Pyrolysis of Waste Plastics into Fuels. Ph.D. Thesis, University of Canterbury, Christchurch, New Zealand, 2010. Available online: https://ir.canterbury.ac.nz/bitstream/handle/10092/4303/Thesis_fulltext.pdf;sequence=1 (accessed on 10 December 2019).
64. Williams, P.T.; Besler, S. The influence of temperature and heating rate on the slow pyrolysis of biomass. *Renew. Energy* **1996**, *7*, 233–250. [CrossRef]
65. Leal-Quirós, E. Plasma processing of municipal solid waste. *Braz. J. Phys.* **2004**, *34*, 158–1593. [CrossRef]
66. Ducharme, C. Technical and Economic Analysis of Plasma-Assisted Waste-to-Energy Processes. Master's Thesis, Columbia University, New York, NY, USA, 2010. Available online: http://www.seas.columbia.edu/earth/wtert/sofos/ducharme_thesis.pdf (accessed on 22 October 2019).
67. Woodford, C. Plasma Arc Recycling. 2016. Available online: http://www.explainthatstuff.com/plasma-arc-recycling.html (accessed on 20 October 2019).
68. Walker, K. Applications of Thermal Depoly-merization. 2013. Available online: http://www.azocleantech.com/article.aspx? (accessed on 20 October 2019).
69. Lemley, B.; Tony, L. Anything into Oil. *Discov. Mag.* **2003**, *20*, 12–18. Available online: http://discovermagazine.com/2003/may/featoil/ (accessed on 20 March 2020).
70. Burne, J. Is this the Ultimate Recycler? The Guardian. 2003. Available online: https://www.theguardian.com/education/2003/may/22/research. (accessed on 20 October 2019).
71. Mendez, D. *Hydrous Thermal Depolymerization*; Stanford University: Stanford, CA, USA, 2010. Available online: http://large.stanford.edu/courses/2010/ph240/mendez2/ (accessed on 25 November 2019).

72. Adams, T.N.; Appel, B.S. Converting Turkey Offal into Bio-derived Hydrocarbon Oil with the CWT Thermal Process. In Proceedings of the Power-Gen Renewable Energy Conference 2004, Las Vegas, NV, USA, 1–3 March 2004.
73. He, C.; Giannis, A.; Wang, J.Y. Conversion of sewage sludge to clean solid fuel using hydrothermal carbonization: Hydrochar fuel characteristics and combustion behavior. *Appl. Energy* **2013**, *111*, 257–266. [CrossRef]
74. Libra, J.A.; Ro, K.S.; Kammann, C.; Funke, A.; Berge, N.D.; Neubauer, Y.; Titirici, M.; Fühner, C.; Bens, O.; Kern, J.; et al. Hydrothermal car-bonization of biomass residuals: A comparative review of the chemistry, processes and applications of wet and dry pyrolysis. *Biofuels* **2011**, *2*, 89–124. [CrossRef]
75. Ramke, H.G.; Blohse, D.; Lehmann, H.J.; Fettig, J. Hydrothermal Carbonization of Organic Waste. In Proceedings of the CISA, Twelfth International Waste Management and Landfill Symposium, Cagliari, Italy, 5–9 October 2009.
76. Berge, N.; Li, L.; Flora, J.R.V.; Ro, K.S. Assessing the Environmental Impacts of Energy Production from Hydrochar Generated via Hydrothermal Carbonization of Food Waste. *Waste Manag.* **2015**, *43*, 203–217. [CrossRef]
77. Lin, Y.; Ma, X.; Peng, X.; Yu, Z. Forecasting the Byproducts Generated by Hydrothermal Carbonization of Municipal Solid Wastes. *Waste Manag. Res.* **2016**, *35*, 92–100. [CrossRef] [PubMed]
78. Karaj, S.; Rehl, T.; Leis, H.J.; Müller, J. Analysis of Biomass Residues Potential for Electrical Energy Generation in Albania. *Renew. Sustain. Energy Rev.* **2010**, *14*, 493–499. [CrossRef]
79. Xhitoni, A. Renewable Energy Scenarios for Albania. Available online: https://core.ac.uk/display/12980736 (accessed on 17 March 2020).
80. Hoxha, B.B.; Dervishi, D.; Sweeney, K.; Bahja, E. Waste-to-Fuel Technology in Albania—Development System to Support an active Drilling Industry. In Proceedings of the 26th European Biomass Conference and Exhibition, Copenhagen, Denmark, 14–17 May 2018; pp. 1603–1616.
81. Bino, T. A New Path for the Sustainable Development: A Green Economy for Albania. The United National Conference on Sustainable Development, RIO+20. 2012. Available online: https://sustainabledevelopment.un.org/content/documents/1014albanianationalreport.pdf (accessed on 17 March 2020).
82. Alcani, M.; Dorri, A.; Hoxha, A. Some Issues of Municipal Solid Waste Management in Albania and Especially Tirana City. *Int. J. Sci. Technol. Manag.* **2015**, *4*, 16–22.
83. Lico, E.; Vito, S.; Boci, I.; Marku, J. Situation of Plastic Waste Management in Albania Problems and Solutions According to European Directive. *J. Int. Acad. Res. Multidiscip.* **2015**, *3*, 176–184.
84. Tushaj, V. AKBN Annual Report—Analiza Vjetore per Vitin 2016. Available online: http://www.akbn.gov.al/ (accessed on 20 March 2020).
85. Buçpapaj, B. Albania: Tracing the Paths of Biomass Energy. *J. Environ. Sci. Comput. Sci. Eng. Technol.* **2012**, *1*, 316–328.
86. Lushaj, B. Renewable Biomass: Suggestions and Proposed Alternatives. *World J. Environ. Biosci.* **2012**, *1*, 100–110.
87. Jaupaj, O.; Berdufi, I.; Mitrushi, D.; Lushaj, B. Biomass Energy Production—Research Gaps and Trends. In Proceedings of the International Scientific Conference: Hydro Climate Resources—An Important Tool for a Sustainable Development of Albania, Tirana, Albania, 16 June 2011.
88. United Nations Economic Commission for Europe. Report Environmental Performance Reviews—Albania—Second Review. Available online: http://www.unece.org/fileadmin/DAM/env/epr/epr_studies/AlbaniaII.pdf (accessed on 17 March 2020).
89. European Neighbourhood Policy and Enlargement Negotiations. EU Environment & Climate Change. Available online: https://ec.europa.eu/neighbourhood-enlargement/policy/conditions-membership/chapters-of-the-acquis_en (accessed on 17 March 2020).
90. Ministry of Environment, Forestry and Water Administration, Albania. Available online: http://adaptation-undp.org/partners/ministry-environment-forestry-and-water-administration-albania (accessed on 20 March 2020).
91. European Topic Centre on Sustainable Consumption and Production (ETC/SCP). Waste Policies Factsheet for Albania. Available online: http://scp.eionet.europa.eu/facts/factsheets_waste/2011_edition/factsheet?country=AL (accessed on 20 March 2020).

92. Ministry of Public Works, Transport and Telecommunication. Waste Generation Data. Available online: http://www.mppt.gov.al/ (accessed on 22 March 2020).
93. European Environment Agency (EEA). Available online: https://www.eea.europa.eu/countries-and-regions/albania (accessed on 22 March 2020).
94. Environmental Policy in Albania. United Nations Development Programme. Available online: http://www.al.undp.org/content/albania/en/home/ourwork/environmentandenergy/overview.html (accessed on 24 March 2020).
95. National Waste Management Strategy. Available online: http://wastepolicy.environment.gov.za/ (accessed on 24 March 2020).
96. Eurostat. Available online: https://ec.europa.eu/eurostat/data/database (accessed on 22 March 2020).
97. Albania Institute of Statistics. Available online: http://databaza.instat.gov.al/pxweb/en/DST/START__EN (accessed on 22 March 2020).
98. Chifari, R.; Piano, S.L.; Bukkens, S.G.; Giampietro, M. A holistic framework for the integrated assessment of urban waste management systems. *Ecol. Indic.* **2018**, *94*, 24–36. [CrossRef]
99. Nsair, A.; Onen Cinar, S.; Alassali, A.; Abu Qdais, H.; Kuchta, K. Operational Parameters of Biogas Plants: A Review and Evaluation Study. *Energies* **2020**, *13*, 3761. [CrossRef]
100. Grigore, L.S.; Priescu, I.; Joita, D.; Oncioiu, I. The Integration of Collaborative Robot Systems and Their Environmental Impacts. *Processes* **2020**, *8*, 494. [CrossRef]
101. Gug, J.; Cacciola, D.; Sobkowicz, M.J. Processing and properties of a solid energy fuel from municipal solid waste (MSW) and recycled plastics. *Waste Manag.* **2015**, *35*, 283–292. [CrossRef]

Article

Anaerobic Degradation of Environmentally Hazardous Aquatic Plant *Pistia stratiotes* and Soluble Cu(II) Detoxification by Methanogenic Granular Microbial Preparation

Olesia Havryliuk [1], Vira Hovorukha [1], Oleksandr Savitsky [2], Volodymyr Trilis [2], Antonina Kalinichenko [3,*], Agnieszka Dołhańczuk-Śródka [3], Daniel Janecki [3] and Oleksandr Tashyrev [1]

1 Department of Extremophilic Microorganisms Biology, Zabolotny Institute of Microbiology and Virology, National Academy of Sciences of Ukraine, 03143 Kyiv, Ukraine; gav_olesya@ukr.net (O.H.); vira-govorukha@ukr.net (V.H.); tach2007@ukr.net (O.T.)
2 Department of Ichthyology and Hydrobiology of River Systems, Institute of Hydrobiology, National Academy of Sciences of Ukraine, 12 Prosp. Geroiv Stalingradu, 04210 Kyiv, Ukraine; a_savitsky@ukr.net (O.S.); trylis@rambler.ru (V.T.)
3 Institute of Environmental Engineering and Biotechnology, University of Opole, 45-040 Opole, Poland; agna@uni.opole.pl (A.D.-Ś.); zecjan@uni.opole.pl (D.J.)
* Correspondence: akalinichenko@uni.opole.pl; Tel.: +48-787-321-587

Citation: Havryliuk, O.; Hovorukha, V.; Savitsky, O.; Trilis, V.; Kalinichenko, A.; Dołhańczuk-Śródka, A.; Janecki, D.; Tashyrev, O. Anaerobic Degradation of Environmentally Hazardous Aquatic Plant *Pistia stratiotes* and Soluble Cu(II) Detoxification by Methanogenic Granular Microbial Preparation. *Energies* **2021**, *14*, 3849. https://doi.org/10.3390/en14133849

Academic Editor: Gabriele Di Giacomo

Received: 4 June 2021
Accepted: 22 June 2021
Published: 25 June 2021

Abstract: The aquatic plant *Pistia stratiotes* L. is environmentally hazardous and requires effective methods for its utilization. The harmfulness of these plants is determined by their excessive growth in water bodies and degradation of local aquatic ecosystems. Mechanical removal of these plants is widespread but requires fairly resource-intensive technology. However, these aquatic plants are polymer-containing substrates and have a great potential for conversion into bioenergy. The aim of the work was to determine the main patterns of *Pistia stratiotes* L. degradation via granular microbial preparation (GMP) to obtain biomethane gas while simultaneously detoxifying toxic copper compounds. The composition of the gas phase was determined via gas chromatography. The pH and redox potential parameters were determined potentiometrically, and Cu(II) concentration photocolorimetrically. Applying the preparation, high efficiency of biomethane fermentation of aquatic plants and Cu(II) detoxification were achieved. Biomethane yield reached 68.0 ± 11.1 L/kg VS of *Pistia stratiotes* L. biomass. The plants' weight was decreased by 9 times. The Cu(II) was completely removed after 3 and 10 days of fermentation from initial concentrations of 100 ppm and 200 ppm, respectively. The result confirms the possibility of using the GMP to obtain biomethane from environmentally hazardous substrates and detoxify copper-contaminated fluids.

Keywords: biomethane; *Pistia stratiotes* L. plants; copper bioremoval; anaerobic degradation of hazardous plants; environmental biotechnology; bioremediation; biomethane production

1. Introduction

Today, our planet suffers from a number of environmental problems that require globally effective approaches to solve them [1]. They include global climate change [2], including rising temperatures, accumulation of organic waste [3,4], and pollution by toxic metals [5]. These factors have a serious impact on many species of plants, animals, and microorganisms, as well as on the relationships between their populations. In the past, the change of the habitat of a particular species of animal or plant was rare. Now, such changes are happening very quickly and unsystematically under the influence of significant man-made load and global environmental changes [6]. This has become one of the main reasons for the change in the species composition of biota, in particular the reduction of aboriginal species [7]. In this regard, such phenomena as uncontrolled outbreaks of growth of various alien monospecies populations of aquatic macrophytes have become more frequent in the aquatic ecosystems of Ukraine [8]. *Pistia stratiotes* L. is one of the ecologically dangerous alien aquatic macrophytes.

Pistia stratiotes L. is a tropical aquatic plant that floats on the water surface. It is widespread in tropical and subtropical areas and in some places reaches mass development, forming a continuous carpet on the surface of water bodies [9]. The species *P. stratiotes*, also known as *Jalkumbhi*, is native to South America, is not a frost-resistant species, and is widely found in ponds, streams, and free-flowing rivers, mainly in tropical and subtropical regions of Asia, Africa, and America [10]. Within the traditional range, the growth of pistia is constrained by its natural enemies and local environmental conditions. But as the global climate warmed, pistia began to spread to countries with a temperate climate. Here, pistia have no natural enemies [11]. In such places, the massive growth of pistia can become uncontrolled, causing the degradation of local aquatic ecosystems and significant economic damage. The harmfulness of these plants is determined by their excessive growth in water bodies, which interferes with the photosynthesis and respiration of aquatic organisms and critically degrades water quality [12]. *P. stratiotes* L. is a free-floating plant with no odor and a bitter taste. This plant usually forms a dense mat on the surface of the water [13]. *P. stratiotes* is widespread as an ornamental plant, is often used in aquariums and garden ponds [14], is often used as a biofertilizer, and is also known as a medicinal plant [15]. However, *P. stratiotes* is able to block navigation channels, prevent fishing and transportation of boats, block the flow of water in irrigation canals, and disrupt the production of hydropower [14]. The dense plant mass of *P. stratiotes* above the water surface impairs the penetration of light and oxygen supply into water bodies and, thus, destroys their biodiversity [9]. Mechanical removal of these plants is widespread [9], but later they begin to rot on the riverbank and again contaminate ecosystems with toxic microbial exometabolites such as fatty acids an alcohols. However, the biomass of pistia is a natural carbohydrate-containing polymer [16]. Therefore, we assumed that the polymers of pistia, like cellulose and starch, could be fermented via anaerobic microorganisms with the formation of CH_4 [16]. The use of other polymeric substrates to produce biomethane is also widespread. Substrates used include meat processing waste [17], wastewater sludge [18], food waste [19], etc. However, anaerobic fermentation of aquatic plant materials is less described.

Hence, it follows that a double-positive result can be achieved with anaerobic degradation of the biomass of pistia: fast and efficient degradation of the ecologically hazardous biomass of pistia and the synthesis of the energy carrier CH_4. Pistia appeared in the water bodies of Europe in the second half of the 20th century. It was first recorded in the canals of the Netherlands in 1973 [20,21], where a periodic active growth of plants in the summer was detected in subsequent years [22]. In 1998, a massive proliferation of pistia in the waters of northern Italy was recorded [23,24]. Now, it is widespread throughout Italy [23]. In Europe, pistia is found in the Czech Republic [25], Slovenia [20], Serbia [26], and in some other countries. In recent years, pistia has been found in the Great Lakes region of North America [27]. The current distribution of this plant is so global that in some EU countries it is even forbidden to use it as an aquarium plant [28]. The intensive spread of pistia in the water reservoirs of Ukraine is greatly facilitated by its popularity among aquarists and landscape designers [29]. The rapid spread of *Pistia stratiotes* has also been observed in Serbia, especially in the Bega River, which has connections with a fairly large part of the country's river network. It is believed that the appearance of *P. stratiotes* is associated with its spontaneous spread from neighboring Romania, where the plant has been recorded for about ten years [26]. In Egypt, *P. stratiotes* occurs in the slow-flowing canals of the northern Nile Delta and reaches the city of Embaba near Cairo, as well as in calm and still waters, especially around the city of Fariskur. It was recently recorded in Lakes Mariut and Mansala in the north of the Nile Delta [30].

Although its specimens usually do not survive the winter and do not create a stable population, the situation may change due to climate change. In Ukraine, the first findings of pistia were registered near Kyiv in 2005. Since then, it has repeatedly occurred here in natural reservoirs, but did not reach mass development [31]. In 2013–2014, a mass outbreak of pistia was registered on the Seversky Donets River in the vicinity of Kharkiv [32]. This outbreak caused economic problems (restrictions on shipping, recreation, and fishing), and more than 6 million UAH were allocated from the regional budget to fight pistia alone. The problem of spontaneous distribution of invasive aquatic macrophytes, in particular *P. stratiotes*, is relevant in Ukraine. This is due to climate change and rising temperatures in the winter [8]. Rapid distribution of the species in the lowlands of the Dnister River and in the Dnipro and Siversky Donets Rivers has been reported [33]. Single specimens of the plant were found both in the city hydrographic network and in the shallow water of the Dnipro River near Kyiv [34]. It has been experimentally confirmed that after the removal of invasive freshwater plants, including *P. stratiotes*, the biodiversity of water bodies is restored [35]. Various methods are used to control these weeds in aquatic environments (e.g., preventive, mechanical, biological, and chemical). The most widespread method is mechanical collection. However, mechanical removal of plants leads to the accumulation of a huge amount of wet weed biomass. In most cases, it is disposed of in wastelands or burned after drying. An alternative is the use of waste in other industrial processes [9].

Since the significant growth of pistia has led to a number of economic and environmental problems, the issue of combating this phenomenon, as well as its consequences, is relevant. World experience shows the mechanical removal of pistia from the reservoir to be the most effective method (Global Invasive Species Database). This method requires fairly resource-intensive technology [36]. Pistia biomass, however, can be seen as a valuable raw material, the effective use of which can not only reduce the cost of combating this invasive species, but also bring profits. Microbial degradation to produce biogas is one of the useful applications of plant biomass [16,37], which can be used on farms. Its incineration to convert into electricity is another application, but a more hazardous one [38]. Fermentation of aquatic plant material can be accompanied by the formation of biohydrogen or biomethane. The biomethane yield was quite high at 103–262 NmL CH_4/g VS during the fermentation of beach-cast seagrass wrack [39]. Aqueous hyacinth is also a hazardous plant that can be successfully used to produce biomethane [40]. A huge CH_4 yield was obtained—337 NL/kg VS_{added}—via its cofermentation with household waste [41]. The aquatic plant *Landoltia punctate* is a substrate for the production of biohydrogen, the yield of which can reach 2.14 mol H_2/mole of reduced sugar [42]. Thus, we consider aquatic plants to be promising renewable substrate for fermentation with biogas obtaining.

Another serious environmental problem in agricultural and industrialized countries is the pollution of natural biogeocenoses with toxic metals [43]. Copper compounds are one of the most common and environmentally hazardous pollutants [44]. The main sources of copper pollution are the uncontrolled use of copper-containing fungicides and pesticides [45], deposits and industrial mining [46], as well as industrial wastewater [47]. Copper in trace concentrations is a necessary trace element for the functioning of living organisms [48]. However, when its concentration increases, it is a very toxic metal. Copper, as a catalytically active metal, is able to influence numerous metabolic, functional, and regulatory pathways in living organisms. The mechanism of the negative action of the metal is to inhibit the functioning of a number of vital enzymes—acetylcholinesterase, succinate dehydrogenase proteases, lipases, and glucosidases [49]. The gradual accumulation of copper in agricultural soils leads to an increase in their phytotoxicity and absorption copper in agricultural crops. Widely used methods of purification of contaminated ecosystems from copper compounds include physicochemical methods: adsorption [50], membrane filtration [51], cementation [52], photocatalysis, and electrodialysis [53]. These methods are cost-effective but, in some cases, do not guarantee complete remediation of contaminated ecosystems [54]. In recent decades, there has been considerable interest in methods of bioremediation of contaminated ecosystems that are based on the use of microorganisms as major biotechnological agents [55]. The microbial pathways of copper removal are based on their accumulation, precipitation (to $Cu(OH)_2\downarrow$, $CuCO_3\downarrow$), as well as reduction to insoluble and nontoxic compounds ($Cu_2O\downarrow$) [56]. Transformation reactions of divalent copper compounds occur due to the release of microbial exometabolites both inside and outside the cells [57,58]. These associated reactions are clearly described in the method section. A number of studies have shown that microorganisms have the genetic potential to remove heavy metals from the environment [59]. The ability to transform copper compounds is determined by genetic determinants [58]. Thus, *cop* genes, chaperones, transporters, and sequestering molecules together encode copper removal mechanisms in different microorganisms [60]. Microbial biotechnology is economically viable because cheap organic substrates, including environmentally hazardous organic waste, which also includes the aquatic plant *Pistia stratiotes* L., can be used as a substrate for the growth of microorganisms [16]. Such promising integrated biotechnological approaches based on the simultaneous solution of several environmental problems are poorly studied and require significant fundamental and applied research. Thus, in this work we considered the possibility of using microbial preparations to provide a double-positive environmental effect—the degradation of the environmentally hazardous plant *Pistia stratiotes* L. and detoxification of hazardous copper compounds.

2. Materials and Methods

2.1. Production of the Granular Microbial Preparation

The method used to produce the granular microbial preparation (GMP) was as follows. The biomass of biomethane-synthesizing microorganisms was sampled in the first stage. Digested sludge from methane tanks at the Bortnytsia aeration station in Kyiv, Ukraine, was collected to be used as a source of methanogenic microorganisms. One liter of digested sludge was then mixed with the starting substrates and regulators of microbial metabolism to manufacture the granular microbial preparation in the second stage. The granules were made by extrusion using a miniextruder.

After forming, the granules were dried in a ventilated laboratory electric furnace at 40 °C to preserve the cells of methanogenic microorganisms. The GMP2 modification containing living methane-synthesizing microorganisms was then ready to use (Figure 1).

Figure 1. General view of the granular microbial preparation.

Preparation of the granules of hydrogen-synthesizing microorganisms (GMP1) is described in our previous work [61]. The preparation procedure for this type of GMP was similar to GMP2. The main difference was drying in a ventilated laboratory electric furnace at 105 °C after forming the granules. This temperature was used to destroy methanogens and select only spore-forming hydrogen-synthesizing microorganisms.

2.2. Sampling and Preparation of Substrate for Anaerobic Fermentation

Plants of *Pistia stratiotes* L. were used as a substrate for degradation via hydrogen-synthesizing (GMP1) and methanogenic (GMP2) microbial preparations.

Samples of aquatic plants were collected from Zoloche Lake (Kyiv region, Ukraine) in October 2020 (Figure 2A). For selection, the generally accepted technique of floristic and geobotanical research of features of formations of the higher water vegetation on reservoirs was used: establishment of features of overgrowth of various sites of reservoirs with the subsequent mapping of a vegetative cover of a reservoir. The biomass was collected during the expedition trip using the trial plots method and the ecological-coenotic profile method [62,63]. Plants were selected only with a satisfactory physiological condition without signs of slowing of vegetation or death. The plants were dried at 105 °C to a constant weight. Plants were dried to kill the most viable microorganisms to avoid the effect of their own microbiome on the experimental fermentation process and distortion of the results. The dried plants were ground (the size was 0.5–2.0 cm) before loading in sealed glass jars to ferment (Figure 2B).

Figure 2. The view of the *Pistia stratiotes* L. plants on the sampling site (**A**) and after drying (**B**) before loading into the sealed glass jars.

2.3. Description of the Fermentation Process

The fermentation of plants was carried out under three conditions: without GMP and with the methanogenic and hydrogen-synthesizing types of GMP to compare the effectiveness of the fermentation process.

To study the dynamics of pistia fermentation, 5 g of samples dried at 105 °C and 400 mL of boiled tap water were loaded into sealed glass jars with a total volume of 500 mL (Figure 3). The initial gas phase was air.

In one case, no GMP was added to the sealed glass jars to check the possibility of spontaneous fermentation of plants by pistia's own microbiome (Figure 3A).

In the other two cases, the samples were dried and treated with 85–90 °C tap water and cooled. Afterward, 2 g of the hydrogen-synthesizing GMP or methanogenic microbiome GMP were loaded into sealed glass jars (second (Figure 3B) and third (Figure 3C) cases, respectively). The bioreactor was closed with rubber stoppers with fittings to sample the aliquots of culture fluid and gas and to remove the synthesized gas. The synthesized gas flowed through the gas controller to the gas holder. The fermentation cycle took place over 76 days at 30 °C.

Figure 3. General view of the sealed glass jars with a total volume of 500 mL: (**A**) control variant without preparations; (**B**) with GMP1 modification containing living hydrogen-synthesizing microorganisms; (**C**) with GMP2 modification containing living methane-synthesizing microorganisms.

The following parameters were monitored: pH, redox potential (Eh), gas volume [64], and the concentrations of H_2, O_2, N_2, CO_2, and CH_4 in the gas phase, as well as total carbon [65] and ammonium ion (NH^{4+}) concentration [66] in the culture fluid.

The completion of the process was evidenced by the stabilization of pH and increase in the redox potential of the medium, termination of gas synthesis, and visual degradation of solid waste particles [67]. The experiment was carried out in triplicate (n = 3). Standard deviation (SD) and average values (x) were calculated.

2.4. Measurement of the Dynamic of Toxic Copper Bioremoval during the Fermentation of *P. stratiotes* L. Plants

The most efficient fermentation process was selected after completion of all fermentation cycles with H_2-producing and CH_4-producing types of GMP to determine the possibility of toxic Cu^{2+} microbial removal. A variant with GMP2 based on a methanogenic microbiome was used. The substrate preparation methods were identical to those described in Section 2.3.

A solution of 60,000 ppm Cu^{2+} was used as the stock. It was prepared by $CuSO_4 \cdot 5H_2O$ dissolving in distilled water in a volumetric flask. Citrate was used as a chelator. The Cu(II) concentration was determined by using the colorimetric method with 4-(2-pyridylazo)resorcinol (PAR) as well as the method of substituent titration with PAR and ethylenediaminetetraacetic acid (EDTA) in the ranges of 0.5–7.0 ppm and 25.0–3000.0 ppm Cu^{2+}, respectively. The methods are based on the property of PAR to form colored red complexes with cations of bivalent heavy metals, including Cu^{2+} [68].

The copper(II) citrate solution was added to the sealed glass jars when microorganisms reached the stage of active metabolic activity (49 day into the fermentation). This was estimated by the increase of biomethane synthesis (concentration of CH_4 in the gas phase of bioreactor increased to 60%) and the decrease of the Eh (in this experiment, Eh was decreased to −30 and −70 mV). Afterward, the copper solution was added to the bioreactor with the final concentrations 100 ppm and 200 ppm Cu^{2+} in the citrate forms, respectively. The experiment was carried out in triplicate (n = 3). Standard deviation (SD) and average values (x) were calculated.

2.5. Measurement of the Fermentation Parameters

The redox potential (Eh) and pHof the medium were measured potentiometrically. For this purpose, measurement of pH and Eh was performed using the EZODO MP-103 universal ionomer with the remote electrodes and temperature sensor. To measure pH and Eh, we used the combined Ezodo ceramic chloride electrodes with BNC connectors—PY41 and PO50 models, respectively. Before the measurement, the electrodes were tested by standard buffer solutions. For the pH check, standard pH buffer solutions were used:

a solution of $KHC_2O_4H_2C_2O_4{\cdot}2H_2O$ (pH= 1.68), a mixture of NaH_2PO_4 and K_2HPO_4 (pH = 6.86), and $Na_2B_4O_7{\cdot}10H_2O$ (pH = 9.18). Standard pH buffers were prepared according to the producer's manual (OJSC "Kiev plant RIAP"). For the validation of Eh measurements, three redox buffer solutions were used. First—ferricyanide, with Eh = +273 mV (13.5 g/L $K_3[Fe(CN)_6]$ and 3.8 g/L $K_4[Fe(CN)_6]{\cdot}3H_2O$), second—Fe(II) citrate (10.0 g/L, with Eh = −150 mV), and third—Ti(III) citrate (15.0 g/L, with Eh = −440 mV) [69].

The gas synthesis was determined by the volume of water squeezed from the gas holder into the intake manifold under the pressure of the synthesized gas.

The H_2 and CH_4 concentration was determined by using the standard gas chromatography method [64]. The chromatograph was equipped with two steel columns: I—for the analysis of H_2, O_2, N_2 and CH_4 and II—for the analysis of CO_2. Column parameters were: I—l = 3 m, d = 3 mm, with molecular sieve 13X (NaX); II—l = 2 m, d = 3 mm, with Porapak Q carrier; column temperature +60 °C, evaporator temperature +75 °C, detector temperature (60 °C), detector current—50 mA. The carrier gas was argon and the flow rate of the gas was 30 cm^3/min. The concentration of H_2 was calculated by the peak squares of its components.

To determine the concentration of soluble organic compounds by the content of the total carbon in the medium, the permanganate method was used [65]. It is based on the oxidation of carbon compounds by the potassium permanganate in strong acidic conditions. The mixture of an aliquot of the culture fluid (1 mL) with 0.1 mL of concentrated sulfuric acid was heated in a boiling water bath. The aliquots (0.1 mL) of 0.1% potassium permanganate solution were gradually added until a steady light pink color appeared. This indicated the completion of the process of carbon compound oxidation by MnO_4^-. Carbon concentration was determined according to standard calibration curves.

2.6. Calculation of the Fermentation Parameters of the Efficiency of Fermentation

The evaluation of the efficiency of the process of the fermentation of the *Pistia stratiotes* L. plants was determined by the following parameters:

- waste degradation time (T, days)—defined as the duration of the process from the moment of the fermentation start until its termination (the termination of gas synthesis, etc.),
- molecular hydrogen yield—calculated as the amount of H_2 (L) synthesized from 1 kg of waste counting to the volatile solids (VS),
- biomethane yield—calculated as the amount of CH_4 (L) synthesized from 1 kg of waste counting to the volatile solids (VS),
- coefficient of waste degradation (Kd)—the degree of waste digestion. For pistia plants, it was calculated as the ratio of initial and final weight of waste counting to the VS: Kd = m_1:m_2 (where m_1 is the initial weight of dry waste; m_2 is the weight of dry detritus). To determine the initial weight of waste, it was dried to a constant weight at 105 °C and weighed. To determine the weight of detritus, the non-fermented residues after fermentation were washed in distilled water, in the solution of weak organic acid, and again in distilled water. The washed residues were dried to a constant weight at 105 °C and weighed.

2.7. Theory/Calculation

Thermodynamic prediction was applied as the theoretical background to predict the possibility of citrate complex of Cu^{2+} removal from the solution by anaerobic microorganisms. The thermodynamic prediction allows for theoretically substantiating the most effective mechanisms of soluble toxic copper(II) ion detoxification by microbial reduction to insoluble copper(I) in the form of $Cu_2O{\downarrow}$ or precipitation of Cu^{2+} in the forms of $Cu(OH)_2{\downarrow}$, $CuCO_3{\downarrow}$, etc. [70].

In accordance with our prediction, microbial metabolism is possible only in the zone of thermodynamic stability of water, i.e., in the range of values of the standard redox potential (E_o') from −414 to +814 MB [71]. The standard redox potential of the reaction

$2Cu^{2+} + 2e + H_2O = Cu_2O\downarrow + 2H^+$ is equal to +380 mV at low concentrations (equal or less than 0.01 M Cu^{2+}). Consequently, the redox potential of this reaction is located inside the zone of thermodynamic stability of water (from −414 to +814 mV). Therefore, reaction of Cu^{2+} reduction to insoluble copper oxide(I) is theoretically premised to be carried out by microorganisms. We consider metal reduction by microorganisms as a binary redox reaction. In this reaction, metabolically active microorganisms are the donor system, and high-potential compounds of copper(II) are the acceptor system. It is obvious that the efficiency of metal reduction is proportional to the potential difference between the acceptor and donor systems. It follows that microorganisms with the lowest values of the redox potential must reduce copper(II) with maximum efficiency. Strict anaerobic bacteria are known to create the lowest redox potential (E_o' = −414 mV). The redox potential of the Cu^{2+} reduction $Cu^{2+} + H_2O + 2e = Cu_2O\downarrow + 2H^+$ is +380 mV (pH = 4.6). It is obvious that the potential difference in 794 mV between the acceptor and donor systems will ensure the fastest possible reduction of soluble copper(II) compounds to insoluble $Cu_2O\downarrow$. From this background it follows that only low-potential strict anaerobes, including biohydrogen- and biomethane-producing microbial communities, are the most effective in removing copper compounds from solutions.

Microbial removal of Cu^{2+} compounds from a solution is possible not only due to the reduction reaction but also as a result of substitution (precipitation) reactions. Firstly, nontoxic insoluble copper hydroxide $Cu(OH)_2\downarrow$ can precipitate, owing to the biologically mediated increase of the medium pH up to 4.5–5.0. Secondly, the Cu^{2+} cation can interact with the $CO_3{}^{2-}$ anion and form insoluble copper carbonate $CuCO_3\downarrow$ in neutral and slightly alkaline conditions: $Cu^{2+} + CO_3{}^{2-} = CuCO_3\downarrow$ [56]. Thus, all metabolic pathways of microorganisms interacting with copper(II) can be theoretically justified via the thermodynamic prediction method to provide the environmental effect of the purification of contaminated solutions and industrial sewage from soluble toxic Cu(II) compounds. Methanogenic microorganisms as the components of the microbial communities of natural and man-made ecosystems are strict anaerobes. In natural microbiocenoses, they function simultaneously with sulfate reducers, which are able to provide them with a source of reduced sulfur (S^{2-}). Many species of methanogenic microorganisms do not have the enzyme dissimilation sulfate reductase, and some are able to synthesize hydrogen sulfide (H_2S) via dissimilation sulfur reduction in the presence of free sulfur S^0 as an electron acceptor and conventional energy substrates (H_2 or methanol) as electron donors. Thus, in the process of biomethane fermentation, a significant amount of hydrogen sulfide is released, which precipitates sulfides of divalent metals, in particular Cu^{2+}, due to the reaction $2Cu^{2+} + H_2S^{2-} \rightarrow Cu^2 + S^{2-}\downarrow + 2H^+$ [72]. That is why the metabolic pathway of biomethane fermentation of organic substrates can be a highly efficient pathway of microbial removal of toxic and soluble copper compounds from solutions.

3. Results

We registered a significant growth of pistia near Kyiv, in the system of lakes and canals connected to the discharge channel Bortnytska aeration station. At the end of summer, a pistia carpet covered a number of reservoirs, in particular, Lake Zoloche near the village of Vyshenky, the mirror coverage of which reached 90%. The pistia biomass in the cluster reached 10 kg/m^2, the total pistia biomass in the lake was about 2000 tons.

Three variants of *P. stratiotes* L. plant fermentation were investigated. The process of fermentation without the addition of GMP was simulated in the first case. In the second and third cases, the plants were fermented with granular microbial preparations as inoculum (biohydrogen and biomethane-synthesizing microorganisms, respectively). The basic fermentation parameters of plant degradation were obtained (Figure 4). The initial values of pH, redox potential, concentration of organic compounds and ammonium ions were approximately the same. Thus, the initial pH in all variants of the experiment ranged from 7.14 ± 0.4 to 7.18 ± 0.3 and Eh from +303 ± 11 to +329 ± 18.2 mV. The concentration of $NH4^+$ ranged from 17.2 ± 8.3 to 23.1 ± 11.5 ppm and the concentration of soluble organic compounds was 30 ± 14.5 to 41 ± 12.1 ppm (Figure 4). In all vials at the beginning of fermentation, the gas phase was air and consisted of O_2 (from 20 ± 2% to 22 ± 2%), N_2 (from 77 ± 7% to 80 ± 5%), and CO_2 (from 0.15 ± 0.05% to 0.22 ± 0.1%) (Figure 5).

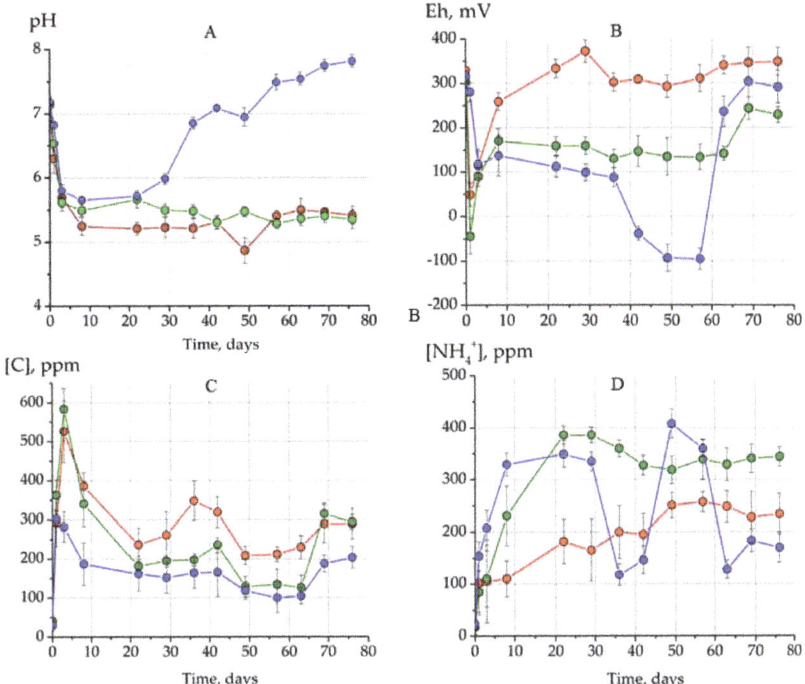

Figure 4. The main metabolic parameters of *P. stratiotes* L. plant fermentation: pH (**A**); Eh (**B**); concentration of total carbon (**C**); concentration of ammonium ions (**D**). The graph shows the dynamics of changes in metabolic parameters in control conditions without GMP (red lines), in the presence of GMP1 based on hydrogen-synthesizing microorganisms (green lines), and in the presence of GMP2 based on methane-synthesizing microorganisms (blue lines).

In the control variant of the experiment, the degradation of the plant was not observed, and the metabolic parameters of the culture fluid indicated the absence of the fermentation process (biohydrogen and biomethane were not synthesized, pH and Eh did not change significantly) (Figures 4 and 5). Thus, in the variant of the experiment without granular microbial preparation (indicated by red lines on the graphs), high values of redox potential were observed during the whole fermentation process, which ranged from 258.6 ± 20.0 mV to 348.6 ± 31.0 mV at the 8th and 76th day of fermentation, respectively (Figure 4B). After treating the fragmented pistia with boiled water, most of the aboriginal microorganisms died. However, due to the residual microbiome there was a decrease in pH from 7.2 ± 0.1 to 5.2 ± 0.2 for 8 days (Figure 4A). The presence of viable microorganisms in the control version of the experiment was also evidenced by the increase in the concentration of total carbon and NH_4^+ (Figure 4C,D). This indicates the presence of processes of hydrolysis of pistia polymers and the accumulation of liquid organic compounds (products of hydrolysis) in the culture medium. Despite this, the fermentation process was insignificant, the synthesis of excess biohydrogen and biomethane was not observed, and the basis of the gas phase was CO_2 and N_2 (Figure 5B,C). The initial gas composition of the sealed glass jars was air. Air does not inhibit the anaerobic degradation of the plants. The presence of O_2 in the air, on the contrary, stimulates the growth of aerobic microorganisms. Aerobic microorganisms in a sealed jar completely consume O_2 and reduce the redox potential in a liquid medium. Such conditions are suitable for the growth of strict anaerobic microorganisms.

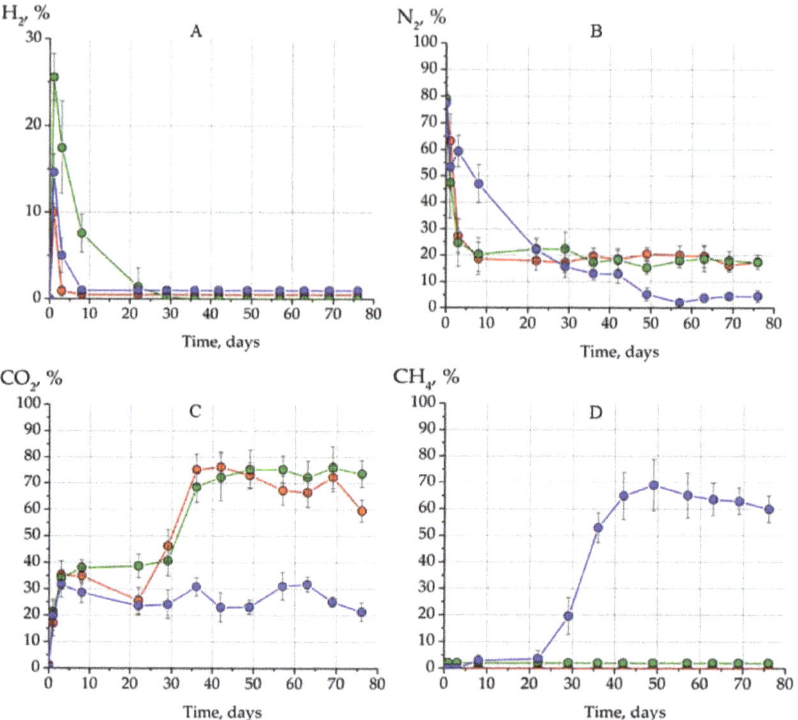

Figure 5. The composition of the gas phase ((**A**)—H_2; (**B**)—N_2; (**C**)—CO_2; (**D**)—CH_4) during fermentation of *P. stratiotes* L. plants in control conditions without GMP (red lines), in the presence of GMP1 based on hydrogen-synthesizing microorganisms (green lines), and in the presence of GMP2 based on methane-synthesizing microorganisms (blue lines).

Thus, the concentration of CO_2 in the gas phase reached 75.2 ± 5.8%, while the N_2 content was 19.7 ± 3.0% on day 36 of fermentation (Figure 5B,C). Hydrogen appeared in the gas phase on day 1 of cultivation (10.0 ± 1.3%), but later its concentration decreased to absolute zero (Figure 5A). Oxygen was completely transformed by microorganisms in all variants of the experiment within 1 day of fermentation, so its illustration on the graphs was impractical. In the first case, pistia was inaccessible to aboriginal microorganisms, and the fermentation process was inefficient. As a consequence, after 76 days of fermentation *Kd* was 1.3, and CO_2 yield was 2.7 ± 0.4 L/kg of plants.

Patterns similar to the control conditions (without GMP) were observed in the second case of the experiment (in the presence of GMP1). Thus, pistia plants were also poorly accessible for degradation via hydrogen-synthesizing microorganisms. The main metabolic parameters of the second variant of the experiment are presented in the form of green lines in Figures 4 and 5. The process of degradation under such conditions was also inefficient (*Kd* = 1.5). The H_2 yield was only 0.4 ± 0.06 L/kg. As in the control experiment, the pH decreased from 7.1 ±0.5 to 5.6 ± 0.1 during the first three days of fermentation (Figure 4A). The Eh decreased to −44.3 ± 19.9 mV on the first day of fermentation, but subsequently increased to positive values in the range of +170.6 ± 27.7 to +243 ± 25.6 mV (Figure 4B). However, in contrast to the control experiment, in the first three days there was a significant increase of the concentration of H_2 in the gas phase of the bioreactor.

Thus, the concentration of H_2 on the first day of fermentation was 25 ± 2.73%, but over time it decreased rapidly—up to 17.5 ± 5.31% on the 3rd day and up to 7.6 ± 2.2% on the 8th day (Figure 5A). During the next two weeks, the hydrogen concentration was at a level of 1.4 ± 0.9 to 2.0 ± 0.7% and then disappeared altogether. Excess gas synthesis was not observed in either the control version without GMP or in the presence of hydrogen-synthesizing granular microbial preparation. The CO_2 concentration increased moderately during the first month of fermentation and subsequently doubled sharply from 40.7 ± 5.8 (29 days of fermentation) to 72.3 ± 9.1% (42 days of fermentation) (Figure 5C). Active hydrogen synthesis on the first day of cultivation and subsequent cessation of its synthesis and the fermentation process in general may indicate rapid depletion of available substrate for hydrogen-synthesizing microorganisms during the first days of cultivation and the absence of enzyme systems for hydrolysis of *P. stratiotes* L. plants in GMP1.

The granular microbial preparation based on methanogenic microorganisms (GMP2) was used as an inoculum in the third case. In this variant, an efficient methane fermentation of pistia and a high level of its degradation were observed ($Kd = 9$, CH_4 and CO_2 yields were 68.0 ± 11.1 L/kg VS and 43.0 ± 6.7 L/kg VS of plants, respectively). The dynamics of the fermentation process are shown in Figures 4 and 5 by blue lines. Thus, the degradation of pistia plants began with the participation of aerobic microorganisms, as evidenced by the decrease in the concentration of O_2 in the gas phase from 21.4 ± 3.8 to 0.05 ± 0.01% within one day. Subsequently, the degradation of polymers occurred via hydrogen fermentation with the formation of gaseous metabolites such as H_2 and CO_2. Hydrogen synthesis began on day 1 of fermentation and its concentration in the gas phase was 14.6 ± 2.1%. On the second day, there was a decrease in the H_2 synthesis (5.0 ± 1.1%), and on the third day synthesis stopped (Figure 5A). The methanogenic microbial preparation was able to autoregulate the metabolic parameters of the medium to optimize for its growth. As a consequence, the pH level decreased only to 5.8 ± 0.8 on the third day of fermentation, but on day 29 it was 6.0 ± 1.1 and on day 42 it increased to 7.1 ± 1.4. At the end of the fermentation process, which lasted 76 days, the pH was 7.8 ± 0.1 (Figure 4A). The decrease of the redox potential from 316.3 ± 32.5 to −93.3 ± 30.5 mV occurred on the 49th day of cultivation and correlated with the increase of the concentration of CH_4 in the gas phase of the bioreactor (Figures 4B and 5D). The concentration of biomethane on the 50th day of fermentation was very high—69 ± 9.8%. This indicates that GMP2 contained a diversified microbial community based on both methanogens and other types of microorganisms—destructors of plant polymers. CO_2 synthesis occurred throughout the experiment and did not exceed 31.8 ± 2.8% (Figure 5C). At the same time, the content of soluble organic compounds (determined by total carbon) and the content of $NH4^+$ (Figure 4C,D) in the culture fluid increased, which also indicates the degradation of plant polymers to liquid compounds (Figure 4C,D).

Usually during the degradation of plant polymer compounds (for example, starch) there is an intensive synthesis of gas, increasing its volume and increasing the pressure in the hermetic bioreactor to 1.0–1.5 atm. However, for pistia fermentation, an increase in gas volume was registered only in the GMP2 case. This was apparently due to the low content of available carbon-containing compounds in pistia tissues for the aboriginal pistia microbiome of GMP1 and the high availability of plant polymers for GMP2. Microorganisms of GMP2 rapidly and effectively destroyed the pistia. Visually, it was the degradation of the plants was noticeable on the 3rd day and appeared to end completely on the 50th day. However, the fermentation of available substrates occurred for another month after complete visual degradation of the pistia. The decrease in pistia weight was determined by the degradation coefficient *Kd*, which is equal to the ratio of the initial and final mass in terms of volatile solids. The initial dry weight of pistia was 5.0 g, and after fermentation it was 0.55 g, so $Kd = 9$. Thus, it was shown that pistia degradation was carried out with the production of the energy carrier biomethane. This preparation can be used for bioremediation of waters overgrown with alien pistia plants. At the first stage—removal and collection of pistia biomass from reservoirs, and at the second—its fermentation in a bioreactor with effective weight reduction and obtaining the high energy carrier CH_4.

The last stage of the work was the study of the dynamics of the removal of soluble compounds of Cu^{2+} in citrate form by GMP2 based on methanogenic microorganisms. The third variant of the experiment (with GMP2) was chosen to model the process of copper detoxification, as only this variant of the experiment was highly effective in the degradation of pistia plants. This part of the research was carried out for several purposes. The first was to confirm the thermodynamic position that methanogens are able to remove copper compounds from solutions. The second was consideration of the degradation of an aquatic plant as an integrated process that can be used for both obtaining biofuel and detoxification of common metals. The copper solution in the form of Cu^{2+} citrate was added in the active phase of fermentation at 49 h of cultivation to final concentrations of 100 ppm and 200 ppm, respectively (Figure 6A,B). In both variants, after the addition of the copper solution there was a sharp increase in the redox potential. The Eh ranged from -86 ± 31.7 mV to $+286 \pm 28.5$ mV after the addition of 100 ppm Cu^{2+} (Figure 6A) and from -16.3 ± 10.1 mV to $+290 \pm 22.2$ mV after the addition of 200 ppm of Cu^{2+} (Figure 6B).

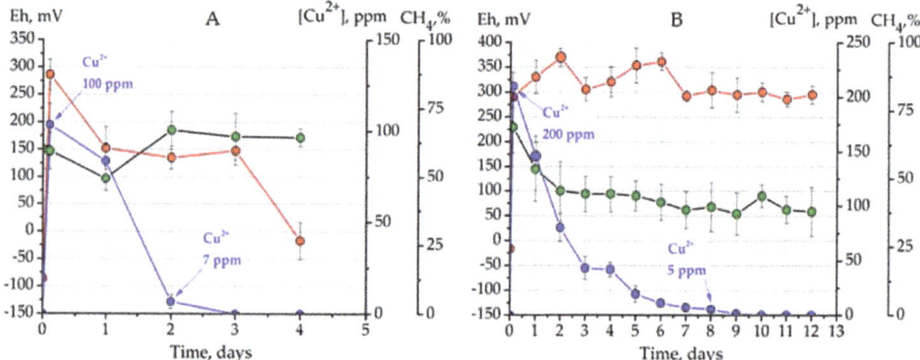

Figure 6. Effective removal of toxic copper compounds during the fermentation of pistia plants at the initial concentrations of 100 ppm (**A**) and 200 ppm (**B**) Cu^{2+}.

Removal of 100 ppm Cu^{2+} took only three days and correlated with a slight decrease in redox potential in the medium. During the first day, there was an inhibition of biomethane synthesis, which manifested itself in a decrease in concentration from $60.0 \pm 7.0\%$ to $49.4 \pm 4.5\%$, but when the concentration of Cu^{2+} decreased after 2 days to 7.0 ± 3.4 ppm, biomethane synthesis again increased to $67.0 \pm 6.9\%$ (Figure 6A). The redox potential also gradually decreased to a more optimal level for the growth of methanogens and on day 4 was -17.0 ± 23.1 mV (Figure 6A). Copper at the concentration of 200 ppm inhibited the process of biomethane fermentation. This was manifested in the inhibition of biomethane synthesis, increasing Eh to very high and suboptimal values for methanogens in the range $+290 \pm 30.2$ to $+370 \pm 17.9$ mV (Figure 6B). Despite the suppression of the metabolic process by soluble copper(II), it was completely removed from the medium within 10 days of fermentation. The metabolic parameters did not return to baseline, and the biomethane concentration finally decreased from $69.3 \pm 7.3\%$ to $37.1 \pm 7.71\%$ (Figure 6B).

Thus, the application of GMP2 for the degradation of pistia plants reduced their concentration by 9 times, as evidenced by the coefficient of degradation of *Kd*. The duration of the cycle was 76 days, after which gas synthesis was terminated. The yield of biomethane was 68.0 ± 1.1 L CH_4/kg VS of plants. The application of GMP1 was accompanied by an inefficient process and did not contribute to the degradation of pistia plants (*Kd* = 1.5).

4. Discussion

Obtaining biogas during the fermentation of macrophytes is one of the promising areas considered in a number of studies. [16,73]. The main advantages of using aquatic plant biomass are lack of competition with food crops for arable land, high growth rates, low fractions of lignin, which reduces the need for energy-intensive pretreatment, and compatibility with the introduction of an approach to bioprocessing [74]. Aquatic *P. stratiotes* plants contain a lignocellulosic biomass, which is a mixture of cellulose, hemicellulose, and lignin, and therefore have great potential for conversion into bioenergy. The anaerobic fermentation of *P. stratiotes* via activated sludge waste as inoculum was investigated. The highest biogas yield previously obtained was 321 mL/g of solids with a biomethane content of 72.5% [74]. In this case, the high yield of biomethane may have been caused by the presence of additional substrates for methanogens in the activated sludge. In our case, the microbial preparation was used as an inoculum, not as an additional substrate.

Obtaining biogas from aquatic plants has been gaining popularity in recent decades. In the case of [75], pistachios, aqueous hyacinths, and channel grasses were used as substrates. The biogas production was only 15.4–23.65 L/kg dry weight after a 21-day fermentation period. Thus, in the absence of additional cosubstrates, the biogas yield was quite low. In our experiment, any additional substrates are absent, and the yield of biomethane was high (68.0 ± 11.1 L/kg VS).

Another promising area of bioremediation of ecosystems is the use of microorganisms, in particular anaerobics, to remove toxic metals from solutions. For example, a microbial community of sulfate-reducing microorganisms (SRM) was able to precipitate $CuS\downarrow$ in an anaerobic sulfidogenic reactor used to treat industrial wastewater from metal compounds. Sulfate-reducing microorganisms reduce SO_4^{2-} in the process of dissimilatory sulfate reduction and emit hydrogen sulfide H_2S^{2-}, which interacts with Cu^{2+} and precipitates it in the form of the insoluble sulfide $CuS\downarrow$ [76]. The possibility of using Cu^{2+} as a terminal acceptor of electrons in the process of anaerobic respiration by microorganisms of the genus *Desulfuromonas* was proved. The efficiency of the immobilization of Cu^{2+} in the form of $CuS\downarrow$ using H_2S synthesized by microorganisms reached 97.3–100.0% in the presence of free sulfur S^0 [77]. Reduction of Cu^{2+} can also occur via interaction with microbial exometabolites. The role of extracellular polymeric substances (EPS) of bacterial strains of *Shewanella oneidensis* MR-1, *Bacillus subtilis* and yeast *Saccharomyces cerevisiae* and their redox state in the reduction of Cu^{2+} under anaerobic conditions was studied. It was found that the reduction occurred for about 10 min with the participation of cytochrome *c* and EPS (substances with phenolic and amide groups) of the studied microorganisms [78]. The possibility of the reduction of Cu^{2+} to Cu^{+} by autotrophic acidophiles of *A. ferrooxidans*. and *A. caldus* was demonstrated. They were also cultivated in the presence of elemental sulfur S^0 [79]. More than 23 ppm of Cu^{2+} was reduced over 24 h at an initial concentration in the medium of 100 ppm by *Pseudomonas putida* NA. Thus, the bioremoval efficiency was only 23%, which is not an effective result. A mixed bacterial culture containing microorganisms of the genera *Thiobacillus* and *Clostridium* was capable to immobilize Cu^{2+} via reduction mechanisms. The reduction efficiency was 89–92% Cu^{2+} in the temperature range 20–35 °C [80]. Thus, it was shown that in most cases complete 100% removal of copper compounds was not achieved. Copper has been found to be one of the most toxic to methanogenic microorganisms. According to recent studies, heavy metals are toxic to methanogenes *Methanospirillum hungatei* GP1 [81], *Methanospirillum hungatei* JF1, *Methanosarcina barkeri* MS, *Methanothermobacter marburgensis* and *Methanobacterium formicicum* [82] and inhibit their vital functions. However, microbial community analysis showed the presence of microorganisms closely related with *Methanobacterium* and *Methanospirillum*, suggesting that methanogenesis can coexist with sulphate reduction and metal precipitation. The presence of sulphate and heavy metals in wastewaters can affect the performance of methanogens and therefore impact energy recovery (in the form of biogas) from organic materials. We obtained similar results and experimentally proved that the methanogenic microbial preparation is able to completely remove toxic copper compounds from solutions. It is theoretically substantiated that one of the most important mechanisms was the precipitation of copper sulfide due to the sulfate reduction associated with methanogenesis. Despite the real evidence of high efficiency of the metal removal process during methane fermentation by precipitation in the form of sulfides, these processes still remain poorly studied and require further investigations.

5. Conclusions

The obtained results showed the general opportunity to effectively use GMP2 for methane fermentation of environmentally hazardous *P. stratiotes* L. plants. GMP2 was shown to be promising for high efficiency rapid degradation of pistia and the synthesis of high energy carrier biomethane and also complete removal of toxic Cu^{2+} from solutions where it is high in concentration. Methanogenic microbiome analysis, design of industrial plants, and economic calculations will be the next important steps to develop the optimal and profitable path of simultaneous hazardous aquatic plant degradation and obtaining of biofuel.

Author Contributions: Investigation, V.H., O.H. and A.K.; conceptualization, V.H., O.T., D.J. and A.D.-Ś.; resources, V.H., A.K., O.H., O.T., D.J. and A.D.-Ś.; methodology, O.S., V.T. and O.T.; supervision, O.T.; writing—original draft preparation, V.H., A.K., O.H., O.T., O.S., A.D.-Ś., D.J. and V.T.; writing—reviewing and editing, V.H. and A.K. All authors have read and agreed to the published version of the manuscript.

Funding: This research was funded under the fundamental research theme of the National Academy of Sciences of Ukraine on the subject #06.80 "Properties of extremophilic microorganisms and their biotechnological potential" from 2016–2020.

Institutional Review Board Statement: Not applicable.

Informed Consent Statement: Not applicable.

Acknowledgments: The authors are thankful to Anastasiia Lapa (student of the Faculty of Natural Sciences and Technologies, University of Opole, Poland) for preparation of the literature review section and analysis of the methods for determining the concentration of copper and ammonium ions in solutions, as well as for mathematical calculations of *P. stratiotes* aquatic plants.

Conflicts of Interest: The authors declare no conflict of interest.

References

1. Wang, A.; Hu, S.; Lin, B. Can Environmental Regulation Solve Pollution Problems? Theoretical Model and Empirical Research Based on the Skill Premium. *Energy Econ.* **2021**, *94*, 105068. [CrossRef]
2. Gernaat, D.E.H.J.; de Boer, H.S.; Daioglou, V.; Yalew, S.G.; Müller, C.; van Vuuren, D.P. Climate Change Impacts on Renewable Energy Supply. *Nat. Clim. Chang.* **2021**, *11*, 119–125. [CrossRef]
3. Lu, H.R.; Qu, X.; Hanandeh, A.E. Towards a Better Environment—the Municipal Organic Waste Management in Brisbane: Environmental Life Cycle and Cost Perspective. *J. Clean. Prod.* **2020**, *258*, 120756. [CrossRef]
4. Schanes, K.; Dobernig, K.; Gözet, B. Food Waste Matters–A Systematic Review of Household Food Waste Practices and Their Policy Implications. *J. Clean. Prod.* **2018**, *182*, 978–991. [CrossRef]
5. Ali, H.; Khan, E.; Ilahi, I. Environmental Chemistry and Ecotoxicology of Hazardous Heavy Metals: Environmental Persistence, Toxicity, and Bioaccumulation. *J. Chem.* **2019**, *2019*, 6730305. [CrossRef]
6. Wessely, J.; Hülber, K.; Gattringer, A.; Kuttner, M.; Moser, D.; Rabitsch, W.; Schindler, S.; Dullinger, S.; Essl, F. Habitat-Based Conservation Strategies Cannot Compensate for Climate-Change-Induced Range Loss. *Nat. Clim. Chang.* **2017**, *7*, 823–827. [CrossRef]
7. Powers, R.P.; Jetz, W. Global Habitat Loss and Extinction Risk of Terrestrial Vertebrates under Future Land-Use-Change Scenarios. *Nat. Clim. Chang.* **2019**, *9*, 323–329. [CrossRef]
8. Olkhovich, O.P.; Taran, N.Y.; Svetlova, N.B.; Batsmanova, L.M.; Aleksiyenko, M.V.; Kovalenko, M.S. Assessment of the Influence of the Invasive Species *Pistia stratiotes* (Araceae) on Some Species of Submerged Macrophytes of Natural Water Bodies of Ukraine. *Hydrobiol. J.* **2017**, *53*, 75–84. [CrossRef]
9. Gusain, R.; Suthar, S. Potential of Aquatic Weeds (*Lemna gibba*, *Lemna minor*, *Pistia stratiotes* and *Eichhornia* sp.) in Biofuel Production. *Process Saf. Environ. Prot.* **2017**, *109*, 233–241. [CrossRef]
10. Ružičková, J.; Lehotská, B.; Takáčová, A.; Semerád, M. Morphometry of Alien Species *Pistia stratiotes* L. in Natural Conditions of the Slovak Republic. *Biologia* **2020**, *75*, 1–10. [CrossRef]
11. Shrestha, B. Climate Change Amplifies Plant Invasion Hotspots in Nepal. *Divers. Distrib.* **2019**, *25*, 1599–1612. [CrossRef]

12. Galal, T.M.; Dakhil, M.A.; Hassan, L.M.; Eid, E.M. Population Dynamics of *Pistia stratiotes* L. *Rend. Lincei Sci. Fis. e Nat.* **2019**, *30*, 367–378. [CrossRef]
13. Tripathi, P.; Kumar, R.; Sharma, A.; Mishra, A.; Gupta, R. *Pistia stratiotes* (Jalkumbhi). *Pharm. Rev.* **2010**, *4*, 153–160. [CrossRef]
14. Jaklič, M.; Koren, Š.; Jogan, N. Alien Water Lettuce (*Pistia stratiotes* L.) Outcompeted Native Macrophytes and Altered the Ecological Conditions of a Sava Oxbow Lake (SE Slovenia). *Acta Bot. Croat.* **2020**, *79*, 35–42. [CrossRef]
15. Achola, K.J.; Indalo, A.A.; Munenge, R.W. Pharmacologic activities of pistia stratiotes. *Int. J. Pharm.* **1997**, *35*, 329–333. [CrossRef]
16. Güngören Madenoğlu, T.; Jalilnejad Falizi, N.; Kabay, N.; Güneş, A.; Kumar, R.; Pek, T.; Yüksel, M. Kinetic Analysis of Methane Production from Anaerobic Digestion of Water Lettuce (*Pistia stratiotes* L.) with Waste Sludge. *J. Chem. Technol. Biotechnol.* **2019**, *94*, 1893–1903. [CrossRef]
17. Okoro, O.V.; Sun, Z.; Birch, J. Prognostic Assessment of the Viability of Hydrothermal Liquefaction as a Post-Resource Recovery Step after Enhanced Biomethane Generation Using Co-Digestion Technologies. *Appl. Sci.* **2018**, *8*, 2290. [CrossRef]
18. Chow, W.L.; Chong, S.; Lim, J.W.; Chan, Y.J.; Chong, M.F.; Tiong, T.J.; Chin, J.K.; Pan, G.-T. Anaerobic Co-Digestion of Wastewater Sludge: A Review of Potential Co-Substrates and Operating Factors for Improved Methane Yield. *Processes* **2020**, *8*, 39. [CrossRef]
19. Morales-Polo, C.; Cledera-Castro, M.D.M.; Moratilla Soria, B.Y. Reviewing the Anaerobic Digestion of Food Waste: From Waste Generation and Anaerobic Process to Its Perspectives. *Appl. Sci.* **2018**, *8*, 1804. [CrossRef]
20. Sajna, N.; Haler, M.; Skornik, S.; Kaligarič, M. Survival and Expansion of *Pistia stratiotes* L. in a Thermal Stream in Slovenia. *Aquat. Bot.* **2007**, *87*, 75–79. [CrossRef]
21. Chapman, D.; Coetzee, J.; Hill, M.; Hussner, A.; Netherland, M.; Pescott, O.; Stiers, I.; van Valkenburg, J.; Tanner, R. *Pistia stratiotes* L. *Bull. OEPP/EPPO Bull.* **2017**, *47*. [CrossRef]
22. Venema, P. Snelle Uitbreiding van Watersla (*Pistia stratiotes* L.) Rond Meppel. *Gorteria* **2001**, *27*, 133–135.
23. Brundu, G.; Stinca, A.; Angius, L.; Bonanomi, G.; Celesti-Grapow, L.; D'Auria, G.; Griffo, R.; Migliozzi, A.; Motti, R. *Pistia stratiotes* L. and Eichhornia Crassipes (Mart.) Solms.: Emerging Invasive Alien Hydrophytes in Campania and Sardinia (Italy). *Bull. OEPP/EPPO Bull.* **2012**, *42*, 568–579. [CrossRef]
24. D'Auria, G. *Indagine Sui Bodri Della Provincia di Cremona: Giovanni D'Auria, Franco Zavagno*; Provincia di Cremona: Cremona, Italy, 2011.
25. Pyšek, P.; Sádlo, J.; Mandák, B. Catalogue of Alien Plants of the Czech Republic Preslia. *Preslia* **2002**, *74*, 97–186.
26. Zivkovic, M. The Beginnings of *Pistia stratiotes* L. Invasion in the Lower Danube Delta: The First Record for the Province of Vojvodina (Serbia). *BioInvasions Rec.* **2019**, *8*, 218–229. [CrossRef]
27. Adebayo, A.; Briski, E.; Kalaci, O.; Hernandez, M.; Ghabooli, S.; Beric, B.; Chan, F.; Zhan, A.; Fifield, E.; Leadley, T.; et al. Water Hyacinth (Eichhornia Crassipes) and Water Lettuce (*Pistia stratiotes*) in the Great Lakes: Playing with Fire? *Aquat. Invasion* **2011**, *6*, 91–96. [CrossRef]
28. Neuenschwander, P.; Julien, M.H.; Center, T.D.; Hill, M.P. *Pistia stratiotes* L. (Araceae). In *Biological Control of Tropical Weeds Using Arthropods*; Muniappan, R., Reddy, G.V.P., Raman, A., Eds.; Cambridge University Press: Cambridge, UK, 2009; pp. 332–352.
29. Olkhovych, O.; Taran, N.; Hrechyshkina, S.; Musienko, M. Influence of Alien Species *Pistia stratiotes* L. 1753 on Representative Species of Genus *Salvinia* in Ukraine. *Transylv. Rev. Syst. Ecol. Res.* **2020**, *22*, 43–56. [CrossRef]
30. Eid, E.M.; Galal, T.M.; Dakhil, M.A.; Hassan, L.M. Modeling the Growth Dynamics of *Pistia stratiotes* L. Populations along the Water Courses of South Nile Delta, Egypt. *Rend. Lincei* **2016**, *27*, 375–382. [CrossRef]
31. Afanasyev, S.; Savitsky, A. Finding of Pistia Stratioides in the Kaniv Reservoir (the Dnieper River, Ukraine) and Assessment of Risk of Its Naturalization. *Hydrobiol. J.* **2016**, *52*, 50–57. [CrossRef]
32. Mosiakin, A.; Kazarinova, H.O. Potential Invasive Range Modeling of Pistia Stratiotes (Araceae) Based on GIS-Analysis of Ecoclimatic Factors. [In Ukrainian: Моделювання інвазійного поширенняpistia stratiotes (araceae) на основі гіс-аналізу кліматичніх факторів]. *Ukr. Bot. J.* **2014**, *71*, 549–557. [CrossRef]
33. Konishchuk, V.V.; Solomakha, I.; Mudrak, O.; Mudrak, H.; Khodyn, O.B. Ecological Impact of Phytoinvasions in Ukraine. *Ukr. J. Ecol.* **2020**, *3*, 69–75. [CrossRef]
34. Zub, L.M.; Prokopuk, M.S. The Features of Macrophyte Invasions in Aquatic Ecosystems of the Middle Dnieper Region (Ukraine). *Russ. J. Biol. Invasions* **2020**, *11*, 108–117. [CrossRef]
35. Coetzee, J.A.; Langa, S.D.F.; Motitsoe, S.N.; Hill, M.P. Biological Control of Water Lettuce, *Pistia stratiotes* L., Facilitates Macroinvertebrate Biodiversity Recovery: A Mesocosm Study. *Hydrobiologia* **2020**, *847*, 3917–3929. [CrossRef]
36. Champion, P. Knowledge to Action on Aquatic Invasive Species: Island Biosecurity—the New Zealand and South Pacific Story. *Manag. Biol. Invasions* **2018**, *9*, 383–394. [CrossRef]
37. Mthethwa, N.P.; Nasr, M.; Bux, F.; Kumari, S. Utilization of *Pistia stratiotes* (Aquatic Weed) for Fermentative Biohydrogen: Electron-Equivalent Balance, Stoichiometry, and Cost Estimation. *Int. J. Hydrogen Energy* **2018**, *43*, 8243–8255. [CrossRef]
38. Fusi, A.; Bacenetti, J.; Fiala, M.; Azapagic, A. Life Cycle Environmental Impacts of Electricity from Biogas Produced by Anaerobic Digestion. *Front. Bioeng. Biotechnol.* **2016**, *4*, 26. [CrossRef] [PubMed]
39. Misson, G.; Mainardis, M.; Incerti, G.; Goi, D.; Peressotti, A. Preliminary Evaluation of Potential Methane Production from Anaerobic Digestion of Beach-Cast Seagrass Wrack: The Case Study of High-Adriatic Coast. *J. Clean. Prod.* **2020**, *254*, 120131. [CrossRef]
40. Bote, M.A.; Naik, V.R.; Jagdeeshgouda, K.B. Production of Biogas with Aquatic Weed Water Hyacinth and Development of Briquette Making Machine. *Mater. Sci. Energy Technol.* **2020**, *3*, 64–71. [CrossRef]

41. León, L.A.R.D.; Diez, P.Q.; Gálvez, L.R.T.; Perea, L.A.; Barragán, C.A.L.; Rodríguez, C.A.G.; León, A.R. Biochemical Methane Potential of Water Hyacinth and the Organic Fraction of Municipal Solid Waste Using Leachate from Mexico City's Bordo Poniente Composting Plant as Inoculum. *Fuel* **2021**, *285*, 119132. [CrossRef]
42. Miranda, A.F.; Kumar, N.R.; Spangenberg, G.; Subudhi, S.; Lal, B.; Mouradov, A. Aquatic Plants, Landoltia Punctata, and Azolla Filiculoides as Bio-Converters of Wastewater to Biofuel. *Plants* **2020**, *9*, 437. [CrossRef]
43. Sparks, D.L. Toxic Metals in the Environment: The Role of Surfaces. *Elements* **2005**, *1*, 193–197. [CrossRef]
44. Flemming, C.A.; Trevors, J.T. Copper Toxicity and Chemistry in the Environment: A Review. *Water Air Soil Pollut.* **1989**, *44*, 143–158. [CrossRef]
45. Husak, V. Copper and Copper-Containing Pesticides: Metabolism, Toxicity and Oxidative Stress. *JPNU* **2015**, *2*, 38–50. [CrossRef]
46. Sikamo, J. Copper Mining in Zambia—History and Future. *J. S. Afr. Inst. Min. Metall.* **2016**, *116*, 491–496. [CrossRef]
47. Al-Saydeh, S.A.; El-Naas, M.H.; Zaidi, S.J. Copper Removal from Industrial Wastewater: A Comprehensive Review. *J. Ind. Eng. Chem.* **2017**, *56*, 35–44. [CrossRef]
48. Tapiero, H.; Townsend, D.M.; Tew, K.D. Trace Elements in Human Physiology and Pathology. Copper. *Biomed. Pharm.* **2003**, *57*, 386–398. [CrossRef]
49. Vincent, M.; Duval, R.E.; Hartemann, P.; Engels-Deutsch, M. Contact Killing and Antimicrobial Properties of Copper. *J. Appl. Microbiol.* **2018**, *124*, 1032–1046. [CrossRef] [PubMed]
50. Andrejkovičová, S.; Sudagar, A.; Rocha, J.; Patinha, C.; Hajjaji, W.; Silva, E.F.; da Velosa, A.; Rocha, F. The Effect of Natural Zeolite on Microstructure, Mechanical and Heavy Metals Adsorption Properties of Metakaolin Based Geopolymers. *Appl. Clay Sci.* **2016**, *126*, 141–152. [CrossRef]
51. Urbina, L.; Guaresti, O.; Requies, J.; Gabilondo, N.; Eceiza, A.; Corcuera, M.A.; Retegi, A. Design of Reusable Novel Membranes Based on Bacterial Cellulose and Chitosan for the Filtration of Copper in Wastewaters. *Carbohydr. Polym.* **2018**, *193*, 362–372. [CrossRef] [PubMed]
52. Nassef, E. Removal of Copper from Wastewater By Cementation From Simulated Leach Liquors. *J. Chem. Eng. Process Technol.* **2015**, *6*, 1. [CrossRef]
53. Dong, Y.; Liu, J.; Sui, M.; Qu, Y.; Ambuchi, J.J.; Wang, H.; Feng, Y. A Combined Microbial Desalination Cell and Electrodialysis System for Copper-Containing Wastewater Treatment and High-Salinity-Water Desalination. *J. Hazard. Mater.* **2017**, *321*, 307–315. [CrossRef]
54. Wang, G.; Zhang, S.; Xu, X.; Zhong, Q.; Zhang, C.; Jia, Y.; Li, T.; Deng, O.; Li, Y. Heavy Metal Removal by GLDA Washing: Optimization, Redistribution, Recycling, and Changes in Soil Fertility. *Sci. Total Environ.* **2016**, *569–570*, 557–568. [CrossRef] [PubMed]
55. Dhaliwal, S.S.; Singh, J.; Taneja, P.K.; Mandal, A. Remediation Techniques for Removal of Heavy Metals from the Soil Contaminated through Different Sources: A Review. *Environ. Sci. Pollut. Res.* **2020**, *27*, 1319–1333. [CrossRef] [PubMed]
56. Hovorukha, V.; Havryliuk, O.; Tashyreva, H.; Tashyrev, O.; Sioma, I. Thermodynamic Substantiation of Integral Mechanisms of Microbial Interaction with Metals. *J. Ecol. Eng. Environ. Prot.* **2018**, *2*, 55–63. [CrossRef]
57. Raab, A.; Feldmann, J. Microbial Transformation of Metals and Metalloids. *Sci. Prog.* **2003**, *86*, 179–202. [CrossRef]
58. Solioz, M. *Copper and Bacteria: Evolution, Homeostasis and Toxicity*; SpringerBriefs in Molecular Science; Springer International Publishing: Cham, Switzerland, 2018; ISBN 978-3-319-94438-8.
59. Cornu, J.-Y.; Huguenot, D.; Jézéquel, K.; Lollier, M.; Lebeau, T. Bioremediation of Copper-Contaminated Soils by Bacteria. *World J. Microbiol. Biotechnol.* **2017**, *33*, 26. [CrossRef]
60. Trevors, J.T.; Cotter, C.M. Copper Toxicity and Uptake in Microorganisms. *J. Ind. Microbiol.* **1990**, *6*, 77–84. [CrossRef]
61. Hovorukha, V.; Havryliuk, O.; Gladka, G.; Tashyrev, O.; Kalinichenko, A.; Sporek, M.; Dołhańczuk-Śródka, A. Hydrogen Dark Fermentation for Degradation of Solid and Liquid Food Waste. *Energies* **2021**, *14*, 1831. [CrossRef]
62. Shirokova, V.A.; Khutorova, A.O.; Shirokov, R.S.; Yurova, Y.D. Hydroecological assessment of surface waters of the Osetr river basin. *Int. Multidiscip. Sci. Geoconf. SGEM* **2019**, *19*, 323–330.
63. Optasyuk, O.; Korotchenko, I. Ecological, Coenotic and Chorological Features of Synanthropic Species of the Genus Linum in the Flora of Ukraine. *Ukr. Bot. J.* **2014**, *71*, 703–707. [CrossRef]
64. Acree, W.E. Basic Gas Chromatography (McNair, Harold M.; Miller, James M.). *J. Chem. Educ.* **1998**, *75*, 1094. [CrossRef]
65. Suslova, O.; Govorukha, V.; Brovarskaya, O.; Matveeva, N.; Tashyreva, H.; Tashyrev, O. Method for Determining Organic Compound Concentration in Biological Systems by Permanganate Redox Titration. *Int. J. Bioautom.* **2014**, *18*, 45–52.
66. Qiu, X.; Liu, G.-P.; Zhu, Y.-Q. Determination of Water-Soluble Ammonium Ion in Soil by Spectrophotometry. *Analyst* **1987**, *112*, 909–911. [CrossRef]
67. Tashyreva, H.; Tashyrev, O.; Hovorukha, V.; Havryliuk, O. The effect of mixing modes on biohydrogen yield and spatial ph gradient at dark fermentation of solid food waste. *J. Ecol. Eng. Environ. Prot.* **2017**, *2*, 53–62. [CrossRef]
68. Prekrasna, Ie.P.; Tashyrev, O.B. Copper Resistant Strain Candida Tropicalis RomCu5 Interaction with Soluble and Insoluble Copper Compounds. *Biotechnol. Acta* **2015**, *8*, 93–102. [CrossRef]
69. Zehnder, A.J.B.; Wuhrmann, K. Titanium (III) Citrate as a Nontoxic Oxidation-Reduction Buffering System for the Culture of Obligate Anaerobes Vertebrate Central Nervous System: Same Neurons Mediate Both Electrical and Chemical Inhibitions. *Science* **1975**, *194*, 1165–1166. [CrossRef] [PubMed]

70. Havryliuk, O.; Hovorukha, V.; Gladka, G.; Tashyrev, O. Bioremoval of copper(II) via hydrogen fermentation of ecologically hazardous multicomponent food waste. *J. Ecol. Eng. Environ. Prot.* **2020**, *2*, 5–14. [CrossRef]
71. Tashyrev, O.; Hovorukha, V.; Suslova, O.; Tashyreva, H.; Havryliuk, O. Thermodynamic prediction for development of novel environmental biotechnologies and valuable products from waste obtaining. *J. Ecol. Eng. Environ. Prot.* **2018**, *1*, 24–35. [CrossRef]
72. Paulo, L.M.; Stams, A.J.M.; Sousa, D.Z. Methanogens, Sulphate and Heavy Metals: A Complex System. *Rev. Environ. Sci. Bio/Technol.* **2015**, *14*, 537–553. [CrossRef]
73. Ramaraj, R.; Dussadee, N. Biological Purification Processes for Biogas Using Algae Cultures: A Review. *Int. J. Sustain. Green Energy* **2015**, *4*, 20–32.
74. Montingelli, M.E.; Tedesco, S.; Olabi, A.G. Biogas Production from Algal Biomass: A Review. *Renew. Sustain. Energy Rev.* **2015**, *43*, 961–972. [CrossRef]
75. Singhal, V.; Rai, J.P.N. Biogas Production from Water Hyacinth and Channel Grass Used for Phytoremediation of Industrial Effluents. *Bioresour. Technol.* **2003**, *86*, 221–225. [CrossRef]
76. Tyagi, S.; Malik, W.; Annachhatre, A.P. Heavy Metal Precipitation from Sulfide Produced from Anaerobic Sulfidogenic Reactor. *Mater. Today Proc.* **2020**, *32*, 936–942. [CrossRef]
77. Hnatush, S.O.; Moroz, O.M.; Yavorska, G.V.; Borsukevych, B.M. Sulfidogenic and Metal Reducing Activities of Desulfuromonas Genus Bacteria under the Influence of Copper Chloride. *Biosyst. Divers.* **2018**, *26*, 218–226. [CrossRef]
78. Xu, H.; He, E.; Peijnenburg, W.J.G.M.; Song, L.; Zhao, L.; Xu, X.; Cao, X.; Qiu, H. Contribution of Pristine and Reduced Microbial Extracellular Polymeric Substances of Different Sources to Cu(II) Reduction. *J. Hazard. Mater.* **2021**, *415*, 125616. [CrossRef]
79. Johnson, D.B.; Hedrich, S.; Pakostova, E. Indirect Redox Transformations of Iron, Copper, and Chromium Catalyzed by Extremely Acidophilic Bacteria. *Front. Microbiol.* **2017**, *8*, 211. [CrossRef] [PubMed]
80. Vernans, A.K.R.; Iswanto, B.; Rinanti, A. Removal of Heavy Metal (Cu^{2+}) by *Thiobacillus* sp. and *Clostridium* sp. at Various Temperatures and Concentration of Pollutant in Liquid Media. *J. Phys. Conf. Ser.* **2019**, *1402*, 022102. [CrossRef]
81. Pankhania, I.P.; Robinson, J.P. Heavy Metal Inhibition of Methanogenesis by Methanospirillum Hungatei GP1. *FEMS Microbiol. Lett.* **1984**, *22*, 277–281. [CrossRef]
82. Jarrell, K.F.; Saulnier, M.; Ley, A. Inhibition of Methanogenesis in Pure Cultures by Ammonia, Fatty Acids, and Heavy Metals, and Protection against Heavy Metal Toxicity by Sewage Sludge. *Can. J. Microbiol.* **1987**, *33*, 551–554. [CrossRef]

Article

Influence of Valorization of Sewage Sludge on Energy Consumption in the Drying Process

Ewa Siedlecka [1,*] and Jarosław Siedlecki [2]

1 Faculty of Infrastructure and Environment, Czestochowa University of Technology, Brzeźnicka St. 60a, 42-200 Częstochowa, Poland

2 Department of Mathematics, Czestochowa University of Technology, Armii Krajowej St. 21, 42-200 Częstochowa, Poland; jaroslaw.siedlecki@pcz.pl

* Correspondence: ewa.siedlecka@pcz.pl; Tel.: +48-34-32-50-957

Abstract: Valorization of digested sewage sludge generated in a medium-sized sewage treatment plant and the effect of valorization on energy consumption during sludge drying used for energy recovery are presented. Anaerobic digestion of sewage sludge reduces dry matter content compared to raw sludge. This lowers its calorific value leading to the lower interest of consumers in using it as fuel. The aim of the study was to valorize digested sewage sludge prior to drying with high-energy waste with low moisture content. The procedure led to the reduction in moisture content by about 50% in the substrate supplied for solidification and drying. The calorific value of digested sewage sludge increased by 50–80%, and the energy consumption of the drying process decreased by about 50%. Physical and chemical properties of sewage sludge and moisture content of substrates and mixtures after valorization were determined. The heat of combustion of valorized sewage sludge mixtures, their elemental composition, and ash content is investigated. Their calorific value in the analytical and working states of 10% H_2O was calculated. The highest calorific value was obtained for the mixture of sewage sludge valorized with waste plastics or combined with wood dust, averaging 23 MJ/kg. A mathematical approximation of sewage sludge valorization is presented.

Keywords: sewage sludge; organic matter; valorization; energy saving; alternative fuel

Citation: Siedlecka, E.; Siedlecki, J. Influence of Valorization of Sewage Sludge on Energy Consumption in the Drying Process. *Energies* **2021**, *14*, 4511. https://doi.org/10.3390/en14154511

Academic Editor: Gabriele Di Giacomo

Received: 30 June 2021
Accepted: 23 July 2021
Published: 26 July 2021

1. Introduction

Municipal sewage sludge is a by-product of the treatment of municipal and industrial wastewater [1,2]. Sewage sludge constitutes only 1–2 vol.% of treated wastewater, but its management is very complicated, and treatment costs amount to 20–60% of wastewater treatment plants' (WWTPs') total operating costs [3]. Sewage sludge from sewage treatment plants (STPs) is characterized by a high content of organic and inorganic substances, including microbial biomass, pathogens, nutrients N and P, and metals. In addition, they have a very heterogeneous composition [4] and a high water content up to 95–99% [5]. The water content varies according to the type of sewage sludge (Table 1).

Table 1. The content of water in sewage sludge according to different authors.

Content of Water	Raw Sludge	Thickened Sewage Sludge	Digested and Dewatered Sewage Sludge		
			Centrifuge	Belt Filter Press	Chamber Filter Press
(%)	99–95	95–90	80–78	75–70	70–68
Ref.	[5]	[6]	[7]	[8]	[9]

Sewage sludge is one of the substances that are difficult to dewater. Without pretreatment, the dewatering effects and separation rates of sewage sludge are very low [10]. High-molecular-weight polyelectrolytes are most commonly used for conditioning. Their consumption depends on the dewatering equipment used and amounts to 5–12 g/kg d.m. [11].

Sludge treatment in large and medium sewage treatment plants includes such processes as thickening, anaerobic stabilization, mechanical dewatering, and drying. Anaerobic stabilization (methane production) changes the properties of sewage sludge. In this process, ca. 50% of the organic matter contained in the water is decomposed. Each kilogram of decomposed organic matter can yield 0.49–0.75 Nm^3/kg d.m. of biogas [12,13]. The gas produced in the process lowers the calorific value of the sludge by approximately 2–4 MJ/kg (Figure 1) [9,12].

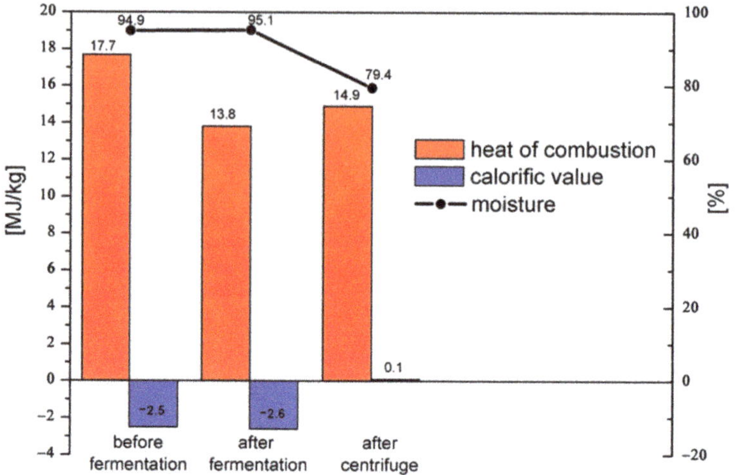

Figure 1. Heat of combustion, calorific value, and moisture of sewage sludge.

Water content in sewage sludge after digestion and dewatering is 68–80% H_2O. The content of heavy metals in the sludge increases [14,15]. Siebielec and Stuczyński [16] studied sewage sludge from 43 wastewater treatment plants in Poland and presented statistical analyzes of the results (Table 2). The authors showed a large variation in the content of heavy metals in sewage sludge. The content of cadmium, chromium, and nickel was the most variable in the tested samples. According to the authors, it is related to the runoff of surface water to the sewage system from metallurgical waste dumps, which are located in the vicinity of the sewage treatment plant.

Table 2. The range and coefficients of variation of selected parameters of sewage sludge [16].

Parameter	Average (Median)	Range	Coefficient of Variation, %
Organic matter, %	42.2 (43.3)	13.6–65.1	33
Nitrogen, %	2.61 (2.48)	0.55–5.64	49
Phosphorus, %	1.83 (1.75)	0.14–4.08	47
Potassium, %	0.25 (0.21)	0.09–0.87	57
Calcium, %	3.93 (3.82)	0.81–19.9	62
Magnesium, %	0.58 (0.55)	0.01–1.7	73
Cu, mg/kg d.m.	184 (154)	41–449	55
Zn, mg/kg d.m.	2135 (1760)	541–9824	68
Cd, mg/kg d.m.	10.5 (4.95)	1.1–149.1	198
Ni, mg/kg d.m.	69.2 (39.1)	18–1172	214
Pb, mg/kg d.m.	173 (132)	45–953	85
Cr, mg/kg d.m	320 (69.9)	24–7544	315

Polycyclic aromatic hydrocarbons (PAHs), which have an affinity for solid sludge particulate matter, are accumulated in sewage sludge [17]. Contamination of sludge with PAHs was confirmed for both raw and digested sludge [18–20]. The development of analytical techniques has led to the emergence of new problems concerning the use of sewage sludge in agriculture and nature, as new contaminants such as hormones, drug residues, flame retardants, and increased amounts of organic pollutants have been identified [21]. There is an ongoing discussion in EU countries regarding the maximum amounts of residual polymers in sewage sludge. This is planned to be taken into account when new legislation on sewage sludge management is drafted.

Sewage sludge contaminated with heavy metals, xenobiotics, and synthetic polymers cannot be used in agriculture or nature. The presence of these contaminants in sewage sludge represents a potential source of contamination of soils and waters. There is growing skepticism in the wider public about the use of sewage sludge in agriculture and soil reclamation [22].

Attempts have been made to use sewage sludge in the production of construction materials [23,24]. Rezaee et al. [25] showed that sewage sludge can be used up to 15% in the production of eco-cement, which has mineral components similar to traditional Portland cement. These applications offer alternative methods for sludge recycling, but the amounts of sludge used are only a small fraction of total sludge production, while sludge-based production is generally of lower quality and causes environmental concerns such as leaching of heavy metals [5].

Therefore, a sludge treatment scheme consisting of such processes as thickening, dewatering, drying, and energy recovery is enforced. A product obtained by drying sewage sludge can be used in thermal processes that allow energy recovery [26].

The energy consumption of drying depends mostly on the water content of sewage sludge and its structure. In sludge with very high water content, the amount of energy required to dry it may exceed the amount of energy recovered during thermal use. According to the work of [9], 60–70% more energy is needed to dry 1 Mg of sewage sludge containing 20% to 90% d.m. than to dry sludge containing 35% to 90% d.m. The theoretical energy requirement needed to evaporate 1 kg H_2O at normal pressure is 0.627 kW/kg H_2O. Depending on design solutions and installation parameters, the average energy demand is 0.6–1.2 kW/kg [27]. Lossman [28] presented a comparison of the energy intensity of sludge drying equipment derived from laboratory tests, with heat consumption coefficients higher than the values presented by Fukas-Płonka and Janik [27] (Table 3). According to the work of [29], the consumption of thermal energy for drying of sewage sludge in order to obtain the appropriate fuel-grade quality is between 1.8 and 2.2 kWh_{th}/kg_{DM}. Additionally, the electricity consumption of the drying equipment is between 0.10 and 0.30 kWh_{el}/kg_{DM}. The authors state that there is questionable energy efficiency between recovery of waste heat from sewage sludge incineration and thermal drying.

Table 3. Energy consumption of sewage sludge dryers.

Energy Consumption	Drum Dryer (DDS)	Belt Dryer (BDS)	Fluidized Bed Dryer (FDS)	Solar Dryer with Underfloor Heating
kW/kg H_2O	1.4–1.6	1.45 [1]–1.75	1.2	1.6

[1] with recuperation of heat.

Waste transformed into products that can be used for energy purposes is referred to as alternative or secondary fuel. It is produced by separating combustible fractions in the processes of sorting, multi-stage crushing, homogenization, and briquetting. The composition of the components that form an alternative fuel varies. They affect the final parameters such as lower heating value (LHV), moisture content, and ash content (Table 4).

Variable properties of secondary fuels reduce their use for energy production. Recipients of such fuels prefer fuels with unchanging parameters, as required by power equipment.

Table 4. Parameters of alternative fuels according to different authors.

LHV, MJ/kg	17.0	21.2	15.0	19.8
Moisture content, %	19.2	15.0	12.0	0.0
Ash content, %	11.0	10.9	16.0	3.4
Ref.	[30]	[31]	[32]	[33]

The paper presents research aimed at reducing the amount of energy needed to dry digested sewage sludge. The aim was achieved by the valorization of the sewage sludge before the actual drying. Waste of vegetable origins such as waste paper, wood dust, and plastic waste was used. Waste materials are characterized by low moisture content and calorific values higher than digested sludge. The resulting mixtures are characterized by plastic consistency, which enables briquetting or pelletizing of the final product. The valorization of sewage sludge offers substantial advantages: it increases the dry matter content of the sludge to be dried and at the same time reduces the energy consumption required for drying, increases the calorific value of the sludge, and contributes to an improvement in the CO_2 emissions toward minimum values. Borzooei et al. [34] showed research on the reduction in carbon footprint in a wastewater treatment plant. They have made an assessment of dynamic sludge thickener, as well as hybrid thermo-alkali pretreatment of waste-activated sludge to enhance the biogas production in the WWTP. To upgrade the produced biogas in sludge treatment units to biomethane with an average efficiency of 98.6%, the selective membranes were studied. The authors also proposed experimental microalgae technology for CO_2 fixation, which significantly reduces the carbon footprint of the wastewater treatment plant.

It is assumed that during the combustion, the amount of CO_2 emitted into the atmosphere equals the amount of the gas absorbed from the environment by the plant material (wood dust, waste paper).

In the present study, it was important to develop a method enabling quantitative and qualitative optimization of fuel components. This will allow for the production of goods with the assumed stable quality parameters, which is possible based on rational fuel valorization using analytical methods.

2. Materials and Methods

2.1. Characterization of Sewage Sludge

Digested sewage sludge (SS) generated in a Warta Sewage Treatment Plant in Częstochowa with a capacity of 46,000 m^3/d was used in the study. The wastewater treatment plant uses the activated sludge method with nitrification, denitrification, dephosphatation, and stabilization by methane fermentation. Stabilized sludge is mechanically dewatered on a belt press. Sludge samples were placed in sealed containers. The moisture content was measured three times, and the content of heavy metals was determined in the examined sludge. The particular physical and chemical analyses were carried out as follows:

- The moisture content of the hydrated sludge was measured using a Radwag HAC 110/NH weighing machine with a reading accuracy of 0.001%;
- Analysis of heavy metals was performed by the emission spectroscopy method with inductively coupled plasma (ICP-OES SPECTRO ARCOS). For this purpose, part of the sewage sludge was dried at 105 °C to a constant mass, ground in a ring mill, and mineralized with aqua regia (mineralization time 2 h, temperature 120 °C).

2.2. Sewage Sludge Valorization

Hydrated sewage sludge (SS) samples were used for valorization examinations. The sludge was valorized with waste paper (P), wood dust (D), and mixed waste plastics of polyethylene terephthalate and polypropylene (PPT). Polyvinyl alcohol PVA and calcium oxide CaO were used as binding additives. A total of 13 mixtures consisting of sewage sludge and selected wastes in appropriate proportions were prepared. Waste paper and waste plastics were shredded before valorization. Mixing was performed in a ribbon mixer. The particular physical and chemical analyses were carried out as follows:

- The moisture contents of waste paper, plastics, and wood dust and mixtures were measured using a Radwag HAC 110/NH moisture analyzer with a reading accuracy of 0.001%. Then, the mixtures were placed in a steel die and pressed under pressure. A total of 13 briquette samples were obtained;
- Solidification of samples in the form of briquettes was carried out on a hydraulic press 250 C Żywiec, using pressures of 100 and 150 MPa;
- Elemental analysis of selected briquettes (C, H, N, S) was performed using a LECO TruSpec CHNS analyzer;
- The higher heating value (HHV) was determined in an IKA C2000 Basic isoperibolic calorimeter;
- The lower heating value (LHV) was calculated using the Boie formula [35];
- The bulk density of selected briquettes was determined according to the work of [36]. The mass of individual briquettes was determined using a laboratory balance. The volume was determined by a geometric method by measuring the diameter and height of cylindrical briquettes. The volume calculation was performed using the formula: Volume = $\Pi \cdot$(briquette radius)$^2 \cdot$height. The bulk density of the briquette was calculated according to the formula: Bulk density = mass of briquette/volume of briquette. Ash content was determined according to the work of [37].

2.3. Approximation of Sewage Sludge Valorization

Taking into account the percentages of components in the 13 tested samples and the calculated calorific values, the mean square approximation of the data set was determined. For this, the following linear multivariable function is proposed:

$$z_{approx} = F(u_1, u_2, \ldots, u_k, p_1, p_2, \ldots, p_k) := \sum_{l=1}^{k} p_l u_l = p_1 u_1 + p_2 u_2 + \ldots + p_k u_k, \tag{1}$$

in which the coefficients p_l, $l = 1, 2 \ldots, k$, are the parameters of the function searched for, and the variables u_l, $l = 1, 2, \ldots, k$, are the arguments of the function. In order to determine the values of the coefficients, the least-squares criterion is defined, which takes the form:

$$S(p_1, p_2, \ldots, p_k) = \sum_{i=1}^{n} \left(\sum_{l=1}^{k} p_l \cdot (u_l)_i - z_i \right)^2 = \min, \tag{2}$$

The partial derivatives of the function S with respect to the particular parameters p_l, $l = 1, 2 \ldots, k$ are:

$$\frac{\partial S(p_1, p_2, \ldots, p_k)}{\partial p_j} = 2\sum_{i=1}^{n} \left(\sum_{l=1}^{k} p_l \cdot (u_l)_i - z_i \right) (u_j)_i \tag{3}$$

The necessary condition for the existence of an extreme of a multivariable function leads to a linear system of equations in the form:

$$\sum_{i=1}^{n} \left(\sum_{l=1}^{k} p_l \cdot (u_l)_i - z_i \right) (u_j)_i = 0 \text{ for } j = 1, 2, \ldots, k, \tag{4}$$

And after its transformation, we obtain the following form:

$$\sum_{l=1}^{k} p_l \sum_{i=1}^{n} (u_l)_i (u_j)_i = \sum_{i=1}^{n} z_i (u_j)_i \text{ for } j = 1, 2, \ldots, k, \tag{5}$$

The above system of equation can also be written in the matrix form:

$$\mathbf{U} \cdot \mathbf{P} = \mathbf{Z}, \tag{6}$$

where

$$\mathbf{U} = \begin{bmatrix} \sum_{i=1}^{n} (u_1)_i(u_1)_i & \sum_{i=1}^{n} (u_2)_i(u_1)_i & \cdots & \sum_{i=1}^{n} (u_k)_i(u_1)_i \\ \sum_{i=1}^{n} (u_1)_i(u_2)_i & \sum_{i=1}^{n} (u_2)_i(u_2)_i & \cdots & \sum_{i=1}^{n} (u_k)_i(u_2)_i \\ \vdots & \vdots & \ddots & \vdots \\ \sum_{i=1}^{n} (u_1)_i(u_k)_i & \sum_{i=1}^{n} (u_2)_i(u_k)_i & \cdots & \sum_{i=1}^{n} (u_k)_i(u_k)_i \end{bmatrix}, \mathbf{P} = \begin{bmatrix} p_1 \\ p_2 \\ \vdots \\ p_k \end{bmatrix}, \mathbf{Z} = \begin{bmatrix} \sum_{i=1}^{n} z_i(u_1)_i \\ \sum_{i=1}^{n} z_i(u_2)_i \\ \vdots \\ \sum_{i=1}^{n} z_i(u_k)_i \end{bmatrix} \tag{7}$$

From the solution of the system of equations (6) $\mathbf{P} = \mathbf{U}^{-1} \cdot \mathbf{Z}$, we obtain the values of parameters p_l, $l = 1, 2 \ldots, k$. The determined parameters p_l allow us to determine the calorific value for any share of particular components of the sample using the approximation Formula (1) [38].

3. Results and Discussion

3.1. Characterization of Sewage Sludge

The main component of sewage sludge is dry organic matter, which constitutes over 64% (Table 5). The sludge is characterized by a high water content of above 77%. The high content of biogenic elements indicates the predominance of municipal sewage inflow to the treatment plant. The presence of heavy metals with the highest fraction of zinc also indicates the content of industrial sewage in the stream flowing into the treatment plant.

Table 5. Characterization of sewage sludge (SS).

Parameter	Unit	Value
Cd	mg/kg d.m.	8.0
Ni	mg/kg d.m.	111.0
Cu	mg/kg d.m.	254.0
Cr	mg/kg d.m.	405.0
Pb	mg/kg d.m.	101.0
Zn	mg/kg d.m.	2950.0
Co	mg/kg d.m.	4.8
K	mg/kg d.m.	2835.0
P	mg/kg d.m.	3.15
C	mg/kg d.m.	40.0
H	mg/kg d.m.	5.0
Organic matter	%	64.5
Mineral matter	%	35.5
Moisture content	%	77.26

The high water content hinders the efficient energy recovery from sewage sludge, which lowers its calorific value [39]. It is important that the sewage sludge sent for drying should have the highest possible dry matter content. In practice, treatment plants use equipment with low dewatering efficiency (centrifuges, belt presses), with a water content of sludge after dewatering of 70–80%. Consequently, sludge with high water content is sent for drying. The removal of 700–800 kg H_2O from 1 Mg of sewage sludge generates 30–50% of the costs incurred by the treatment plant. Drying sewage sludge is an energy-intensive process, which, with the expected increase in energy consumption, will result in higher prices for water supply and sewage treatment for both individual and business customers.

3.2. Sewage Sludge Valorization

The purpose of the valorization of digested sewage sludge was to reduce its moisture content and increase its calorific value. The waste selected for this purpose, i.e., waste paper (P), wood dust (W), and waste plastics (PPT), is characterized by low moisture content and calorific value higher than in sewage sludge (Table 6). The range of calorific value for individual substrates is, respectively: for wood dust 17 MJ/kg [40], waste paper 11–26 MJ/kg, waste plastics 35 MJ/kg on average [41], sewage sludge 9–13 MJ/kg [42]. The change in the structure of sewage sludge resulting from the valorization made it possible to obtain mixtures of appropriate consistency, facilitating the solidification of briquettes. The results of the substrate moisture content measurements are shown in Figures 2–5.

Table 6. Moisture content of the substrates.

Substrates	Moisture Content, %
Sewage sludge (SS)	77.26
Wood dust (D)	6.82
Waste paper (P)	6.58
Waste plastics (PPT)	29.80

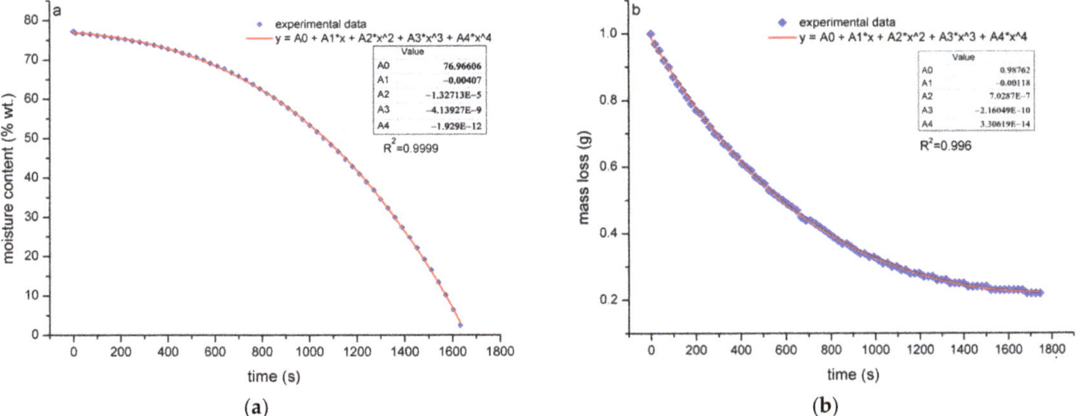

(**a**) (**b**)

Figure 2. Relation between: (**a**) moisture content over time of SS; (**b**) mass loss over time of SS.

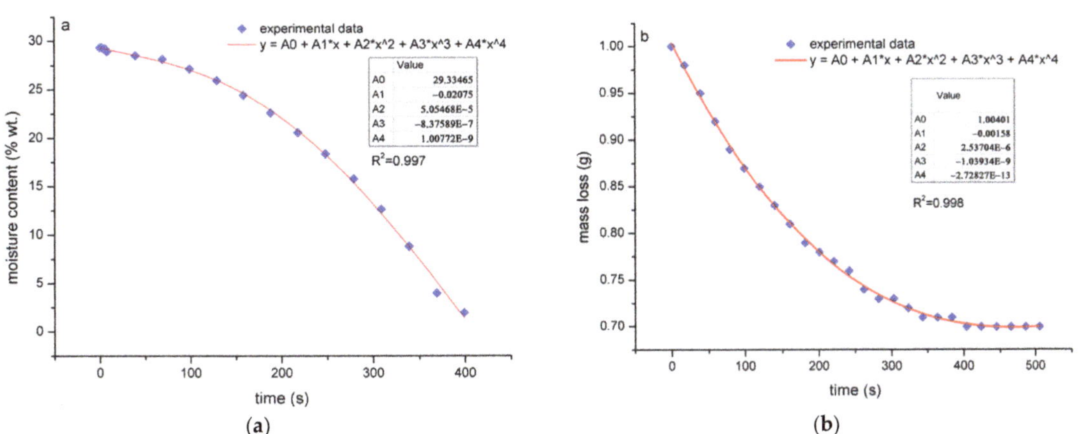

(**a**) (**b**)

Figure 3. Relation between: (**a**) moisture content over time of PPT; (**b**) mass loss over time of PPT.

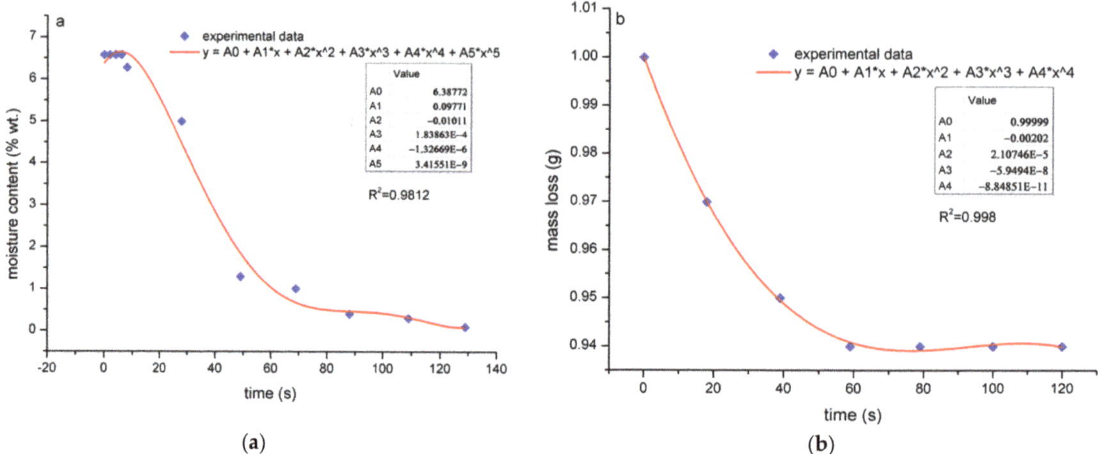

(**a**) (**b**)

Figure 4. Relation between: (**a**) moisture content over time of P; (**b**) mass loss over time of P.

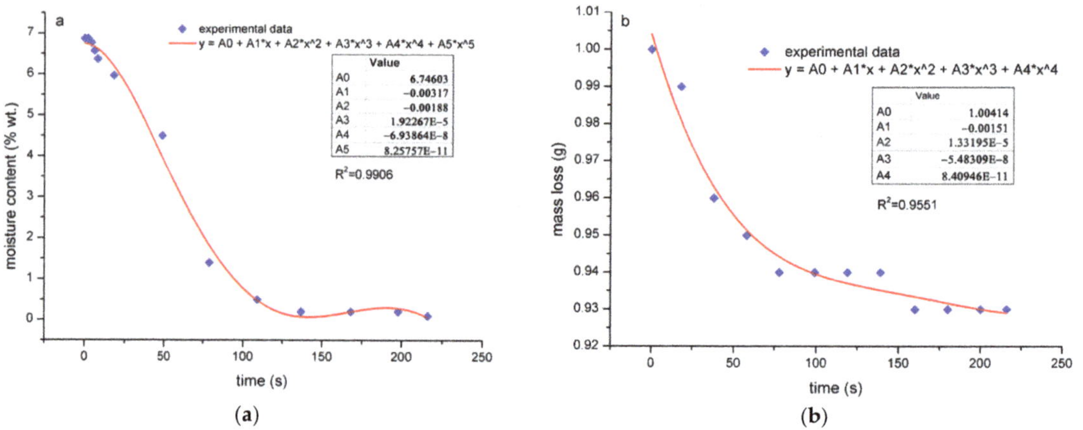

(**a**) (**b**)

Figure 5. Relation between: (**a**) moisture content over time of D; (**b**) mass loss over time of D.

The moisture content is 10 times lower in wood dust and waste paper and 2.6 times lower in waste plastics with respect to sewage sludge (Table 6). Drying of sewage sludge with a moisture content of 77.26% to analytical moisture took 1600 s. It took four times less time to dry PPT waste plastics (400 s) and 10 times less time to dry wood dust and waste paper (140 and 115 s, respectively). Therefore, it was assumed that the valorization of sewage sludge with these substrates would reduce its moisture content and drying time, thus contributing to a reduction in energy consumption. Thirteen mixtures were obtained as a result of valorization of sewage sludge with waste paper, wood dust, waste plastics, and PVA and CaO binding additives, with their composition and moisture content presented in Table 7.

Table 7. Composition and moisture content of the samples.

No	Mixture	Moisture Content (%)	Composition of the Mixtures, %wt.				
			Sewage Sludge (SS)	Wood Dust (D)	Waste Paper (P)	Waste Plastics (PPT)	PVA CaO (+)
1	SSD	42.04	50.0	50.0	-	-	-
2	SSD1+	41.20	49.0	49.0	-		2.0
3	SSD2+	35.57	41.0	57.0	-	-	2.0
4	SSD3+	32.75	37.0	61.0	-	-	2.0
5	SSP1	51.10	63.0	-	37.0	-	-
6	SSP2	41.92	50.0	-	50.0	-	-
7	SSP1+	50.27	62.0	-	36.0	-	2.0
8	SSP2+	41.08	49.0	-	49.0	-	2.0
9	SSP3+	33.31	38.0	-	60.0	-	2.0
10	SSPPT	53.52	50.0	-	-	50.0	-
11	SSPPTD	37.93	33.3	33.3	-	33.3	-
12	SSPPT+	52.46	49.0	-	-	49.0	2.0
13	SSPPTD+	37.16	32.7	32.7	-	32.7	2.0

The moistúre content data for mixtures 1–13 indicate that this content in sewage sludge may be altered by valorization with other types of low-moisture waste. The proportions of low-moisture waste had a significant effect on the moisture content. The higher the content of low-moisture waste, the lower the moisture content of the resulting mixture (Figures 6–8). The lowest moisture content was found for SSD2+ and SSD3+ mixtures, whereas its highest values were observed in SSP1+, SSPPT, SSPPT+, and SSP1. The presence of binders in the mixtures has no significant effect on their moisture content (Table 7).

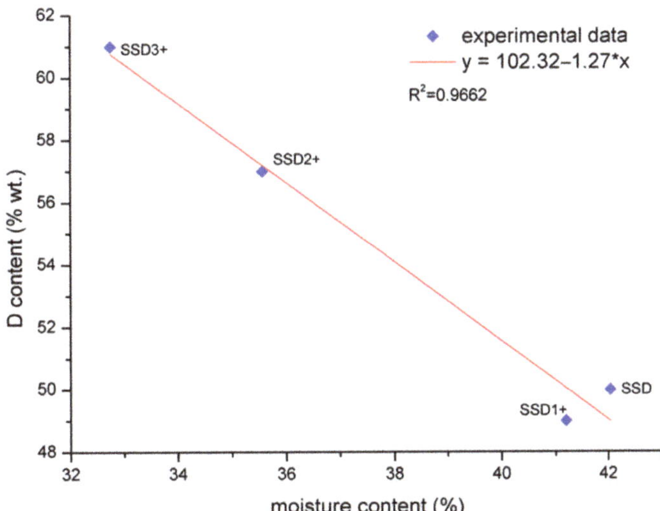

Figure 6. Correlation between percentage of D and moisture content for SSD, SSD1+, SSD2+, and SSD3+.

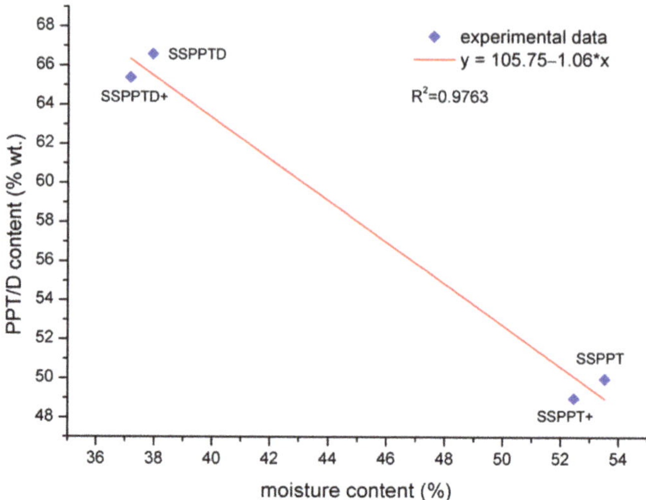

Figure 7. Correlation between percentage of PPT/D and moisture content for SSPPTD, SSPPTD+, SSPPT, and SSPPT+.

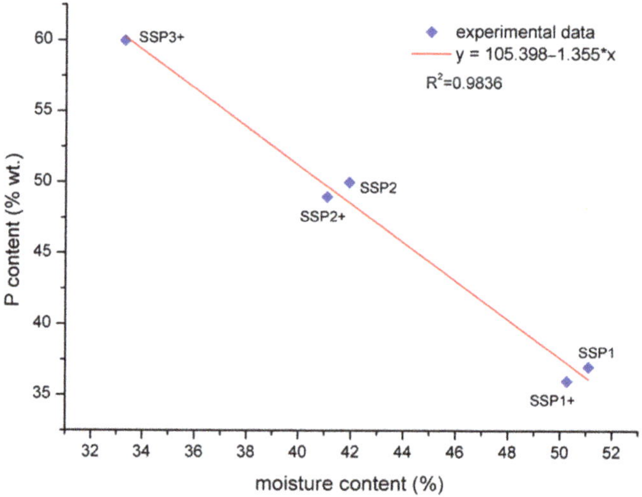

Figure 8. Correlation between percentage of P and moisture content for SSP1, SSP2, SSP1+, SSP2+, and SSP3+.

Mixtures with a similar moisture content of 37–41% were selected for further studies: SSD1+, SSP2+, and SSPPTD+. These mixtures are characterized by a low bulk weight of 0.09 g/cm^3 for SSP2+ and 0.28 g/cm^3 for SSPPTD+ and a low energy density (Table 8). A hydraulic press was used to compress the selected mixtures. As a result, cylindrical briquettes with a diameter of 25 mm, a height of 10–15 mm, and a bulk density of 0.9–1.6 g/cm^3 were obtained (Table 8).

Table 8. Bulk density of the mixtures/briquettes.

Mixture /Briquette	Bulk Density of the Mixtures, g/cm^3	Bulk Density of the Briquettes, g/cm^3	
		Pressure	
		100 MPa	150 MPa
SSPPTD+	0.28	0.9	0.84
SSD1+	0.24	0.91	0.81
SSP2+	0.09	1.59	1.61

SSD1+ and SSPPTD+ briquettes have a suitable degree of solidification and hardness. Transverse cracks that weakened compactness were noticed in briquettes higher than 15 mm (Figure 9). In both cases, smaller amounts of material should be applied to the die to obtain briquettes of sufficient height (10–15 mm) and better durability. The delamination in briquettes with a height greater than 15 mm may be due to the phenomenon of sample decompression after taking it out of the mold and the content of plastics fractions of bigger dimensions, which, after pressing, return to the previous size. According to the work of [43], many factors can affect the results of sample pressing, including the amount of material pressed, pressure, moisture content in the mixture, fineness of the material, binding agents, and many others. Therefore, a wide range of additional research is required for an accurate understanding of the briquette manufacturing process.

Figure 9. Briquettes of SSP2+, SSD1+, and SSPPTD+ (pressure 100–150 MPa).

The analysis of moisture loss as a function of time for selected mixtures clearly indicates the energy gain associated with drying of the valorized sludge (Figure 10). The drying time to working value for the SSD1+ mixture was reduced by 50%, SSP2+ by 54%, and SSPPTD+ by 60% compared to the drying time of sludge alone.

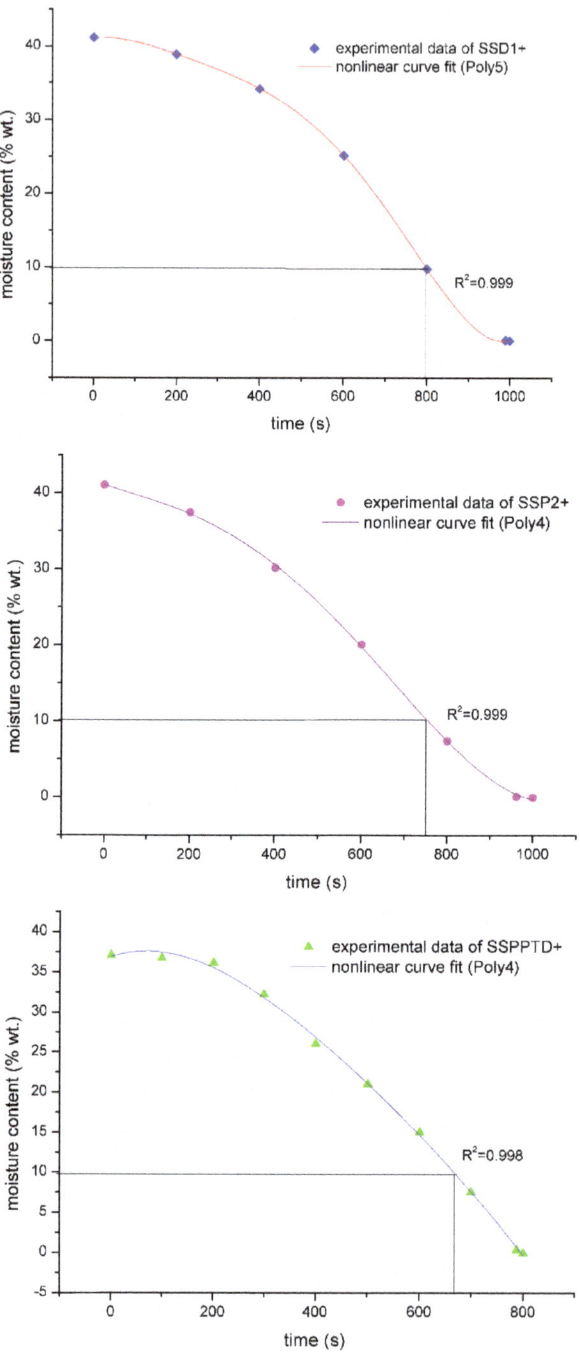

Figure 10. Moisture content as a function of time of SSD1+, SSP2+, and SSPPTD+.

According to design assumptions, sewage sludge in belt dryers should have a moisture content of approximately 60%. In order to reduce the moisture content from 80% to 60%, it is assumed to introduce 0.3 Mg of dried sludge per 1 Mg of wet sludge to avoid the phenomenon of sludge "sticking". Dry sludge is poorly wettable, which does not completely reduce this problem. The proposed valorization of digested sewage sludge before the drying process changes the structure of the sludge and significantly reduces its moisture content. It improves heat convection and facilitates moisture outflow. Using the low-moisture waste and higher calorific values for valorization shortens the time of sludge drying by 50–60% while reducing the energy consumption of the process by the respective value. According to the work of [44], the energy needed to produce pellets using sewage sludge and biomass (50 + 50%) was 50% lower than the energy needed to produce pellets from pure biomass. The phenomenon of sludge "sticking" is beneficial in the proposed solution by facilitating the formation of the final product in the form of briquettes and increasing their compactness and durability. Kubonova et al. [45] showed that sewage sludge is a very suitable binder for individual components of the alternative fuel, at the same time indicating that the moisture content of sewage sludge (about 72 wt.%) was an advantage for mixing with other types of waste.

The next stage of the research was to carry out elemental analysis of the mixtures, to determine the combustion heat, and to calculate the calorific value in the analytical state and in the working state for moisture content of 10%. Depending on the type of substrate and its percentage in the valorized sewage sludge mixtures, the results obtained varied (Table 9). The heat of combustion is directly proportional to the content of C and H and inversely proportional to the ash content. Sewage sludge valorized with wood dust (SSD1+, SSD2+, and SSD3+) had a calorific value ranging from 15.8 to 16.6 MJ/kg. Mixtures consisting of sludge and waste paper reached values ranging from 14.2 MJ/kg for SSP1+ to 15.2 MJ/kg for SSP3. The mixtures of sewage sludge valorized with waste plastic and wood dust have the highest calorific value, with 22.2 MJ/kg for SSPPTD+ and 23.9 MJ/kg for SSPET+. Chen et al. [46] showed that the fuel made from sewage sludge and wood dust in a proportion of 10:1, with a moisture content of 14.2–18.5%, is characterized by a calorific value of 21.8–23.4 MJ·kg^{-1}. The fuel obtained by the authors met all the requirements of the Taiwanese company Taipower. According to the work of [47], the calorific value of pellets consisting of sewage sludge and fir (CFSP) was 17.54 MJ/kg. These results are comparable to SSD1+, SSD2+, and SSD3+, respectively. Park et al. [48] showed that if 50% of waste plastics is included in the mixture, it is possible to achieve a calorific value of about 23 MJ/kg for the RDF. Valorization of sewage sludge in the proportions 1:1:1 (sewage sludge: waste plastics: wood dust) allowed to obtain a comparable calorific value.

Table 9. Elemental analysis, HHV and LHV of briquettes.

No	Symbol of Briquette	Element				Ash Content	HHV	LHV	LHV (Moisture 10%)
		C	H	S	N				
		%				%	MJ/kg	MJ/kg	MJ/kg
1	SSD	44.50	5.70	0.400	1.36	24.00	17.40	16.18	15.9
2	SSD1+	44.62	5.75	0.397	1.34	23.90	17.30	16.09	15.8
3	SSD2+	45.68	5.80	0.360	1.41	23.56	17.90	16.60	16.4
4	SSD3+	46.17	5.81	0.344	1.38	23.08	18.07	16.77	16.6
5	SSP1	39.85	5.87	0.397	1.31	26.30	15.95	14.63	14.4
6	SSP2	40.50	5.95	0.325	1.33	25.00	16.30	14.90	14.8
7	SSP1+	39.02	5.90	0.343	1.40	26.80	15.70	14.40	14.2
8	SSP2+	40.75	6.03	0.312	1.29	25.50	16.40	15.08	14.8
9	SSP3+	41.24	6.19	0.258	1.35	24.40	16.80	15.40	15.2
10	SSPPT	64.00	6.75	0.300	1.36	22.50	25.38	23.86	23.7
11	SSPPTD	59.65	6.79	0.266	1.37	21.00	24.00	22.38	22.3
12	SSPPT+	63.70	7.20	0.294	1.37	23.05	25.80	24.17	23.9
13	SSPPTD+	59.52	6.84	0.264	1.41	21.61	23.90	22.38	22.2

According to the requirements for alternative fuels specified in the PN-EN-15359:2012 standard [49], there is a division into fuel classes in terms of the net calorific value. Class 2 comprises fuels with a calorific value of ≥20 MJ/kg, including sewage sludge mixtures valorized with waste plastic (SSPPT+) or in combination with wood dust (SSPPTD+) with the calorific value of 23 MJ/kg. Class 3, according to PN-EN-15359:2012 [49], with a calorific value above 15 MJ/kg, includes sewage sludge mixtures valorized with wood dust (SSD, SSD1+, SSD2+, and SSD3+). The mean calorific value, in this case, is 16.2 MJ/kg. Class 4, with a calorific value above 10 MJ/kg, includes all sewage sludge mixtures valorized with waste paper, with a mean calorific value of 14.7 MJ/kg.

3.3. Approximation of Sewage Sludge Valorization

The values presented in Table 10 can be treated as a set of numerical data that can be written in general form as:

$$\{(u_1)_1, (u_2)_1, \ldots, (u_k)_1, z_1\}, \{(u_1)_2, (u_2)_2, \ldots, (u_k)_2, z_2\}, \ldots, \{(u_1)_n, (u_2)_n, \ldots, (u_k)_n, z_n\} \quad (8)$$

where $(u_l)_j$, $l = 1, \ldots, k$, $j = l, \ldots n$, are the shares of particular components in sample j (total number of sample is $n = 13$). Each sample is described by the set of components k (here $k = 5$), and the particular indexes denote: 1-sewage sludge (SS), 2-wood dust (D), 3-waste paper (P), 4-mixed waste plastics (PPT), and 5-polyvinyl alcohol and calcium oxide (+). Values of z_j, $j = 1, \ldots n$, denote the "calorific values" of the sample j and are determined experimentally.

Table 10. Set of numerical data and approximation of sewage sludge valorization.

No	SS	D	P	PPT	+	z	z_{approx}	$\lvert z_{approx}\text{-}z \rvert$	$(z_{approx}\text{-}z)^2$
	u_1	u_2	u_3	u_4	u_5				
1	0.5	0.5	0	0	0	15.7	15.61	0.09	0.0081
2	0.49	0.49	0	0	0.02	15.6	15.65	0.05	0.0025
3	0.41	0.57	0	0	0.02	16.1	16.12	0.02	0.0004
4	0.37	0.61	0	0	0.02	16.3	16.35	0.05	0.0025
5	0.63	0	0.37	0	0	14.2	14.02	0.18	0.0324
6	0.5	0	0.5	0	0	14.5	14.49	0.01	0.0001
7	0.62	0	0.36	0	0.02	13.9	14.09	0.19	0.0361
8	0.49	0	0.49	0	0.02	14.6	14.55	0.05	0.0025
9	0.38	0	0.6	0	0.02	14.9	14.95	0.05	0.0025
10	0.5	0	0	500	0	23.3	23.58	0.28	0.0784
11	0.333	0.333	0	0.333	0	21.9	21.9	0	0
12	0.49	0	0	0.49	20	23.7	23.46	0.24	0.0576
13	327	0.327	0	0.327	0.02	21.9	21.83	0.07	0.0049

On the basis of the given data in Table 10, the numerical values of p_l are equal to: $p_1 = 12.695874$, $p_2 = 18.527441$, $p_3 = 16.282293$, $p_4 = 34.465184$ and $p_5 = 17.626107$. The calculated values of p_l allow us to determine the calorific value for any share of particular components in the sample using the approximation Formula (1). Table 10 presents the calculated values z_{approx} on the basis of Equation (1) for the determined parameters p_l and the component shares (u_l) in each sample. The square error $(z_{approx}\text{-}z)^2$ was determined for each sample j.

4. Discussion

Waste with energetic potential should be used locally all over the world. This is supported by the limitation of the use of conventional energy sources and the limitation of landfilling. Guidelines for this can be found in the Directive of the European Parliament and of the Council 2018/850 from 30 May 2018 [50] amending Directive 1999/31/EC concerning the storage of waste [51].

According to the European Commission's report [52], the term refuse-derived fuel (RDF) refers to waste that has been processed to meet industry requirements, mainly those concerning high calorific value. RDF should include specific fractions of municipal waste, industrial and commercial waste, sewage sludge, industrial hazardous waste, and biomass. Municipal waste, including selectively collected fractions, can be used to produce alternative fuel for energy recovery in the cement kiln installations. It is possible to add waste from other waste groups (without hazardous waste) to achieve the parameters expected by the cement plant. Preferred parameters of alternative fuel are: moisture content < 20%, calorific value > 20 MJ/kg, sulfur content < 1%. Deviations from the required parameters can be agreed upon on a case-by-case basis, taking into account the specific design of the installation [53,54]. Restrictions on the chemical composition of the refuse-derived fuel depend on individual kiln installations. The content of trace components in the raw materials for clinker production in the basic fuel and in the refuse-derived fuel is taken into consideration. Furthermore, it is also important how and where the fuel is collected for the installation. The use of biomass waste for energy purposes makes it possible to meet environmental standards for emissions of CO_2, SO_x, NO_x, dust, dioxins, chlorine, mercury, and heavy metals. Biomass burning reduces the balance of CO_2 emissions because the amount of CO_2 previously collected from the environment is released into the atmosphere. The low nitrogen content of biomass waste reduces the emissions of NO_x compounds into the atmosphere compared to burning coal [55,56].

The diagram (Figure 11) shows a typical solution for a sewage treatment plant (a) and the feasibility of the solution presented in the article (b) that would consist in the use of a "bypass" between the centrifuge and, for example, a sludge belt dryer. The "bypass" would consist of a homogenizing mixer and a press. The supply of waste materials in the form of waste paper (P), wood dust (D), and waste plastics (PPT) would not generate costs as it would burden the supplier.

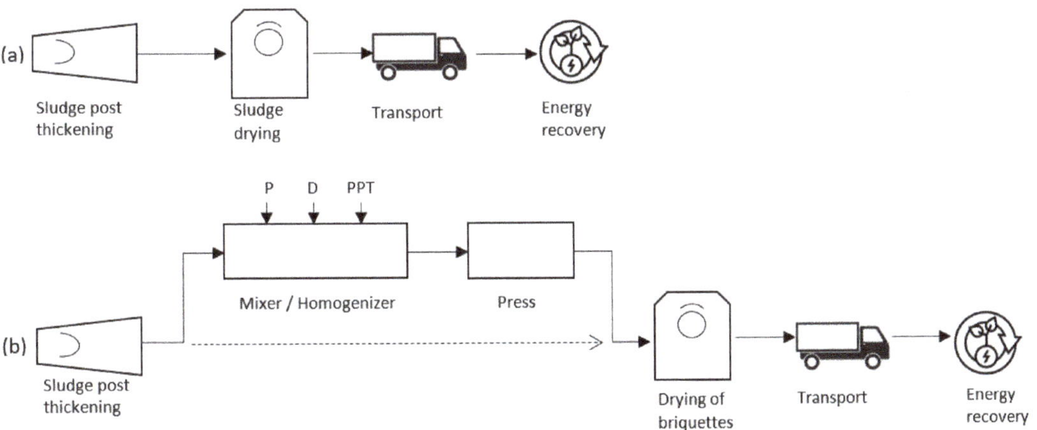

Figure 11. Scheme of (**a**) typical solution for sewage sludge treatment, (**b**) feasibility of sewage sludge valorization.

The difficulties arising from the mixing of waste are related to its variable composition and differences in physical and chemical properties. This requires constant laboratory control of the quality of refuse-derived fuels [57]. Examination of sewage sludge valorization was oriented at obtaining a high calorific value and reducing the carbon footprint related to wood dust and waste paper. Valorization with mixed waste plastics does not contribute to reducing the carbon footprint of the obtained product. However, this does not mean that there is no reduction in the carbon footprint of the obtained SSPPTD+ product, where the proportion of individual substrates is 1:1:1 (sewage sludge: waste plastics: wood dust). The consequences for atmospheric emissions when burning valorized dried sewage sludge will be insignificant because, in the structure of heat from individual fuels used for clinker production, dry sewage sludge constitutes 0.17–3.1% in cement plants in Poland [58,59]. The PP and PET waste contained in the product does not contain chlorine, fluorine, mercury, and heavy metals. In the case of cement plants, flue gas emissions are continuously monitored, which guarantees compliance with the restrictive emission standards for individual compounds (NO_x, SO_x, TOC, CO, HCl, HF, dust). Many other parameters determining the furnace operation are automatically adjusted to the level of recorded emissions. This provides the ability to respond when elevated levels of hazardous compounds occur [60].

Digested and dried sewage sludge is a substitute for fuel used in cement plants. The low calorific value of the fuels (13–15 MJ/kg) limits their substitution in the cement industry to about 5–10% [42]. An increase in the calorific value of the valorized sewage sludge above 20 MJ/kg, at a low moisture content of 10% H_2O, makes it possible to substitute the fuel used in cement plants by up to 50%. In energy and environmental terms, it is an attractive fuel for the consumer. CO_2 emissions are limited for the cement industry. Sewage sludge represents biomass and is neutral in terms of CO_2 emissions [61].

The selected products obtained through sewage sludge valorization are characterized by a calorific value of 22–23 MJ/kg, which makes it possible to use them for co-combustion with coal in the energy industry. This requires additional testing to accept them for use in boiler equipment. Furthermore, attempts will be made to perform mono-combustion of the products at the wastewater treatment plant combined with energy recovery, phosphorus recovery from ash, and use of the mineral residue in the construction industry.

5. Conclusions

Based on the performed analyses, the following conclusions can be formulated:

1. The current degree of substitution of conventional fuels with alternative fuels in the cement industry has exceeded 40% on average in Poland. This has raised expectations for alternative fuels with a required calorific value of 20 MJ/kg;
2. Most biomass fuels have a calorific value of 16–18 MJ/kg. The proposed valorization of sewage sludge increases this value up to 23 MJ/kg;
3. The valorization of biomass waste with the waste with high calorific value ensures the possibility to carry out energy recovery in cement industry installations, in accordance with their technological requirements;
4. The mass fraction of biomass in the valorized sewage sludge ranges from 50% to 60%. This improves the balance of CO_2 emissions, which is limited in the cement industry;
5. The valorization of sewage sludge with low-moisture waste reduces the energy consumption of the drying process in sewage treatment plants by up to 50%;
6. The sludge adhesion to the drying belt observed in the drying process is a favorable parameter in the case of solidification of the product in the form of briquette or granulate. It ensures the durability and compactness of the product.

Author Contributions: Conceptualization, E.S. and J.S.; methodology, E.S. and J.S.; validation, E.S. and J.S.; investigation, E.S.; writing—original draft preparation, E.S.; writing—review and editing, E.S. and J.S.; visualization, E.S. Both authors have read and agreed to the published version of the manuscript.

Funding: The scientific research was funded by the statute subvention of the Czestochowa University of Technology, Faculty of Infrastructure and Environment and Faculty of Mechanical Engineering and Computer Science: Department of Mathematics under BS/PB-1-100-3011/21/P.

Institutional Review Board Statement: Not applicable.

Informed Consent Statement: Not applicable.

Data Availability Statement: The data presented in this study are available on request from the corresponding author.

Conflicts of Interest: The authors declare no conflict of interest. The funders had no role in the design of the study; in the collection, analyses, or interpretation of data; in the writing of the manuscript; or in the decision to publish the results.

References

1. Kocbek, E.; Garcia, H.A.; Hooijmans, C.M.; Mijatović, I.; Lah, B.; Brdjanovic, D. Microwave treatment of municipal sewage sludge: Evaluation of the drying performance and energy demand of a pilot-scale microwave drying system. *Sci. Total Environ.* **2020**, *742*, 140541. [CrossRef]
2. Nielsen, S.; Stefanakis, A.I. Sustainable Dewatering of Industrial Sludges in Sludge Treatment Reed Beds: Experiences from Pilot and Full-Scale Studies under Different Climates. *Appl. Sci.* **2020**, *10*, 7446. [CrossRef]
3. Andreoli, C.V.; Von Sperling, M.; Fernandes, F. Sludge Treatment and Disposal. *Water Intell. Online* **2015**, 4–6. [CrossRef]
4. Zaker, A.; Chen, Z.; Wang, X.; Zhang, Q. Microwave-assisted pyrolysis of sewage sludge: A review. *Fuel Process. Technol.* **2019**, *187*, 84–104. [CrossRef]
5. Chang, Z.; Long, G.; Zhou, J.L.; Ma, C. Valorization of sewage sludge in the fabrication of construction and building materials: A review. *Resour. Conserv. Recycl.* **2020**, *154*, 104606. [CrossRef]
6. Heidrich, Z.; Witkowski, A. *Urządzenia Do Oczyszczania Ścieków*; Wydawnictwo Seidel-Przywecki Sp. z o.o.: Warszawa, Poland, 2005; pp. 201–212.
7. Feitz, A.J.; Guan, J.; Waite, T.D. Size and structure effects on centrifugal dewatering of digested sewage sludge. *Water Sci. Technol.* **2001**, *44*, 427–435. [CrossRef]
8. Chen, G.; Po Lock Yue, P.L.; Mujumdar, A.S. Sludge dewatering and drying. *Dry. Technol.* **2002**, *20*, 883–916. [CrossRef]
9. Bień, J.B.; Bień, J.D.; Matysiak, B. *Gospodarka Odpadami W Oczyszczalniach Ścieków*; Wydawnictwa Politechniki Częstochowskiej: Częstochowa, Poland, 1999.
10. Oladejo, J.; Shi, K.; Luo, X.; Yang, G.; Wu, T. A Review of Sludge-to-Energy Recovery Methods. *Energies* **2019**, *12*, 60. [CrossRef]
11. Kłaczyński, E. Oczyszczalnia ścieków—Stabilizacja osadów ściekowych cz. 2. *Wodociągi I Kanaliz* **2013**, *9*, 40–41.
12. Ostojski, A.; Swinarski, M. Znaczenie potencjału energetycznego osadów ściekowych w aspekcie gospodarki o obiegu zamkniętym—Przykład oczyszczalni w Gdańsku. *Rocz. Ochr. Sr.* **2018**, *20*, 1252–1268.
13. Mortusiewicz, A. Współfermentacja osadów ściekowych i wybranych kosubstratów jako metoda efektywnej biometanizacji. Monografie—Polska Akademia Nauk. *Kom. Inżynierii Sr.* **2012**, *98*, 12–45.
14. Álvarez, E.A.; Callejón Mochón, M.; Jiménez Sánchez, J.C.; Ternero Rodríguez, M. Heavy metal extractable forms in sludge from wastewater treatment plants. *Chemosphere* **2002**, *47*, 765–775. [CrossRef]
15. Wang, C.; Hu, X.; Chen, M.L.; Wu, Y.H. Total concentrations and fractions of Cd, Cr, Pb, Cu, Ni and Zn in sewage sludge from municipal and industrial wastewater treatment plants. *J. Hazard Mater.* **2005**, *119*, 245–249. [CrossRef]
16. Siebielec, G.; Stuczyński, T. Metale śladowe w komunalnych osadach ściekowych wytwarzanych w Polsce. *Proc. ECOpole* **2008**, *2*, 479–484. [CrossRef]
17. Haftka, J.J.H.; Govers, H.A.J.; Parsons, J.R. Influence of temperature and origin of dissolved organic matter on the partitioning behavior of polycyclic aromatic hydrocarbons. *Environ. Sci. Pollut. Res.* **2010**, *17*, 1070–1079. [CrossRef]
18. Pérez, S.; Guillamón, M.; Barceló, D. Quantitative analysis of polycyclic aromatic hydrocarbons in sewage sludge from wastewater treatment plants. *J. Chromatogr. A* **2001**, *938*, 57–65. [CrossRef]
19. Hamzawi, N.; Kennedy, K.J.; McLean, D.D. Anaerobic digestion of co-mingled municipal solid waste and sewage sludge. *Water Sci. Technol.* **1998**, *38*, 127–132. [CrossRef]
20. Wlodarczyk-Makula, M. The loads of PAHs in wastewater and sewage sludge of municipal treatment plant. *Polycl. Aromat. Compd.* **2005**, *25*, 183–194. [CrossRef]
21. Hudcova, H.; Vymazal, J.; Rozkosný, M. Present restrictions of sewage sludge application in agriculture within the European Union. *Soil Water Res.* **2019**, *14*, 104–120. [CrossRef]
22. Schnell, M.; Horst, T.; Quicker, P. Thermal treatment of sewage sludge in Germany: A review. *J. Environ. Manag.* **2020**, *263*, 110367. [CrossRef]
23. Gherghel, A.; Teodosiu, C.; De Gisi, S. A review on wastewater sludge valorisation and its challenges in the context of circular economy. *J. Clean. Prod.* **2019**, *228*, 244–263. [CrossRef]
24. Świerczek, L.; Cieślik, B.M.; Konieczka, P. The potential of raw sewage sludge in construction industry–A review. *J. Clean. Prod.* **2018**, *200*, 342–356. [CrossRef]
25. Rezaee, F.; Danesh, S.; Tavakkolizadeh, M.; Mohammadi-Khatami, M. Investigating chemical, physical and mechanical properties of eco-cement produced using dry sewage sludge and traditional raw materials. *J. Clean. Prod.* **2019**, *214*, 749–757. [CrossRef]
26. Syed-Hassan, S.S.A.; Wang, Y.; Hu, S.; Su, S.; Xiang, J. Thermochemical processing of sewage sludge to energy and fuel: Fundamentals, challenges and considerations. *Renew. Sustain. Energy Rev.* **2017**, *80*, 888–913. [CrossRef]
27. Fukas-Płonka, Ł.; Janik, M. Plusy i minusy suszenia osadów ściekowych. *Forum Eksploatatora* **2008**, *5*, 25–27.
28. Lossman, O. Wybrane metody suszenia osadów ściekowych. *Gaz Woda I Tech. Sanit.* **2014**, *9*, 346–349.
29. Đurđević, D.; Blecich, P.; Jurić, Ž. Energy Recovery from Sewage Sludge: The Case Study of Croatia. *Energies* **2019**, *12*, 1927. [CrossRef]

30. Costa, M.; Massarotti, N.; Mauro, A.; Arpino, F.; Rocco, V. CFD modelling of a RDF incineration plant. *Appl. Therm. Eng.* **2016**, *101*, 710–719. [CrossRef]
31. Rahman, M.G.; Rasul, M.M.K.; Khan, S.; Sharma, S. 2015. Recent development on the uses of alternative fuels in cement manufacturing process. *Fuel* **2015**, *145*, 84–99. [CrossRef]
32. Evangelisti, S.; Tagliaferri, C.; Clift, R.; Lettieri, P.; Taylor, R.; Chapman, C. Life cycle assessment of conventional and two-stage advanced energy-from-waste technologies for municipal solid waste treatment. *J. Clean. Prod.* **2015**, *100*, 212–223. [CrossRef]
33. Edo, M.; Skoglund, N.; Gao, Q.; Persson, P.; Jansson, S. Fate of metals and emissions of organic pollutants from torrefaction of waste wood, MSW, and RDF. *Waste Manag.* **2017**, *68*, 646–652. [CrossRef] [PubMed]
34. Borzooei, S.; Campo, G.; Cerutti, A.; Meucci, L.; Panepinto, D.; Ravina, M.; Riggio, V.; Ruffino, B.; Scibilia, G.; Zanetti, M. Feasibility analysis for reduction of carbon footprint in a wastewater treatment plant. *J. Clean. Prod.* **2020**, *271*, 122526. [CrossRef]
35. Sheng, C.; Azevedo, J.L.T. Estimating the higher heating value of biomass fuels from basic analysis data. *Biomass Bioenergy* **2005**, *28*, 499–507. [CrossRef]
36. CEN/TS 15401:2010 Standard. In *Solid Recovered Fuels-Determination of Bulk Density*; BSI Group: London, UK, 2010.
37. Polish Std. PN-EN 15403:2011. In *Stałe Paliwa Wtórne-Oznaczanie Zawartości Popiołu*; Wydawnictwa Normalizacyjne: Warszawa, Poland, 2011.
38. Majchrzak, E.; Mochnacki, B. *Metody Numeryczne: Podstawy Teoretyczne, Aspekty Praktyczne i Algorytmy Wyd*; Politechniki Śląskiej: Gliwice, Poland, 1998; pp. 146–180.
39. Tic, W.J.; Guziałowska-Tic, J.; Pawlak-Kruczek, H.; Woźnikowski, E.; Zadorożny, A.; Niedzwiecki, Ł.; Wnukowski, M.; Krochmalny, K.; Czerep, M. Novel concept of an installation for sustainable thermal utilization of sewage sludge. *Energies* **2018**, *11*, 748. [CrossRef]
40. Kuczyńska, I. Badania mułów węglowych i odpadów mineralnych w zakresie możliwości otrzymania paliw. *Górnictwo I Geoinżynieria* **2005**, *29*, 83–92.
41. Klojzy-Kaczmarczyk, B.; Staszczak, J. Szacowanie masy frakcji energetycznych w odpadach komunalnych wytwarzanych na obszarach o różnym charakterze zabudowy. *Polityka Energetyczna Energy Policy J.* **2017**, *20*, 143–154.
42. Wzorek, M. Characterisation of the properties of alternative fuels containing sewage sludge. *Fuel Process. Technol.* **2012**, *104*, 80–89. [CrossRef]
43. Jewiarz, M.; Mudryk, K.; Wróbel, M.; Frączek, J.; Dziedzic, K. Parameters Affecting RDF-Based Pellet Quality. *Energies* **2020**, *13*, 910. [CrossRef]
44. Li, H.; Jiang, L.B.; Li, C.Z.; Liang, J.; Yuan, X.Z.; Xiao, Z.H.; Wang, H. Co-pelletization of sewage sludge and biomass: The energy input and properties of pellets. *Fuel Process. Technol.* **2015**, *132*, 55–61. [CrossRef]
45. Kubonova, L.; Janakova, I.; Malikova, P.; Drabinova, S.; Dej, M.; Smelik, R.; Skalny, P.; Heviankova, S. Evaluation of Waste Blends with Sewage Sludge as a Potential Material Input for Pyrolysis. *Appl. Sci.* **2021**, *11*, 1610. [CrossRef]
46. Chen, W.S.; Chang, F.C.; Shen, Y.H.; Tsai, M.S. The characteristics of organic sludge/sawdust derived fuel. *Bioresour. Technol.* **2011**, *102*, 5406–5410. [CrossRef]
47. Jiang, L.; Yuan, X.; Xiao, Z.; Liang, J.; Li, H.; Cao, L.; Wanga, H.; Chen, X.; Zeng, G. A comparative study of biomass pellet and biomass-sludge mixed pellet: Energy input and pellet properties. *Energy Convers. Manag.* **2016**, *126*, 509–515. [CrossRef]
48. Park, K.; Ahn, B.-J.; Shin, H.C. A study on producing Refuse Derived Fuel (RDF) using waste plastic film and sewage sludge: The usability of Raw Materials for RDF. *Eng. Agric. Environ. Food.* **2008**, *1*, 57–62. [CrossRef]
49. Polish Std. PN-EN 15359:2012. In *Solid Recovered Fuels—Specifications and Classes*; Polish Committee for Standardization: Warszawa, Poland, 2013.
50. The European Parliament and the Council of the European Union. Directive 2018/850 of the European Parliament and of the Council 2018/850 from 30 May 2018. *J. Eur. Union* **2018**, *L 150*, 109–140.
51. Czop, M.; Poranek, N.; Czajkowski, A.; Wagstyl, Ł. Fuels from Waste as Renewable Energy in Distributed Generation on the Example of the ORC System. *Recycling* **2019**, *4*, 26. [CrossRef]
52. Gendebien, A.; Leavens, A.; Blackmore, K.; Godley, A.; Lewin, K.; Whiting, K.J. *Refuse Derived Fuel, Current Practice and Perspectives (B4-3040/2000/306517/MAR/E3) Final Report*; European Commission; Water Research Centre (WRc): Swindon, UK, 2003.
53. Sever Akdag, A.; Atımtay, A.; Sanin, F.-D. Comparison of fuel value and combustion characteristics of two different RDF samples. *Waste Manag.* **2016**, *47*, 217–224. [CrossRef]
54. Brunner, P.H.; Rechberger, H. Waste to energy–key element for sustainable waste management. *Waste Manag.* **2015**, *37*, 3–12. [CrossRef] [PubMed]
55. Tripathi, N.; Hills, C.D.; Singh, R.S.; Atkinson, C.J. Biomass waste utilisation in low-carbon products: Harnessing a major potential resource. *NPJ Clim. Atmos. Sci.* **2019**, *2*, 1–10. [CrossRef]
56. Smoliński, A.; Karwot, J.; Bondaruk, J.; Bąk, A. The bioconversion of sewage sludge to bio-fuel: The environmental and economic benefits. *Materials* **2019**, *12*, 2417. [CrossRef]
57. Rajca, P.; Zajemska, M. Ocena możliwości wykorzystania paliwa RDF na cele energetyczne. *Rynek Energii* **2018**, *4*, 29.
58. Głodek-Bucyk, E.; Kalinowski, W. Paliwa z odpadów wyzwaniem środowiskowym dla technologii wypalania klinkieru. *Pr. Inst. Ceram. I Mater. Bud.* **2014**, *7*, 188–197.

59. Siemieniuk, J.; Szatyłowicz, E. Zmniejszenie emisji CO_2 w procesie produkcji cementu. *Bud. I Inżynieria Sr.* **2018**, *9*, 81–87.
60. Chatziaras, N.; Psomopoulos, C.S.; Themelis, N.J. Use of waste derived fuels in cement industry: A review. *Manag. Environ. Qual.* **2016**, *27*, 178–193. [CrossRef]
61. Werle, S. Wielowariantowa analiza możliwości współspalania osadów ściekowych w kotłach energetycznych opalanych węglem. *Arch. Gospod. Odpadami I Ochr. Sr.* **2011**, *13*, 21–38.

Article

Challenges in Sustainable Degradability of Bio-Based and Oxo-Degradable Packaging Materials during Anaerobic Thermophilic Treatment

Magdalena Zaborowska [1,*], Katarzyna Bernat [1], Bartosz Pszczółkowski [2], Irena Wojnowska-Baryła [1] and Dorota Kulikowska [1]

1 Department of Environmental Biotechnology, Faculty of Geoengineering, University of Warmia and Mazury in Olsztyn, 10-719 Olsztyn, Poland; bernat@uwm.edu.pl (K.B.); irka@uwm.edu.pl (I.W.-B.); dorotak@uwm.edu.pl (D.K.)

2 Department of Materials and Machines Technology, University of Warmia and Mazury in Olsztyn, ul. Oczapowskiego 11, 10-719 Olsztyn, Poland; bartosz.pszczolkowski@uwm.edu.pl

* Correspondence: magdalena.zaborowska@uwm.edu.pl; Tel.: +48-89-523-4118; Fax: +48-89-523-4131

Abstract: Although the manufacturers labelled commercially available bio-based products as biodegradable, there are discrepancies concerning the time frame for their sustainable biodegradation and methane production. Starch-based, polylactic acid-based and oxo-degradable foils were anaerobically treated in thermophilic condition (55 °C, 100 days). The effect of alkaline pretreatment on foils degradation was also investigated. To examine changes in their mechanical and physical properties, static tensile tests and microscopic analyses, FTIR and surface roughness analyses were conducted. Despite the thermophilic condition, and the longer retention time compared to that needed for biowaste, a small amount of methane was produced with bio-based foils, even after pretreatment (ca. 30 vs. 50 L/kg VS) and foils only lost functional and mechanical properties. The pieces of bio-based materials had only disintegrated, which means that digestate may become contaminated with fragments of these materials. Thus, providing guidelines for bio-based foil treatment remains a challenge in waste management.

Keywords: biopolymers; starch- and polylactic acid-based material; methane production; surface roughness; tensile strength; FTIR and microscopic analyses

Citation: Zaborowska, M.; Bernat, K.; Pszczółkowski, B.; Wojnowska-Baryła, I.; Kulikowska, D. Challenges in Sustainable Degradability of Bio-Based and Oxo-Degradable Packaging Materials during Anaerobic Thermophilic Treatment. *Energies* **2021**, *14*, 4775. https://doi.org/10.3390/en14164775

Academic Editor: Gabriele Di Giacomo

Received: 1 July 2021
Accepted: 30 July 2021
Published: 5 August 2021

1. Introduction

Bio-based products are defined as entirely or partially biological in origin. Among the bio-based products, biodegradable polymers (biopolymers) play an important role. Biopolymers can be degraded by microorganisms into water and biomass, and depending on whether oxygen is present, into carbon dioxide under aerobic conditions, or into carbon dioxide and methane under anaerobic conditions. In recent years, the share of biopolymers on the market has been increasing and, in 2019, 1.174 million tons were produced at a global level. Further increases in the share of these plastics on the market are expected (i.e., 1.334 million tons by 2024) [1,2].

One of the most known biopolymers is polylactide (PLA), which is produced by the polymerization of lactic acid via the fermentation of dextrose derived from carbohydrate-rich crops, such as potatoes or corn. PLA is widely used because of its properties, including its biodegradability, biocompatibility, high elastic modulus, transparency and mechanical strength [3]. The other commonly used natural polymer (meaning that it is produced directly in nature) is starch, which is, more specifically, a reserve polysaccharide composed of amylopectin (75−80%) and amylose (20−25%). Starch is used due to its abundance, inexpensiveness, and biodegradability. It is known that, on the market, apart from biopolymers, oxo-degradable polymers are still used in some countries [4]. They contain pro-degradants (e.g., d2w® or TDPA (totally degradable plastic additives)), usually produced from cobalt, manganese, and iron compounds, which should ensure the polymer decomposition. Some authors indicate that oxo-degradable polymers may be partially biodegradable, e.g., with mineralization level of 12.4% after three months of incubation with compost [5].

Among the bio-based and oxo-degradable products present on the market, the most popular are materials used for food packaging or waste bags for biowaste collection. Because they are labelled as biodegradable, these products should be separately collected, together with biowaste. Thus, most of the waste from these products is eventually gathered and processed, aerobically or anaerobically, in mechanical–biological treatment plants (MBT installations), or via a combination of the two methods. Because the share of biodegradable polymers in the stream of selectively collected biowaste will likely continue to increase, it is important to better understand the mechanisms and characteristics of their degradation. Most studies have focused on the biodegradation of biopolymers under environmental conditions, i.e., in soil, compost, and aquatic environments [6]. In aerobic conditions, the characteristics of the process depend on the types of polymers involved (i.e., their molecular weight, crystallinity or type of functional groups) and environmental factors (i.e., temperature, moisture, pH, and presence of oxygen) [7,8]. However, knowledge concerning the possibility of polymer degradation under anaerobic conditions is limited, especially in the context of commonly used methods for the treatment of biowaste. Therefore, it is necessary to determine the conditions that facilitate biodegradation of bio-based and oxo-degradable products and the characteristics of this process. Anaerobic processes seem to be especially profitable because they produce methane, which can be used for energy generation. However, because of amount of time needed for methane production from bio-based products [9], pretreatment is needed. The use of pretreatment may increase the surface area of the material, resulting in more rapid degradation and improved methane production [10]. For example, Yu et al. [11] reported over 70% abiotic degradation of PHB at 70 °C in 4 M sodium hydroxide after 4 h of treatment. However, the treatment of PHB in acidic solutions of sulfuric acid (0.05–2 M) at 70 °C for up to 14 h did not result in abiotic degradation. Myung et al. [12] observed the near complete abiotic degradation of poly(3-hydroxybutyrate-co-3-hydroxyvalerate) (PHBV) at 60 °C in 0.1 M sodium hydroxide after 18 h of treatment.

In conclusion, despite many commercial products on the market being labeled as biodegradable, their detailed compositions are not provided by manufacturers. The content of the polymer and various types of additives in the products can differ, but most likely, the composition affects the biodegradability of polymers and the time frame for their biodegradation, especially in the context of their end-of-life considerations in waste management. Thus, in the present study, degradability and methane production (MP) from commercially available bio-based foils based on starch (FS), polylactic acid (PLA), and additionally oxo-degradable product (OXO), was analyzed. Moreover, alkaline pretreatment was used to check the acceleration of the process. Microscopic and FTIR analyses, and determination of the functional properties of the surfaces and the mechanical properties of the materials (static tensile tests) were also performed. Determination of the mechanical properties of material, as another indicator of its degradation, is not a common practice.

2. Materials and Methods

2.1. Bio-Based and Oxo-Degradable Materials Used in the Experiment

In the experiment, commonly available waste bags made of bio-based materials and oxo-degradable material were used. As bio-based materials, BioBag waste bags made from Mater-Bi® material based on starch (FS) and bags made from polylactic acid (PLA) were used. The oxo-degradable waste bag (OXO) was made from petroleum-based polyethylene containing d2w pro-degradant additive. The characteristics of the materials are as follows:

- PLA (the VS content of 99.3 % DM, 71.5 g COD/kg DM, 454.9 mg N_{tot}/kg DM);
- FS (the VS content of 99.9 % DM, 63.9 g COD/kg DM, 408.6 mg N_{tot}/kg DM);

- OXO (the VS content of 93.1 % DM, 3766.8 mg N_{tot}/kg DM (COD was not possible to determine)).

The products were obtained from the commercial public market; thus, they may have contained other additives, improving their functional properties but, unfortunately, the manufacturer did not report their detailed composition. The waste bags were shredded to a particle size of 10 × 10 mm.

Two sets of experiments were carried out. In the first, all materials (PLA, FS, OXO) were used without pretreatment. In the second set, PLA, FS and OXO were pretreated with 0.1 M potassium hydroxide (KOH) for 2 h; pretreatment was performed at ambient temperature (PLA_{KOH}, FS_{KOH}, OXO_{KOH}). After 2 h, the mixture was neutralized with hydrochloric acid. Solubilized PLA ($PLA_{solubilized}$) and FS ($FS_{solubilized}$) were used as control samples. Each material was solubilized with the use of 1 M KOH for 24 h. The OXO material did not solubilize under these conditions.

2.2. Measurement of Methane Production (MP) under Thermophilic Conditions

The anaerobic degradation test of bio-based and oxo-degradable products was carried out in thermophilic conditions (55 °C). The experiment was carried out in a methane potential analysis device (Automatic Methane Potential Test System, AMPTS II, Bioprocess Control, Sweden AB) for 100 days. The device consisted of reactors with a volume of 630 mL. The reactors were mixed at a speed of 80 rpm for 1 min every hour. To each reactor, 200 mL of inoculum and the appropriate amount of material was added to ensure a starting organic load of 4 kg VS/m^3. Anaerobic conditions were achieved by flushing each reactor with pure nitrogen for 2 min. The inoculum was obtained from a closed chamber for the mesophilic anaerobic digestion of sewage sludge. Thermophilic preincubation of the inoculum was carried out at 55 °C for 14 days. The dry mass (DM) content of the inoculum was approximately 1.7%, whereas VS constituted 69.5% of DM.

The anaerobic degradation test was based on determining the volume of methane produced online. The measurement of MP with each material and inoculum, and with inoculum alone was performed in triplicate. In order to not disturb the MP in the AMPTS II, additional glass bottles were prepared for sampling for microscopic analysis, and roughness and static tensile tests. These bottles had the same volume as the reactors used in the AMPTS II and were dosed with the same amount of inoculum and material, and placed in a thermostatic incubator at 55 °C. When the fragments of bio-based and oxo-degradable materials were collected for analysis, they were cleaned with distilled water.

Based on the anaerobic degradation test in the methane potential system, the biodegradation degree (BD) was assumed. The BD of the bio-based materials (PLA, FS, PLA_{KOH} and FS_{KOH}) during anaerobic treatment at 55 °C was calculated as the ratio of the cumulative MP from the material (after pretreatment (treated) or without pretreatment (untreated)) to the cumulative MP from the material after its solubilization. The cumulative MP from the solubilized material was assumed to be the maximal MP from this material under the experimental conditions. For the OXO material, which did not solubilize, the BD was not determinable.

Methane production (MP) followed pseudo first-order kinetics, and can be described with this equation:

$$C_{t,CH4} = C_{CH4} \cdot (1 - e^{kCH4}\ t) + C_{0,CH4}, \quad (1)$$

where $C_{t;CH4}$ (L/kg VS) is the cumulative MP at time t (days) of anaerobic measurement; C_{CH4} (L/kg VS) is the maximal MP; $C_{0,CH4}$ (L/kg VS) is the initial MP; k_{CH4} (d^{-1}) is the kinetic coefficient of MP.

The rate of MP is the product of C_{CH4} and k_{CH4}.

The values of C_{CH4}, $C_{0,CH4}$ and k_{CH4} were obtained by nonlinear regression analysis with Statistica software, version 13.0 (StatSoft, Tulsa, OK, USA).

2.3. Microscopic and FTIR Analysis

During anaerobic degradation tests, pieces of FS, PLA, OXO, PLA_{KOH}, FS_{KOH}, and OXO_{KOH} were removed from the inoculum and cleaned with distilled water for the microscopic analysis, and the roughness and static tensile tests.

The changes on the surface of the bio-based and oxo-degradable materials were analyzed using a Nikon eclipse50i (Nikon, Tokyo, Japan) polarizing microscope at a magnification of 10×. Pictures were taken at 0, 20, 45, 59 and 100 days.

The FTIR spectra (with three replicates for each foil material) were obtained using a Perkin Elmer Spectrum Two (with diamond ATR) (Perkin Elmer, Waltham, MA, USA) brand device having a resolution of 4 cm^{-1} within the wavenumber range of 4000–400 cm^{-1} at room temperature.

2.4. Functional Properties of the Surface

The geometric structure of the surface and the functional properties of the fragments of foils (five replications) were performed with the profile method (PN-ISO 3274) using a Mitutoyo Surftest SJ-210 profilometer (Mitutoyo, Kawasaki, Japan) with a diamond measuring tip with a point angle of 60° and a stylus radius of 2 µm. The measured length for each material consisted of five elementary sections that were 0.8 mm long. The changes in amplitude (arithmetical mean roughness value (Ra), mean square surface profile deviation (Rq)) were determined. The Abbott–Firestone curves [13] were used to determine the reduced peak height without bearing properties (Rpk); the roughness core height (roughness core profile), characterizing the surface bearing properties (Rk); the reduced valleys depth, characterizing the ability of the surface to hold liquids (Rvk); total height of the roughness profile (Rt) and the percentage of the peaks and valleys (Mr1, Mr2, respectively). A detailed description is given in [14].

2.5. Static Tensile Tests

Static tensile tests (tensile strength at break and elongation at break) were performed according to ISO standard (ISO 527-2:2012, ISO 527-3:2019) with the use of a TA.HD.Plus (Stable Microsystems, Surrey, UK). Five replicate samples of foils were used for the measurements until they were too fragile for analysis, and it was impossible to place them in the tensile test device. The samples were subjected to a tensile force at a constant speed until the moment of fracture.

2.6. Analytical Methods

The volatile substance (VS) content of the FS, PLA and OXO materials was determined as loss after ignition at 550 °C. The chemical oxygen demand (COD) of these materials and the content of nitrogen (Kiejdahl nitrogen) were also determined. All physicochemical analyses were performed according to Greenberg et al. [15]. Flash 2000 Organic Elemental Analyzer (Thermo Scientific, Milan, Italy) was used to determine elemental composition of bio-based materials according to the manufacturer's protocol.

3. Results and Discussion

3.1. Methane Production during Anaerobic Test

Biodegradation anaerobic of pieces of waste bags made of PLA, FS, OXO, and also the materials after pretreatment (PLA_{KOH}, FS $_{KOH}$, and OXO_{KOH}) were carried out under thermophilic conditions at 55 °C. The MP analysis tool allowed measurement of the volume of methane (in mL) produced online. Two MP profiles are considered during the analyses of each material: the first profile corresponds to the inoculum alone, and the second, to both the inoculum and the material. By subtracting the values of the first profile from the second, the profile of the methane volume for the material alone is obtained. In the present study, the VS content of the materials was considered to reflect the organic matter content, and the MP was shown as L/kg VS (Figure 1). To improve the readability of the figure, 60 days of measurements are shown, although the time of measurements was ca.

100 days. The OXO material did not degrade; thus, no MP was observed, and the profiles are not provided.

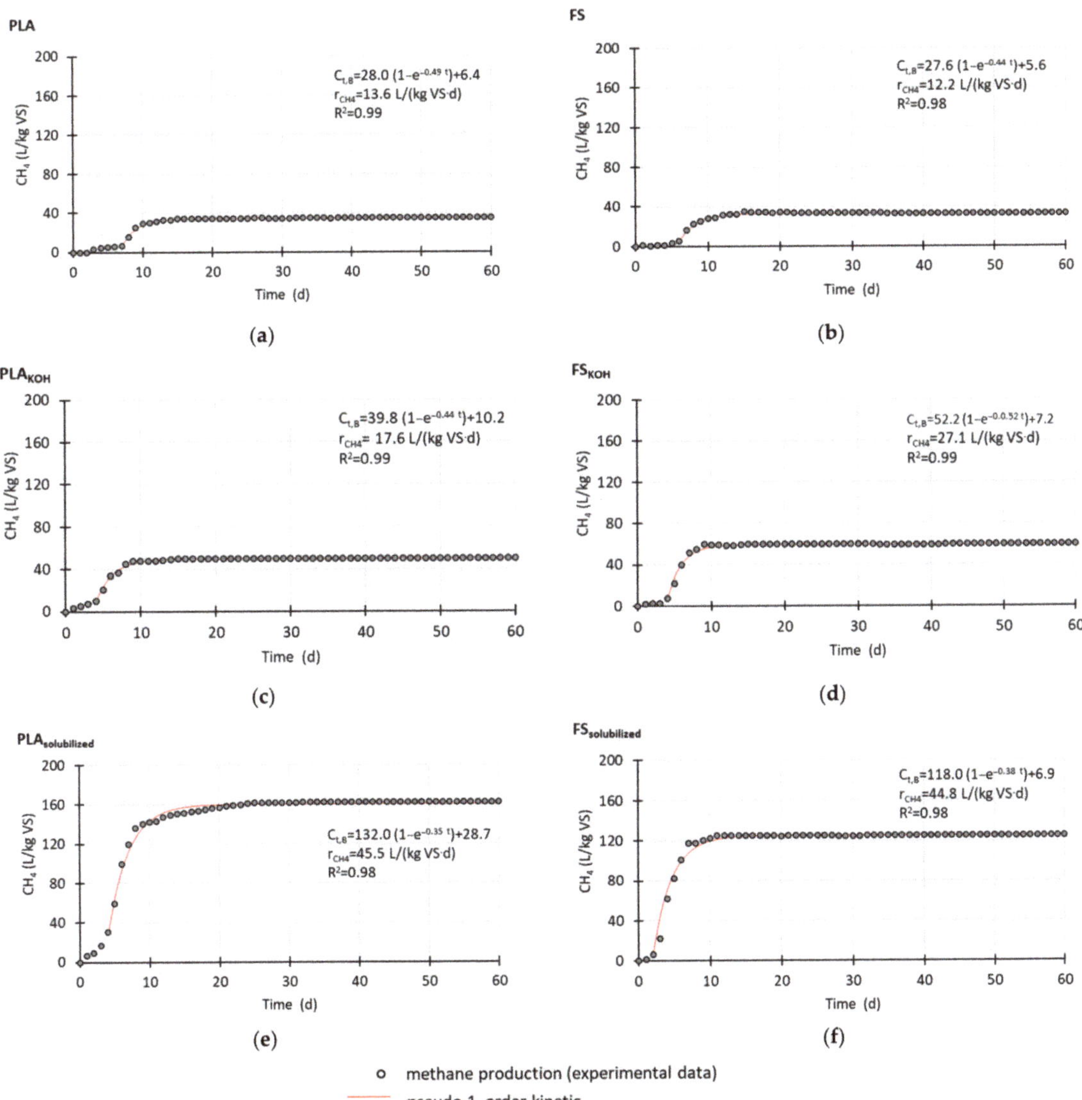

Figure 1. Methane production profiles during an anaerobic degradation test at 55 °C of bio-based materials made of PLA and FS without pretreatment (**a**,**b**), of these materials with pretreatment (**c**,**d**), and solubilized materials (**e**,**f**); to improve the readability of the figure, 60 days of measurements are shown, although the time of measurements was ca. 100 days.

Regarding bio-based materials made of PLA and FS without pretreatment, lag phases in MP were observed. With FS, the lag phase was shorter, lasting 6 days, whereas with PLA, it lasted 7 days. After the lag phase, MP increased until day 15 or 16 of measurements. Then, the methane volumes remained the same until the end of the anaerobic test, which lasted almost 100 days, and the final methane volumes for PLA and FS were similar, 34.5 and 32.7 L/kg VS, respectively.

After pretreatment, the lag phases shortened to 4 days for both PLA_{KOH} and FS_{KOH}, and after this time, intensive MP took place. After the lag phase, during next 6 days of measurement, PLA_{KOH} produced 47.5 L/kg VS, whereas FS_{KOH} produced almost 25% more, 58.6 L/kg VS.

When solubilized materials were investigated in anaerobic degradation tests, it was found that MP started from the beginning of the measurements. After 15 and 12 days of the measurements with $PLA_{solubilized}$ and $FS_{solubilized}$, respectively, MP reached maximum. Despite the fact that PLA needed a longer time to obtain maximum MP, the final value was 1.3 times higher than MP with FS (162 L/kg VS for $PLA_{solubilized}$ vs. 125 L/kg VS for $FS_{solubilized}$).

Cumulative MP after 100 days of anaerobic treatment of PLA and FS, without pretreatment, with pretreatment, and the solubilized materials, is shown in Figure 2a. To summarize, cumulative MP was similar with both materials without pretreatment. After pretreatment, cumulative MP with PLA_{KOH} and FS_{KOH} increased almost 40% and 80%, respectively; with the solubilized materials, it was highest. However, the values were much lower than with pure PLA and pure starch materials.

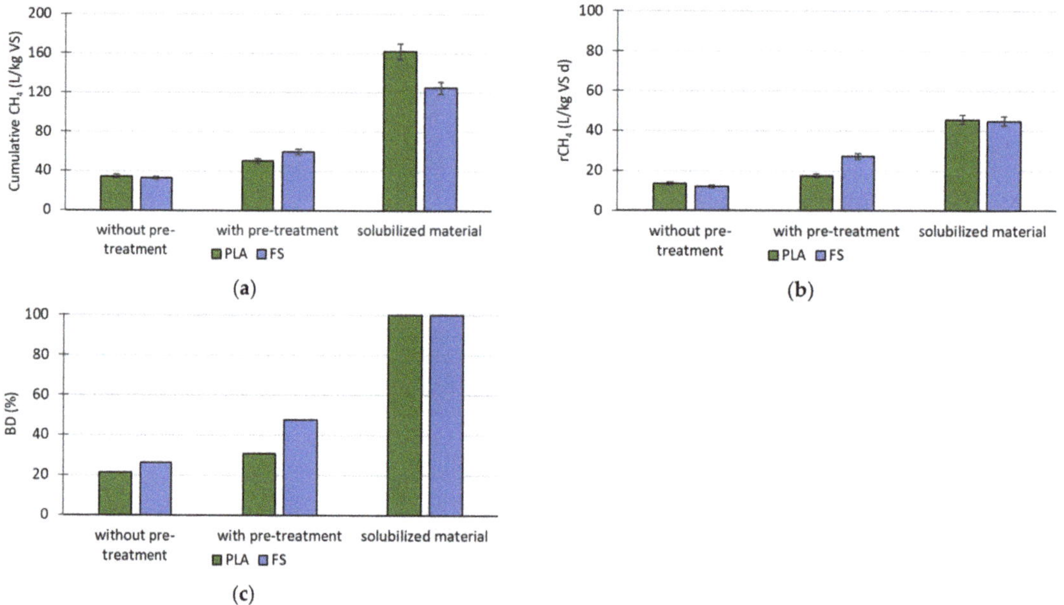

Figure 2. Cumulative methane production after 100 days of anaerobic degradation of PLA and FS (**a**), the rate of methane production (**b**), and the biodegradation degree (BD) (**c**).

Methane production proceeded with first-order kinetics, and the model fit the experimental data well (R^2 = 0.98–0.99). The kinetic models and the initial rates of MP are shown in Figure 1.

The rates of MP, summarized in Figure 2b, corresponded with the cumulative MP with untreated and treated materials. However, the rates of MP with both solubilized materials were similar, despite the difference in cumulative MP.

It was assumed that the solubilized materials ($PLA_{solubilized}$ and $FS_{solubilized}$, i.e., control samples) produced the maximal methane volume, and this equaled 100% of the biodegradation degree (BD) of these materials. For PLA without pretreatment and FS without pretreatment, BD was higher for FS, but for both of the materials, it did not exceed 26% in comparison with solubilize materials. The BD of PLA_{KOH} FS_{KOH} increased to 30 and 47.5%, respectively (Figure 2c).

It should be emphasized that, during the long-term anaerobic biodegradation test, the pieces of the materials were still visible in the inoculum, and they did not disintegrate, as shown by the microscopic analyses.

The elemental compositions of the bio-based products, PLA and FS, were determined (Table 1).

Table 1. Elemental composition, empirical formula, and theoretical methane production (TMP) of foil materials (PLA, FS) and of pure polymers (pure PLA, pure FS).

Characteristics	Pure PLA	PLA	Pure FS	FS
Empirical formula	$C_3H_4O_2$	$C_{4.4}H_{6.31}O_2$	$C_6H_{10}O_5$	$C_{9.8}H_{14.7}O_5$
Elemental composition				
Carbon, % TS	50.00	57.56	44.44	55.36
Hydrogen, % TS	5.56	6.88	6.17	6.90
Oxygen, % TS	44.44	34.86	49.38	37.64
Methane production				
* TMP, L/kg	467	610	413	577
** MP, L/kg	–	162	–	125

* TMP calculated using the Buswell equation [16]; ** cumulative methane production (MP) after 100 days of anaerobic thermophilic treatment of solubilized PLA and FS.

PLA contained more carbon and less oxygen than S. The hydrogen content of both foils was about 6.9%. However, it must be emphasized that both foils (PLA, FS) contained more C and H than pure PLA and pure FS, with lower content of O.

The FS foil material is described by the manufacturer as consisting mainly (as much as 85% or more) of modified thermoplastic starch that is biodegradable and compostable [17]. On the basis of the elementary composition of the foil materials and pure polymers (pure PLA, pure FS), the theoretical amounts of methane (TMP) and carbon dioxide that could be obtained, assuming complete conversion of organics into biogas, were calculated using the Buswell equation [16]:

$$C_aH_bO_c + \left(a - \frac{b}{4} - \frac{c}{2}\right) H_2O \rightarrow \left(\frac{a}{2} - \frac{b}{8} + \frac{c}{4}\right) CO_2 + \left(\frac{a}{2} + \frac{b}{8} - \frac{c}{4}\right) CH_4 \quad (2)$$

The TMP values of pure PLA and pure FS were approximately 1.3–1.4-times lower than those of the PLA and FS foils, due to the lower carbon and hydrogen contents of the pure materials (Table 1). As for the cumulative MP during anaerobic digestion at 55° C, PLA and FS produced more than two-times less methane than $PLA_{solubilized}$ and $FS_{solubilized}$. However, even solubilized materials produced methane in amounts constituted only 27% and 22%, respectively, of the TMP values. The small amount of methane produced by all the bio-based materials is connected with the fact that, until 100 days of anaerobic degradation had passed, pieces of the foils were still visible in the inoculum, and these fragments only disintegrated when they were touched.

The results concerning anaerobic treatment of starch-based material are scarce. For example, Mohee et al. [18] found that the cumulative methane production from starch-based material was of 245 mL over 32 days of batch digestion assays. The value was much higher than that obtained in the present study; however, the authors did not provide the initial doses of the material. Therefore, it is not known what the methane production was per 1 g VS, and the efficiency of the process cannot be estimated.

Studies with PLA are more numerous, however, most of them are carried out with the used of pure polymer, not with commercially available foil materials. Itävaara et al. [19] found that at temperatures of 37 °C during 40 days of measurements PLA degradation was not higher than 5%. They reported 60% of PLA degradation after 100 days of an anaerobic test without providing more specific information. Under thermophilic conditions, about 60% of PLA was also degraded, but after 40 days. Contrarily, the results of Yagi et al. [20] showed that PLA biodegradability was 80–91% in aquatic conditions at a thermophilic temperature. In another study, Yagi et al. [21] found approximately 60% biodegradability of PLA powder of 90% in 60 days under thermophilic conditions. However, these studies focused on the biodegradation degree but without determination of the efficiency of methane production. The recovery of energy from municipal waste one of the main key points of the strategy of circular economy. Thus, it is important to determine the biogas potential of packaging materials (foils) considered to be biodegradable. This was done in the present study.

3.2. Changes in the Structure of Tested Materials—Microscopic Analysis

At the beginning, the foil pieces of the investigated materials looked like typical foil waste bag materials, and all of them were green in color and smooth.

Microscopic images of the structure of untreated and treated foils were present in Figure 3. Regarding the structure of FS, the changes were visible from the 28th day of degradation, whereas changes in the structure of PLA were visible from about the 59th day. When the changes became apparent, there were individual cracks and fissures in the structure. After this time, the cracks and fissures had become deeper and more numerous on the surface. Unfortunately, both of the bio-based materials did not disintegrate completely, and they were still visible until almost 100 days of anaerobic treatment. The use of the pretreatment step did not facilitate the process of disintegration of the bio-based foils. The visible changes, in the form of cracks and fissures, appeared on the surface of foils around the same time as in the variant without pretreatment. However, the pretreatment step caused the foil pieces to become severely weakened and fragile, so that they crumbled when touched after a shorter time of degradation.

As mentioned, the OXO foil did not produce methane, and microscopic analyses did not show any structural changes even after the pretreatment step.

3.3. Changes in the Mechanical Properties of the Materials

During the anaerobic degradation test, the mechanical properties of the foils before and after the pretreatment were investigated. The results of tensile strength are presented in Figure 4a,b. Regarding the untreated OXO material, the tensile strength decreased by about 17.5 ± 6.8% in the first 10 days of the experiment. Then, until the end of the test, the tensile strength remained at a similar level. The initial decrease could result from the presence of the oxo-degradant, which may weaken the mechanical properties of polyethylene (PE) under thermophilic conditions [22,23].

Regarding the bio-based products without pretreatment, a decrease in tensile strength was observed, which indicated that degradation was progressing (Figure 4). It was possible to measure the tensile strength of FS until to the 39th day of the anaerobic degradation test, at which point the tensile strength had decreased by almost three fold. After this time, the foil was too brittle to be subjected to the strength test and it was not possible to place particles of FS in the measuring device. During the first 10 days, the tensile strength decreased sharply, from 16.6 ± 0.8 MPa in the raw material, to 11.1 ± 0.7 MPa. After this time, the tensile strength continued to decrease steadily, and after 39 days, it reached 5.9 ± 0.9 MPa (Figure 4a).

Tensile strength measurements of PLA were possible until the 59th day of the test, which means that, in this regard, the material was more resistant than FS.

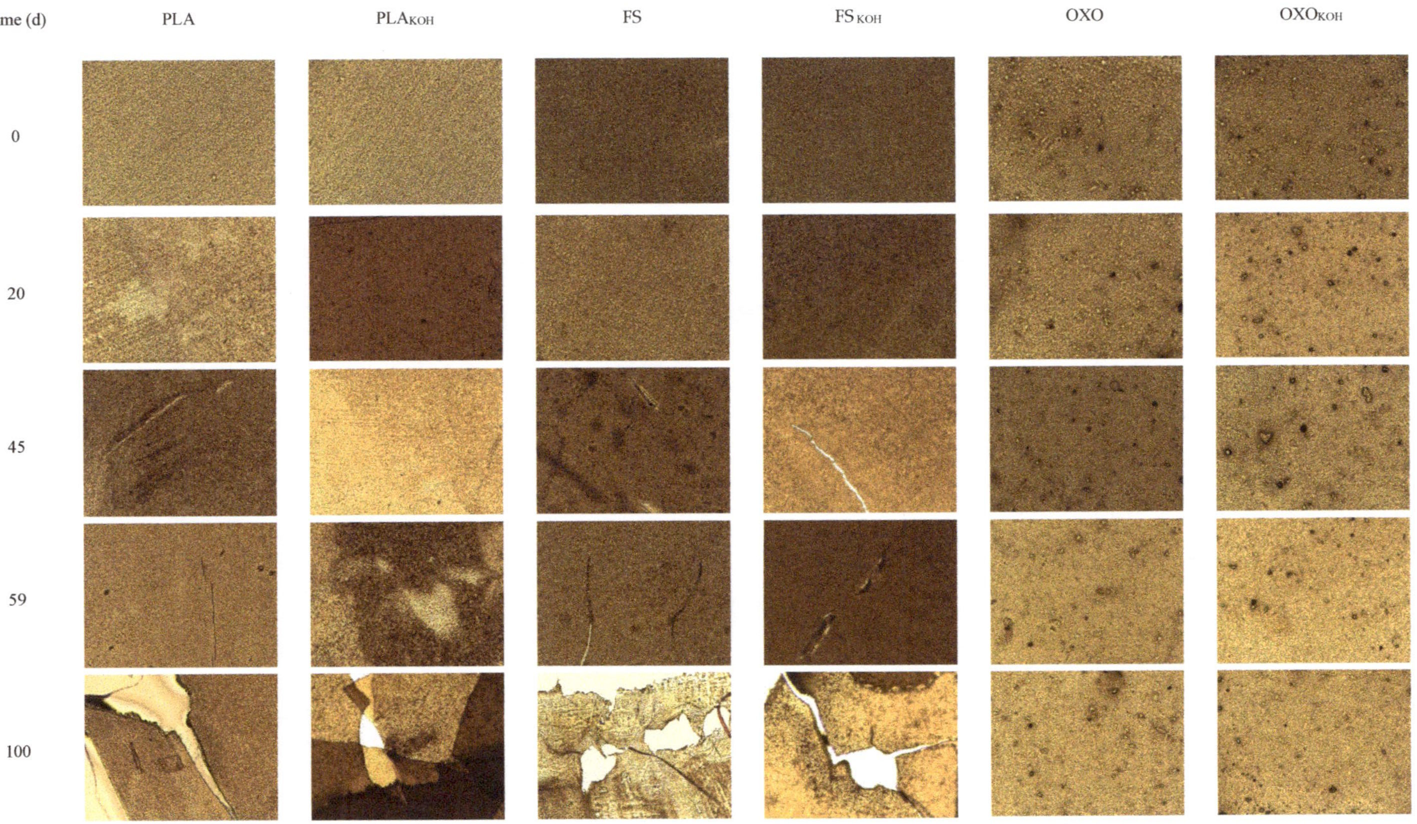

Figure 3. Microscopic images of the structure of untreated and treated foil materials during anaerobic thermophilic degradation test.

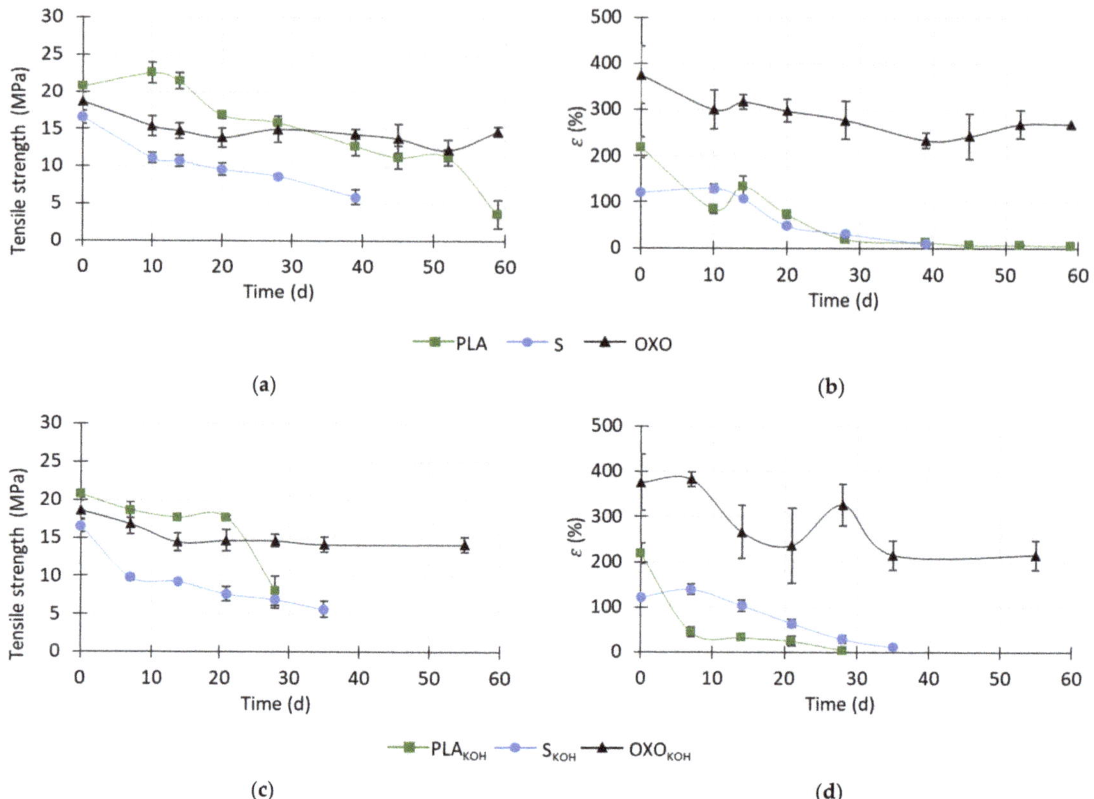

Figure 4. Changes in the mechanical properties (tensile strength and elongation at break ε) during an anaerobic degradation test of untreated foils (**a**,**b**) and foils after pretreatment with KOH (**c**,**d**).

Up to day 14 of the experiment, the tensile strength of PLA increased by about 9% because the material stiffened. Vasile et al. [24] carried out burial soil degradation of pure PLA and found that, after an initial decrease in tensile strength by about 14% on day 50, the material stiffened slightly, corresponding to a tensile strength increase after 100 days. After 150 days, the tensile strength decreased again. Stiffening of PLA may be caused by the water diffusing into the amorphous region of the material and random scission of ester bonds. This causes an increase in the degree of crystallinity during the degradation process. Next, after the degradation of the amorphous regions, degradation proceeds towards the center of the crystalline parts [25,26].

In the present study, the tensile strength of the PLA continuously decreased from day 14 to day 52 of the test, reaching 11.3 ± 1.0 MPa. On the last day, the PLA material was still collectible for the test, and the tensile strength was 3.6 ± 2.0 MPa, which was almost six times lower than in the raw/starting material (Figure 4a).

Rajesh et al. [27] carried out the soil burial degradation of pure PLA for 90 days at room temperature. Before starting degradation, the tensile strength of the material was 50 MPa and, during their study, it decreased by 26.5% (to ca. 38 MPa). Compared to the present study, the reduction in tensile strength was much smaller, but the tensile strength values were much higher. It should be emphasized that those authors conducted their degradation tests under aerobic conditions at room temperature, and they used pure PLA, which has greater tensile strength than the material used in the study presented here. The decrease in the tensile strength of bio-based products during degradation tests may be caused by the diffusion of water into the polymer matrix, causing hydrolysis to begin faster [28].

Then, as a result of breaking the secondary bonds (hydrogen, van der Waals bonds), the polymer is weakened, and the long polymer chains are converted to chains of low molecular weight. Then, the covalent bonds between the monomers are broken, which continues until the polymer is completely degraded [29].

The elongation at break (ε) of PLA, FS and OXO was also determined (Figure 4b,d). The value of ε for OXO material slightly decreased during the first 10 days of the degradation test, but over the next few days, only the small changes were noted. However, it should be emphasized that the initial ε of OXO was the highest of all the tested materials (375%), in contrast to its tensile strength, which was similar to that of the other foil materials.

Generally, the ε value of FS and PLA decreased as anaerobic degradation progressed. Despite the fact that the tensile strengths of both materials were similar, the elongation at the break of PLA was almost two-times higher than that of FS (ca. 220% and ca. 120%, respectively). However, in the case of PLA foil, a rapid initial decrease in ε (to ca. 85%) corresponded to the initial increase in tensile strength (from ca. 21 to ca. 23 MPa). The material may have stiffened due to an increase in its degree of crystallinity, which caused its tensile strength to increase. Then, at ca. 15 days, its elongation at break increased to 135%, which may have been caused by the breakdown of the secondary bonds between the polymer chains. From day 14 to day 45, the ε value decreased, reaching ca. 7%, which was only 3.2% of the initial value. The ε value then remained the same until the 59th day of the test, when the last measurement of this characteristic was possible. Regarding FS, its ε value increased by about 6% (from ca. 122%) during the first 10 days of measurement. This might have been caused by the polymer absorbing water. Then, at day 39, when the last measurement was possible, ε had decreased to ca. 10 $\pm$ 2% (Figure 4b,d).

The tensile strength and ε of KOH-pretreated OXO were virtually the same as those of OXO that had not been pretreated. Initially, the tensile strength slowly decreased, and after the 14th it was 22 $\pm$ 6% lower than the initial value. Then, the tensile strength and ε remained the same until the end of the degradation test. However, it should be emphasized that the KOH-pretreatment of OXO increased the standard deviation of the ε values (Figure 4c,d).

In contrast, pretreatment weakened the mechanical properties of both bio-based materials. Moreover, the time during which it was possible to determine the mechanical properties of these two foils was shorter with pretreatment than without pretreatment (for PLA_{KOH}, reduced from 52 to 28 days; for FS_{KOH}, from 39 to 35 days).

After 7 days of anaerobic degradation of PLA_{KOH}, the tensile strength decreased by about 11 $\pm$ 4% in comparison to the raw material (ca. 21 MPa). No temporary stiffness was observed, unlike the PLA, which was not pretreated. This lack of stiffness was also confirmed by the fact that, during this time, elongation at break decreased rapidly to 45%, which was almost two-times lower than the value obtained during the first days with the untreated PLA. Until 21 days of the test, the tensile strength remained at a similar level, while ε decreased slightly. Then, at 28 days, the tensile strength and ε decreased to ca. 8 MPa and 5%, respectively. With regard to both FS and FS_{KOH}, the tensile strength decreased sharply to ca. 10 MPa within the first 10 days of degradation.

In general, the decrease in elongation at the break of both the pretreated bio-based products and those without pretreatment was caused by an increase in the stiffness and a decrease in the flexibility of those materials. Similarly, Vasile et al. [24] reported that the tensile strength and elongation at break of pure PLA materials and PLA biocomposites decreased during soil burial biodegradation. The authors stated that, as the time of degradation progressed, the stiffness of the investigated materials increased.

3.4. Changes in the Surface Roughness of the Foils

During anaerobic degradation, the surface roughness of the foils decreased. Measurements of roughness were carried out to days 59 and 53 of degradation of the untreated and treated foils, respectively. After this time, the foils were too brittle to be analyzed.

During the first 9 days of anaerobic degradation, the Ra, Rq and Rt values of untreated PLA decreased, corresponding to a decrease in the roughness core (Rk). In addition, the reduced valley depth (Rvk) also decreased. Taken together, the decreases in these values indicated that the roughness of the PLA decreased, which may be due to the diffusion of the inoculum into the polymer matrix and water adsorption by the polymer. From day 9 to day 20, the Ra, Rq, Rt values increased, and on day 28, these values were similar to values noted in the raw material. In contrast, the Rpk value decreased until day 4, and then until day 14, it increased more than 2-times, and then remained at a similar level until the end of the measurements (Figures 5 and 6).

Until day 45, the the Ra, Rq, Rt values for PLA decreased again by about 1.4–1.5 times compared to the raw material, which was probably related to the secondary diffusion of water into the material structure. This also may indicate a gradual decrease in the thickness of the foil. The percentage of the peaks and valleys (Mr1 and Mr2) remained at a same level until day 45, then slightly decreased.

Regarding the FS foil, the Ra, Rq and Rt values increased during the first days of the anaerobic testing, which may indicate that the top layer of the material quickly dissolved. The structure of the material then became more complex and non-homogeneous. The roughness (Ra, Rq, Rt, Rk) of foil FS decreased to the 39th day of the analysis, and after this time, it remained at a constant level. However, the total height of the profile (Rt) increased, which may mean that larger random height peaks occurred in the structure, which are characteristic of damage to the surface structure.

Regarding the OXO foil, the roughness indicators did not display any differences up to day 14, and then, on day 20, all of these values except Mr1 and Mr2 increased 1.7–2.3-times. After day 20, all roughness indicators decreased. However, on the 59th day of degradation, the Ra, Rq, Rt, Rk, Rpk and Rvk values increased, which may have been related to the removal of oxo-additives from the surface structure. During the degradation test, the value of Mr1 decreased slightly, while that of Mr2 remained at a similar level (Figures 5 and 6).

Generally, after the pretreatment with KOH, the roughness of the bio-based materials decreased. This may be due to degradation of the external layers of the foils. However, the MR2 and MR1 values remained at a level similar to that in the untreated materials (Figures 5 and 6).

The pretreatment caused the initial weakening of the surface structure of the PLA. After the pretreatment with KOH, binding of PLA and KOH might have taken place, which would have created a layer that partially blocked the pores in the structure, resulting in the lack of change in the roughness indicators. The slight fluctuations in their values may have resulted from the damage to the surface of the materials after pretreatment. The Ra, Rq, Rt, Rk, Rpk and Rvk values decreased on day 53 of the anaerobic test. This may indicate that particles were rinsed out of the material into the inoculum. Moreover, the lower Rk value indicated a decrease in the roughness core, which could be connected with damage to the top layer of the material. At this time, the material defragmented when it was touched and further determination of the surface roughness was not possible. In contrast, Luo et al. [30] found that, in compost conditions, the neat PLA did not present significant changes on the surface after 5 days biodegradation, but after 20 days, the surface roughness increased. This could be owned to the hydrolysis of PLA and microorganism activities. With increasing incubation time, the cracks and voids became substantially deep and large.

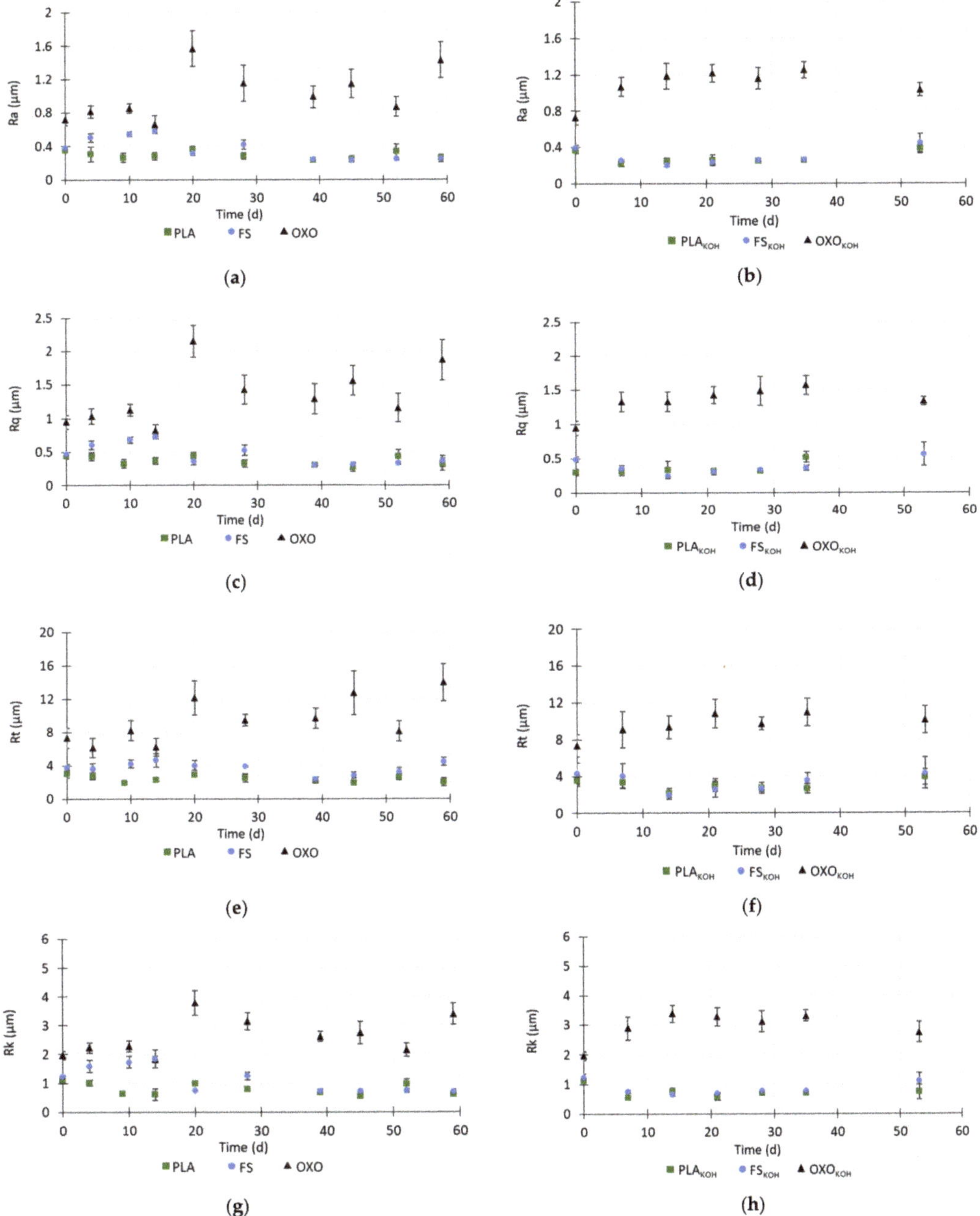

Figure 5. The surface roughness indicators (arithmetical mean roughness value (Ra), mean square surface profile deviation (Rq), the roughness core height (roughness core profile) (Rk), total height of the roughness profile (Rt)) during an anaerobic degradation test of untreated foils (PLA, FS, OXO) (**a**,**c**,**e**,**g**) and foils after pretreatment with KOH (PLA_{KOH}, FS_{KOH}, OXO_{KOH}) (**b**,**d**,**f**,**h**).

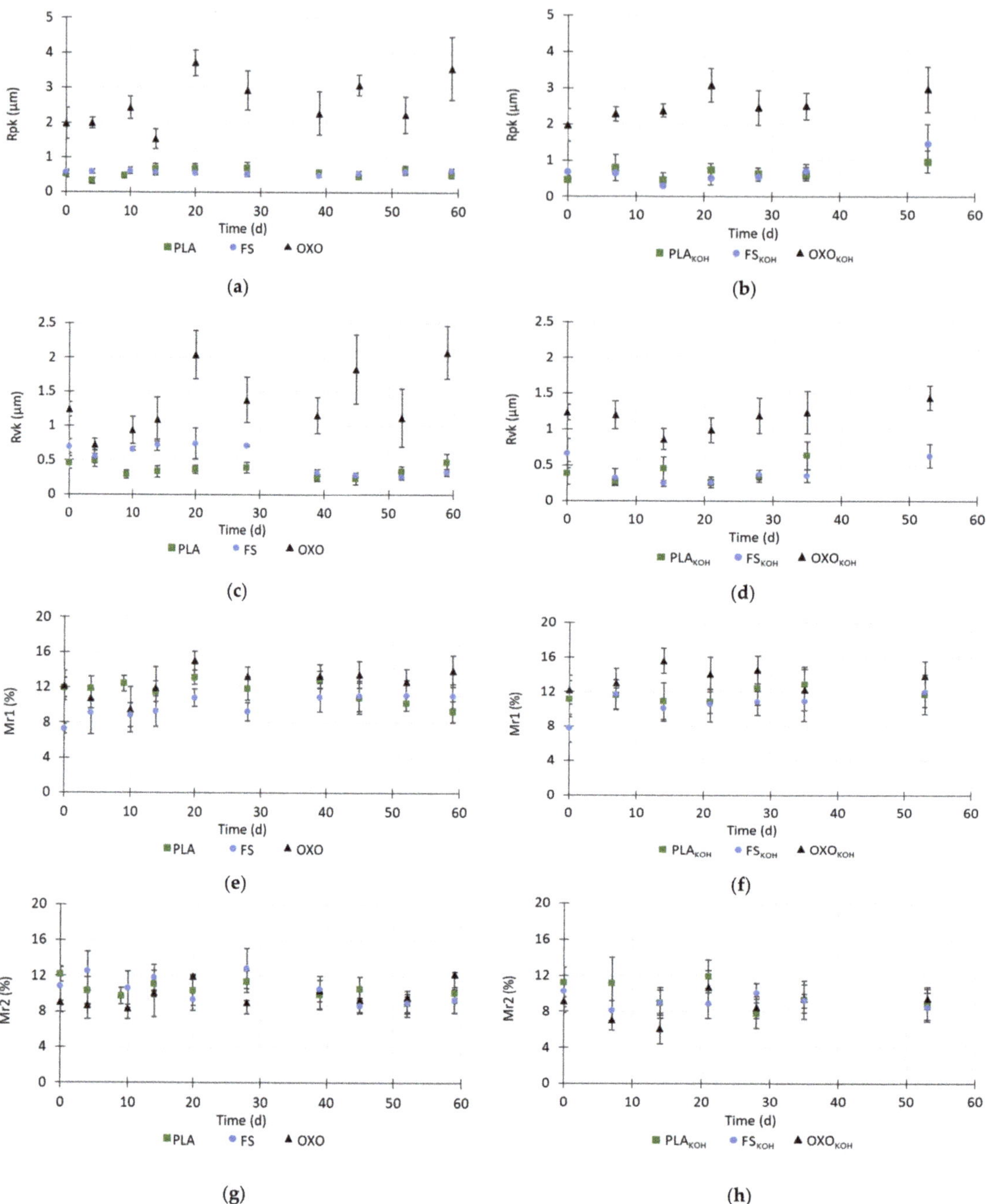

Figure 6. The surface roughness indicators (the reduced peak height without bearing properties (Rpk), the reduced valleys depth (Rvk); the percentage of the peaks and valleys (Mr1, Mr2, respectively)) during an anaerobic degradation test of untreated foils (PLA, FS, OXO) (**a**,**c**,**e**,**g**) and foils after pretreatment with KOH (PLA_{KOH}, FS_{KOH}, OXO_{KOH}) (**b**,**d**,**f**,**h**).

After the KOH pretreatment, the roughness indicators of the FS were lower than those of the untreated material, and they decreased with time until 14 days of the anaerobic test had passed, indicating rapid initial degradation of S. The changes in the values of the roughness indicators corresponded to changes in the MP profile. The lag phase of MP lasted only 4 days and was shorter with pretreated material than with untreated foil, and after this time methane started to be produced rapidly until day 15.

Cumulative MP was higher with treated FS than with the untreated material. Between 15 and 35 days of the test, the roughness indicators did not change much. On day 53, these values differed from what was observed on the other days. This was connected with the non-homogenous structure of S_{KOH} and the presence of small, deep micropores on the surface, which increased the roughness indicator values.

The pretreatment of OXO foil caused the initial values of its roughness indicators to be higher than those of the untreated material. This might have been caused by rapid removal of oxo-additives from the surface structure of OXO. It should be emphasized that the OXO was very rough and heterogenous in its structure. The polymer was not damaged during the anaerobic test, even after the pretreatment, and the changes in its roughness resulted from this heterogenous structure and degradation of the oxo-additive.

3.5. Changes in the Structure of Bio-Based and Oxo-Degradable Foil Materials—FTIR Analysis

Regarding the raw PLA foil material, the FTIR analysis showed typical peaks for pure PLA. Hydrogen-bonded (intra- and intermolecular) -OH groups corresponded to the band at 3371 cm^{-1}. The asymmetric and symmetric vibrations of the -CH_3 groups were represented by the bands located at 3000–2850 cm^{-1}. The strong band at 1713 cm^{-1} indicated asymmetric stretching of carbonyl C=O groups, which may correspond to the amorphous phase of PLA.

In contrast, if a weak band is observed at ~1720 cm^{-1}, it may be assigned to C=O stretching of the crystalline phase of PLA [31]. The bands at 1271, 1104, and 1019 cm^{-1} corresponded to stretching vibrations of the C-O-C and -CH-O- groups. The band at 930 cm^{-1} combined contributions of both -CH_3 rocking and deformation of the carboxylic -OH$_{\ldots}$ H groups [32]. The O-CH-CH_3, esters and/or C-C and rocking vibrations of -CH_3 groups were noted at 874 and 727 cm^{-1}, respectively (Figure 7).

With the untreated PLA foil material, the intensity of all bands increased after ca. 100 days of the anaerobic thermophilic test, except the one corresponding to -OH (at 3371 cm^{-1}), which indicated PLA degradation. For example, the intensity of the peak at 1713 cm^{-1} increased, which corresponds to a larger amount of carbonyl C=O groups, which are terminal groups in PLA chains. This indicates that the molecular weight of the PLA chains decreased. Although high molecular weight PLA is insoluble in water, during the degradation process, water diffuses into the polymer matrix, causing the hydrolysis of the ester groups of the amorphous phase. As a result, short-chain soluble carboxyl-terminated oligomers and monomers are formed and penetrate the aqueous environment. This causes an increase in the number of carboxyl chain ends, which was shown by the greater intensity of the peak corresponding to this group [33,34].

On the contrary, with PLA foil pretreated with KOH, the intensity of all bands decreased. However, in both cases biodegradation took place, which was confirmed by the weakness of the mechanical properties of the material and MP.

However, the time during which it was possible to determine the mechanical properties of PLA_{KOH} was 28 days, which was almost 2 two times shorter than that during which the properties of PLA could be determined. Pretreatment influenced the mechanical properties more than biogas production.

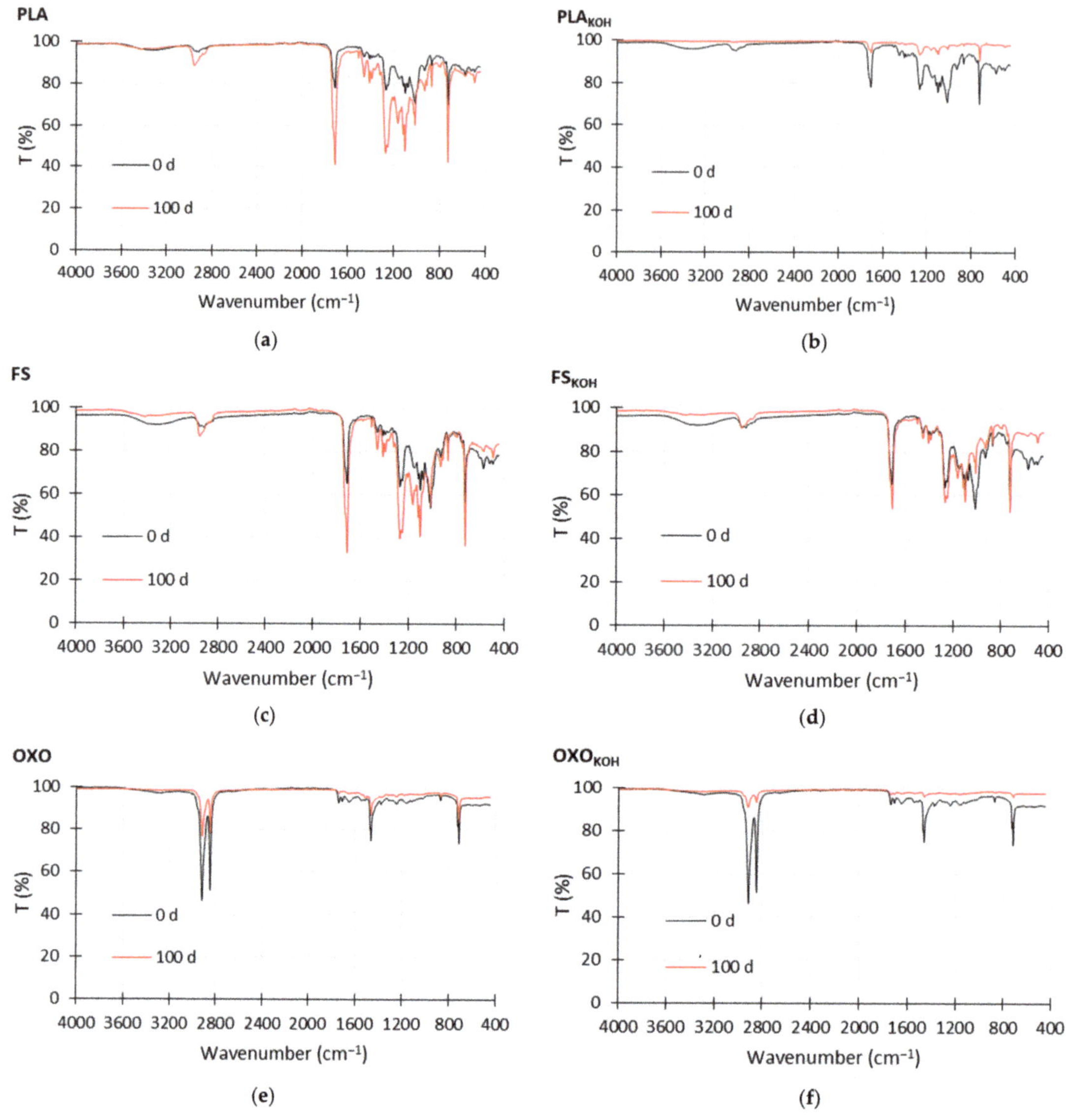

Figure 7. FTIR spectra of untreated foil materials (PLA, FS, OXO) (**a**,**c**,**e**) and foil materials after pretreatment with KOH (PLA_{KOH}, FS_{KOH}, OXO_{KOH}) (**b**,**d**,**f**) before and after 100 days of anaerobic degradation at 55 °C.

Vasile et al. [24] found that, during the degradation of PLA and its composite in soil burial conditions, degradation included the ester groups destroyed by hydrolysis.

In the present study, the intensity of the ATR-FT-IR spectra decreased due to hydrolytic degradation. These results are consistent with those in the literature, Maharana et al. [35] concluded that chain scission of the PLA main chain takes place where the ester bonds are located, leading to the formation of oligomers.

Regarding the raw FS foil material, the bands at 2915 cm^{-1} indicated C–H stretching in aliphatic and aromatic groups (Figure 7). The bands at 1715 and 1272 cm^{-1} corresponded to C=O groups and C–O bonds found in ester linkages. The peak at 1019 cm^{-1} was caused by the stretching of phenylene groups. The band at 727 cm^{-1} indicated four or more adjacent methylene (–CH_2–) groups. The FTIR spectra of FS foil contained bands characteristic of plasticized starch. The peaks between 3558–3100 cm^{-1} and between 1269–900 cm^{-1} corresponded to O–H stretching groups and to C-O stretching and hydrogen bonding peaks, respectively. The peaks recorded at 1140 and 1103 cm^{-1} corresponded to C–O stretching in the C–O–H group. At 1016 cm^{-1}, evidence of C–O stretching in the C–O–C group of the starch anhydroglucose ring could be seen. After 100 days of anaerobic degradation of untreated FS foil material, the bands in the FTIR spectrum were almost identical to those in the spectrum of the raw material, and no new peaks were formed. Only the intensity of the spectra increased.

Regarding the treated FS foil material, the bands in the FTIR spectrum and their intensities were even more similar to those in the spectrum obtained from the raw material. It should be emphasized that pretreatment weakened the mechanical properties of the material more rapidly, and FS_{KOH} produced 80% more methane than FS. Thus, the FTIR spectrums that were obtained after degradation of FS and FS_{KOH} differed.

The FTIR spectra of OXO and OXO_{KOH} were typical of PE (Figure 7). The bands at 2849 and 2846 cm^{-1} corresponded to C–H stretching asymmetric and symmetric vibrations, whereas the band at 1464 cm^{-1} was due to scissoring CH_2 groups. The bands at 720 cm^{-1} and 717 cm^{-1} were caused by CH_2 rocking vibrations. A small band was noted at 1630 cm^{-1}, related to a double bond (or aromatic rings), probably due to some surface antioxidant additive. The spectra obtained from the material after 100 days of anaerobic treatment displayed lower intensity peaks. The band at 1630 cm^{-1} was not observed, which means that the antioxidant additive was washed away or assimilated by bacteria and the spectrum was from pure PE [36]. As a result of the disappearance of the additive, the OXO chains shortened, which slightly weakened the mechanical properties of the material. Degradation of modified polyethylene can proceed via two mechanisms (i) hydro-biodegradation and (ii) oxo-biodegradation, depending on the two additives, starch or pro-oxidant, used in the synthesis polymer [37]. Starch blend polyethylene contains starch inclusions which makes the material hydrophilic and, therefore, microbial amylase enzymes can easily access, attack and remove this part.

Polyethylene with pro-oxidant additive can be photodegraded and chemically degraded, and low molecular weight products are sequentially oxidized [38,39].

4. Conclusions

Because the share of usable polymers has increased in municipal waste streams, it is necessary to the determine the degradability of these materials, which is the current challenge of organic recycling strategies. It was shown that untreated and treated OXO did not degrade; thus, there was no methane production and no visible changes on the surface or in the mechanical properties. Untreated bio-based foils began to lose mechanical properties by day 50 of the anaerobic thermophilic degradation test; the pieces were still visible even after 100 days, and they only disintegrated when touched. Cumulative MP reached only 34 L/kg VS. After pretreatment, cumulative MP increased to 47.5 L/kg VS (PLA_{KOH}) and 58.6 L/kg VS (FS_{KOH}). However, these values constituted only 8–10% of the theoretical MP. Pretreatment of bio-based foils increased surface damage and weakened their mechanical properties (tensile strength, elongation at break). This means that the pieces of bio-based materials may contaminate digestate at municipal waste treatment plant.

Author Contributions: Conceptualization, M.Z. and K.B.; methodology, M.Z., K.B., and B.P.; formal analysis, K.B., I.W.-B., and D.K.; validation, M.Z. and B.P.; investigation, M.Z., B.P., and K.B.; resources, M.Z., K.B., I.W.-B., and D.K.; data curation, M.Z. and K.B.; writing—original draft preparation, M.Z., K.B., B.P., and D.K.; writing—review and editing, M.Z., B.P., K.B., I.W.-B., and D.K.; visualization, M.Z. and B.P.; supervision, K.B.; funding acquisition, K.B., I.W.-B., and D.K. All authors have read and agreed to the published version of the manuscript.

Funding: We are grateful for the financial support of the Ministry of Science and Higher Education, Poland (statutory project No. 29.610.024-110). Publication (Open Access) of this research was financially co-supported by Minister of Science and Higher Education in the range of the program entitled "Regional Initiative of Excellence" for the years 2019–2022, Project No. 010/RID/2018/19, amount of funding 12,000,000 PLN.

Institutional Review Board Statement: Not applicable.

Informed Consent Statement: Not applicable.

Data Availability Statement: Not applicable.

Conflicts of Interest: The authors declare no conflict of interest. The funders had no role in the design of the study; in the collection, analyses, or interpretation of data; in the writing of the manuscript, or in the decision to publish the results.

References

1. Plastics Europe. Plastics—The Facts 2019. An Analysis of European Plastics Production, Demand and Waste Data. Available online: https://www.plasticseurope.org/application/files/9715/7129/9584/FINAL_web_version_Plastics_the_facts2019_14102019.pdf (accessed on 30 June 2021).
2. Zhao, X.; Cornish, K.; Vodovotz, Y. Narrowing the Gap for Bioplastic Use in Food Packaging: An Update. *Environ. Sci. Technol.* **2020**, *54*, 4712–4732. [CrossRef]
3. Mallegni, N.; Phuong, T.V.; Coltelli, M.-B.; Cinelli, P.; Lazzeri, A. Poly(lactic acid) (PLA) Based Tear Resistant and Biodegradable Flexible Films by Blown Film Extrusion. *Materials* **2018**, *11*, 148. [CrossRef]
4. Abdelmoez, W.; Dahab, I.; Ragab, E.M.; Abdelsalam, O.A.; Mustafa, A. Bio-and oxo-degradable plastics: Insights on facts and challenges. *Polym. Adv. Technol.* **2021**, *32*, 1981–1996. [CrossRef]
5. Ojeda, T.F.; Dalmolin, E.; Forte, M.M.; Jacques, R.J.S.; Bento, F.M.; Camargo, F.A. Abiotic and biotic degradation of oxo-biodegradable polyethylenes. *Polym. Degrad. Stab.* **2009**, *94*, 965–970. [CrossRef]
6. Emadian, S.M.; Onay, T.T.; Demirel, B. Biodegradation of bioplastics in natural environments. *Waste Manag.* **2017**, *59*, 526–536. [CrossRef] [PubMed]
7. Artham, T.; Doble, M. Biodegradation of Aliphatic and Aromatic Polycarbonates. *Macromol. Biosci.* **2007**, *8*, 14–24. [CrossRef] [PubMed]
8. Ahmed, T.; Shahid, M.; Azeem, F.; Rasul, I.; Shah, A.A.; Noman, M.; Hameed, A.; Manzoor, N.; Manzoor, I.; Muhammad, S. Biodegradation of plastics: Current scenario and future prospects for environmental safety. *Environ. Sci. Pollut. Res.* **2018**, *25*, 7287–7298. [CrossRef] [PubMed]
9. Bernat, K.; Kulikowska, D.; Wojnowska-Baryła, I.; Zaborowska, M.; Pasieczna-Patkowska, S. Thermophilic and mesophilic biogas production from PLA-based materials: Possibilities and limitations. *Waste Manag.* **2021**, *119*, 295–305. [CrossRef]
10. Benn, N.; Zitomer, D. Pretreatment and Anaerobic Co-digestion of Selected PHB and PLA Bioplastics. *Front. Environ. Sci.* **2018**, *5*, 93. [CrossRef]
11. Yu, J.; Plackett, D.; Chen, L.X. Kinetics and mechanism of the monomeric products from abiotic hydrolysis of poly[(R)-3-hydroxybutyrate] under acidic and alkaline conditions. *Polym. Degrad. Stab.* **2005**, *89*, 289–299. [CrossRef]
12. Myung, J.; Strong, N.I.; Galega, W.M.; Sundstrom, E.R.; Flanagan, J.C.; Woo, S.-G.; Waymouth, R.; Criddle, C.S. Disassembly and reassembly of polyhydroxyalkanoates: Recycling through abiotic depolymerization and biotic repolymerization. *Bioresour. Technol.* **2014**, *170*, 167–174. [CrossRef]
13. Abbott, E.J.; Firestone, F.A. Specifying surface quality: A method based on accurate measurement and comparison. *Spie MS* **1995**, *107*, 63.
14. Pszczółkowski, B.; Bramowicz, M.; Rejmer, W.; Chrostek, T.; Senderowski, C. The influence of the processing temperature of polylactide on geometric structure of the surface using FDM technique. *Arch. Metall. Mater.* **2021**, *66*, 181–186. [CrossRef]
15. Greenberg, A.E.; Clesceri, L.S.; Eaton, A.D. *Standard Methods for the Examination of Water and Wastewater*, 18th ed.; APHA: Washington, DC, USA, 1992.
16. Buswell, A.M.; Mueller, H.F. Mechanism of Methane Fermentation. *Ind. Eng. Chem.* **1952**, *44*, 550–552. [CrossRef]
17. Aldas, M.; Ferri, J.M.; Lopez-Martinez, J.; Samper, M.D.; Arrieta, M.P. Effect of pine resin derivatives on the structural, thermal, and mechanical properties of Mater-Bi type bioplastic. *J. Appl. Polym. Sci.* **2019**, *137*. [CrossRef]
18. Mohee, R.; Unmar, G.; Mudhoo, A.; Khadoo, P. Biodegradability of biodegradable/degradable plastic materials under aerobic and anaerobic conditions. *Waste Manag.* **2008**, *28*, 1624–1629. [CrossRef]
19. Itävaara, M.; Karjomaa, S.; Selin, J.-F. Biodegradation of polylactide in aerobic and anaerobic thermophilic conditions. *Chemosphere* **2002**, *46*, 879–885. [CrossRef]
20. Yagi, H.; Ninomiya, F.; Funabashi, M.; Kunioka, M. Anaerobic biodegradation tests of poly(lactic acid) and polycaprolactone using new evaluation system for methane fermentation in anaerobic sludge. *Polym. Degrad. Stab.* **2009**, *94*, 1397–1404. [CrossRef]

21. Yagi, H.; Ninomiya, F.; Funabashi, M.; Kunioka, M. Anaerobic Biodegradation Tests of Poly(lactic acid) under Mesophilic and Thermophilic Conditions Using a New Evaluation System for Methane Fermentation in Anaerobic Sludge. *Int. J. Mol. Sci.* **2009**, *10*, 3824–3835. [CrossRef]
22. Corti, A.; Muniyasamy, S.; Vitali, M.; Imam, S.H.; Chiellini, E. Oxidation and biodegradation of polyethylene films containing pro-oxidant additives: Synergistic effects of sunlight exposure, thermal aging and fungal biodegradation. *Polym. Degrad. Stab.* **2010**, *95*, 1106–1114. [CrossRef]
23. Yashchuk, O.; Portillo, F.; Hermida, E. Degradation of Polyethylene Film Samples Containing Oxo-Degradable Additives. *Procedia Mater. Sci.* **2012**, *1*, 439–445. [CrossRef]
24. Vasile, C.; Pamfil, D.; Râpă, M.; Darie-Niţă, R.N.; Mitelut, A.C.; Popa, E.E.; Popescu, P.A.; Draghici, M.C.; Popa, M.E. Study of the soil burial degradation of some PLA/CS biocomposites. *Compos. Part B Eng.* **2018**, *142*, 251–262. [CrossRef]
25. De Jong, S.; Arias, E.; Rijkers, D.T.; van Nostrum, C.; Bosch, J.K.-V.D.; Hennink, W. New insights into the hydrolytic degradation of poly(lactic acid): Participation of the alcohol terminus. *Polymer* **2001**, *42*, 2795–2802. [CrossRef]
26. Elsawy, M.; Kim, K.-H.; Park, J.-W.; Deep, A. Hydrolytic degradation of polylactic acid (PLA) and its composites. *Renew. Sustain. Energy Rev.* **2017**, *79*, 1346–1352. [CrossRef]
27. Rajesh, G.; Prasad, A.R.; Gupta, A. Soil Degradation Characteristics of Short Sisal/PLA Composites. *Mater. Today Proc.* **2019**, *18*, 1–7. [CrossRef]
28. Sevim, K.; Pan, J. A model for hydrolytic degradation and erosion of biodegradable polymers. *Acta Biomater.* **2018**, *66*, 192–199. [CrossRef]
29. Lv, S.; Zhang, Y.; Gu, J.; Tan, H. Physicochemical evolutions of starch/poly (lactic acid) composite biodegraded in real soil. *J. Environ. Manag.* **2018**, *228*, 223–231. [CrossRef]
30. Luo, Y.; Lin, Z.; Guo, G. Biodegradation Assessment of Poly (Lactic Acid) Filled with Functionalized Titania Nanoparticles (PLA/TiO2) under Compost Conditions. *Nanoscale Res. Lett.* **2019**, *14*, 56. [CrossRef]
31. Silva, R.D.N.; Oliveira, T.A.; Da Conceição, I.D.; Araque, L.M.; Alves, T.S.; Barbosa, R. Evaluation of hydrolytic degradation of bionanocomposites through fourier transform infrared spectroscopy. *Polimeros* **2018**, *28*, 348–354. [CrossRef]
32. Socrates, G. *Infrared and Raman Characteristic Group Frequencies*, 3rd ed.; John Wiley and Sons, Ltd.: Chichester, UK, 2004; Volume 35, ISBN 978-0-470-09307-8.
33. Gupta, A.; Kumar, V. New emerging trends in synthetic biodegradable polymers—Polylactide: A critique. *Eur. Polym. J.* **2007**, *43*, 4053–4074. [CrossRef]
34. Middleton, J.C.; Tipton, A.J. Synthetic biodegradable polymers as orthopedic devices. *Biomaterials* **2000**, *21*, 2335–2346. [CrossRef]
35. Maharana, T.; Mohanty, B.; Negi, Y. Melt–solid polycondensation of lactic acid and its biodegradability. *Prog. Polym. Sci.* **2009**, *34*, 99–124. [CrossRef]
36. Bandini, F.; Frache, A.; Ferrarini, A.; Taskin, E.; Cocconcelli, P.S.; Puglisi, E. Fate of Biodegradable Polymers Under Industrial Conditions for Anaerobic Digestion and Aerobic Composting of Food Waste. *J. Polym. Environ.* **2020**, *28*, 2539–2550. [CrossRef]
37. Shah, A.A.; Hasan, F.; Hameed, A.; Ahmed, S. Biological degradation of plastics: A comprehensive review. *Biotechnol. Adv.* **2008**, *26*, 246–265. [CrossRef]
38. Bonhomme, S.; Cuer, A.; Delort, A.-M.; Lemaire, J.; Sancelme, M.; Scott, G. Environmental biodegradation of polyethylene. *Polym. Degrad. Stab.* **2003**, *81*, 441–452. [CrossRef]
39. Yamada-Onodera, K.; Mukumoto, H.; Katsuyaya, Y.; Saiganji, A.; Tani, Y. Degradation of polyethylene by a fungus, Penicillium simplicissimum YK. *Polym. Degrad. Stab.* **2001**, *72*, 323–327. [CrossRef]

 MDPI

Article

Environmental and Economic Aspects of Biomethane Production from Organic Waste in Russia

Svetlana Zueva [1,*], Andrey A. Kovalev [2], Yury V. Litti [3], Nicolò M. Ippolito [1], Valentina Innocenzi [1] and Ida De Michelis [1]

1 Department of Industrial and Information Engineering and Economics, University of L'Aquila, 67100 L'Aquila, Italy; nicolomaria.ippolito@univaq.it (N.M.I.); valentina.innocenzi@univaq.it (V.I.); ida.demichelis@univaq.it (I.D.M.)
2 Department of Renewable Energy, Federal Scientific Agroengineering Center VIM, 109428 Moscow, Russia; kovalev_ana@mail.ru
3 Federal Research Centre "Fundamentals of Biotechnology" of the Russian Academy of Sciences, 119071 Moscow, Russia; litty-yuriy@mail.ru
* Correspondence: svetlana.zueva@graduate.univaq.it

Abstract: According to the International Energy Agency (IEA), only a tiny fraction of the full potential of energy from biomass is currently exploited in the world. Biogas is a good source of energy and heat, and a clean fuel. Converting it to biomethane creates a product that combines all the benefits of natural gas with zero greenhouse gas emissions. This is important given that the methane contained in biogas is a more potent greenhouse gas than carbon dioxide (CO_2). The total amount of CO_2 emission avoided due to the installation of biogas plants is around 3380 ton/year, as 1 m^3 of biogas corresponds to 0.70 kg of CO_2 saved. In Russia, despite the huge potential, the development of bioenergy is rather on the periphery, due to the abundance of cheap hydrocarbons and the lack of government support. Based on the data from an agro-industrial plant located in Central Russia, the authors of the article demonstrate that biogas technologies could be successfully used in Russia, provided that the Russian Government adopted Western-type measures of financial incentives.

Keywords: organic waste; biogas; greenhouse gas; economic feasibility

Citation: Zueva, S.; Kovalev, A.A.; Litti, Y.V.; Ippolito, N.M.; Innocenzi, V.; De Michelis, I. Environmental and Economic Aspects of Biomethane Production from Organic Waste in Russia. *Energies* **2021**, *14*, 5244. https://doi.org/10.3390/en14175244

Academic Editor: Wencheng Ma

Received: 14 July 2021
Accepted: 17 August 2021
Published: 24 August 2021

1. Introduction

Electric power generation is the scope of a multi-faceted industry which, all over the world, continues to have a serious impact on the environment.

Among the many ways to produce power, generation from biomass appears to be one of the most virtuous, under the environmental profile [1]. In 2018, the world produced 35 million tons (oil equivalent) of biogas and biomethane, whereas their full potential is 570 and 730 million tons [2], respectively.

A land in which biomass undoubtedly has a bright future is that of the Russian Federation. However, a number of problems are presently hindering the development of this technology in Russia. Legislation creating incentives such as the "green certificates" should be introduced in that country. Another considerable obstacle is the weak exchange rate of the ruble against the euro, worsened by seriously overpriced imports, if at all available, due to the sanctions stemming from the current discrepancies between Russia and the Western bloc in international politics.

In Europe, "green certificates" have existed since 2001. The share of "Green" energy in the European Union reached 18% back in 2018, while in Russia it is only 0.2%, and, according to the Ministry of Energy of Russia, by 2035, the share of renewable energy sources in the energy balance of the Russian Federation will increase to a mere 4%. The Russian government is preparing to launch a national green certificate system by 2022. Under the program, energy retail and large industrial companies will finally be able to sell energy at an increased cost, which will allow firms to "recoup" costs [3].

Today, the main types of renewable energy in Russia are wind and solar power. However, as said above, biomass plants are one of the most attractive areas for investors in Russia in the coming years due to the huge volume of agricultural waste, food industry enterprises, and city sewage treatment plants [4].

In the last few decades, the biogas production from anaerobic digestion of organic waste has become widespread in Europe and has become an important part of the circular economy of many countries [5–7]. Biogas and biomethane are renewable gases, produced by the decomposition of organic matter, which is converted into a combustible biogas rich in methane and a liquid effluent.

The raw material for biogas production can be a wide range of organic waste: solid and liquid waste from the agro-industrial complex, wastewater, and solid household waste. Many studies have been devoted to demonstrating the benefits of co-digestion to increase biogas production of various types of organic waste: sewage sludge with organic waste [8,9], or cattle manure with agricultural waste [10,11]. In the case of processing multiple waste streams in one plant, it becomes possible to increase the production of biogas per unit volume of the digester without compromising the stability of the process.

The biogas technology has promising prospects for further development around the world in the presence of affordable inexpensive raw materials and a wide range of possible options for using biogas [12].

Today, one of the main problems of Russian agricultural and food enterprises is organic waste disposal. Russia generates about 600 million tons of organic waste per year, of which 150 million tons is waste from livestock [13]. Currently, waste is often accumulated near farms, which results in soil acidification, alienation of agricultural land (in Russia, more than 2 million hectares are occupied for storing manure), groundwater pollution, and emissions of greenhouse gas (methane and carbon dioxide) into the atmosphere [14].

From all of the above, it is easy to understand that among the most promising areas of research in Russia today are projects entailing the use of biogas technologies. Investing in these projects is becoming increasingly attractive for many reasons, e.g., (i) production of clean energy (electric or thermal), (ii) production of organic fertilizers from the effluents resulting from the digestion processes, and (iii) last but not least, safe disposal of organic waste [15] with an evident environmental advantage (reduced surface/ground water contamination [16,17] and greenhouse gasses emission [18,19].

There are a number of technical solutions for the use of anaerobically produced biogas. Along with the production of heat and electricity for the needs of the enterprise itself, the biogas produced can be used as a fuel for positive ignition engines or for the production of pure methane and carbon dioxide.

In general, biogas consists of 55% to 75% methane (CH_4) and 25% to 50% carbon dioxide (CO_2) [20–22]. However, depending on the type of organic waste and the parameters of the anaerobic fermentation, small amounts of other gases such as nitrogen (<10%), hydrogen sulfide (<3%), and hydrogen (<1%) may be present. It is the methane component of the biogas that will burn and produce energy [23]. In terms of calorific value, 1 m^3 of biogas is the equivalent of up to 0.8 m^3 natural gas [15].

A typical biogas system consists basically of a manure (or other waste) receiving unit, anaerobic digestion facilities, storage facilities for digester effluent, and gas-handling and gas-use equipment (Figure 1).

The typical structure of an investment in a biomass plant includes (i) project development (technical, legal, and planning consultants; financing, utilities connection; and licensing), (ii) capital investment (equipment and construction), (iii) operation, maintenance, and training costs.

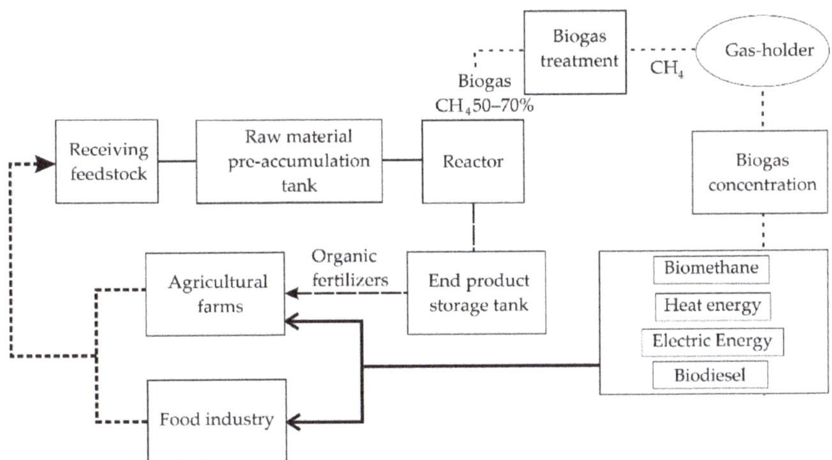

Figure 1. Schematic diagram of the production of renewable energy from organic waste.

Anaerobic digestion is a microbiological process involving a methanogenic community that gradually breaks down complex organic material (OM) to form biogas. A methanogenic community is a biocenosis consisting of anaerobic bacteria and archaea. They are active at four different stages of OM decomposition. The first stage is hydrolysis, which involves hydrolytic bacteria that decompose polymeric compounds into monomers. The second stage is a fermentation process where acidogenic bacteria ferment monomers to organic acids (mainly volatile fatty acids (VFA), alcohols (mainly ethanol), and hydrogen). In the third acetogenic stage, syntrophic bacteria degrade VFA, alcohols, and some other products generated during hydrolysis and fermentation to H_2, CO_2, and acetate. They may also degrade acetate to H_2 and CO_2. Finally, the fourth stage includes methanogenic archaea producing biogas mainly by hydrogenotrophic, acetoclastic, and, to a lesser extent, methylotrophic routes [24]. A schematic diagram of OM decomposition by the methanogenic community is shown in Figure 2.

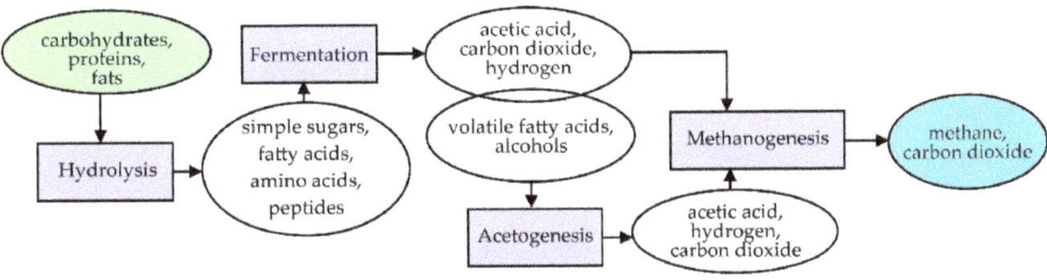

Figure 2. Flow diagram of the anaerobic digestion process.

To obtain biogas rich in methane, constant monitoring of the anaerobic fermentation process is required. For example, the methanogens are very sensitive to changes in to environmental parameters. It is important to maintain the optimal temperature, humidity, pH value, and the composition of organic waste [25].

This article presents the results of a joint Russian–Italian work, in which the authors studied the feasibility of producing biomethane for cogeneration of electricity and heat by treating the organic waste and sludge resulting from the wastewater of a meat processing plant located in the Lipetskaia Region of Russia.

2. Materials and Methods

Calculations were carried out based on data provided by the above-mentioned agro-industrial complex. The calculations were carried out according to the Standard Procedure for Calculating the Economic Efficiency of Biogas Complexes [26]. Waste from 5 farms (manure) and waste from a meat processing plant (manure and sewage sludge) were proposed as biomass for biogas production.

The number of farms (three swine farms plus two dairy farms) and their distance from the proposed biogas plant were taken into account. A centralized plant seemed to be the best choice; it was supposed that it was constructed next to the biggest farm of the selected cluster. It was also supposed that, in addition to the organic waste from farms, this hypothetical biogas plant would receive manure and sewage sludge from a meat processing plant, which is part of the agro-industrial complex (Figure 3).

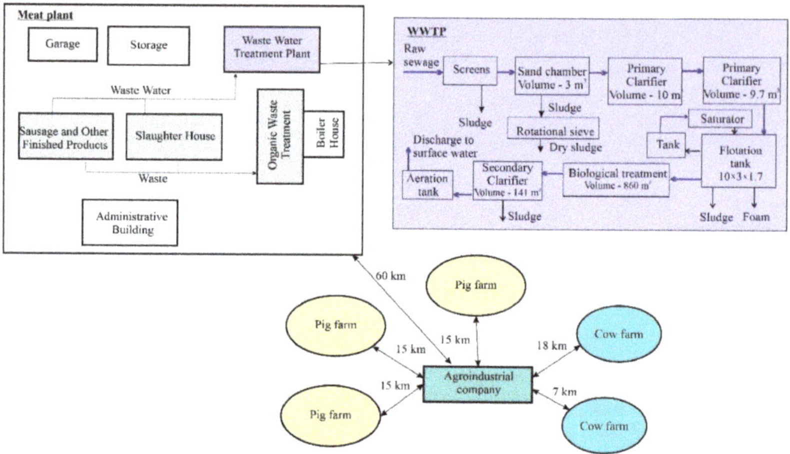

Figure 3. The structure of the agro-industrial complex.

The waste quantities and their relevant main physical/chemical parameters introduced in the calculations are shown in Table 1.

Table 1. Physico-chemical parameters of investigated organic waste.

	Cow Manure	Pig Manure	Sewage Sludge
Total quantity	17,000 ton/year	83,150 ton/year	1.9 ton/year
pH	7.3	6.1	7.4
Dry matter, %	17.6	25.0	5.3
Total Nitrogen, %	1.5	2.5	8.9
Total P_2O_5, %	1.1	2.8	5.8
Total K_2O, %	2.8	1.1	NDA *
Ash, %	12.7	16.6	NDA *
Organic substance, %:			
-converting in carbon	43.6	41.7	NDA *
-converting in organic substance	87.3	83.4	NDA *
Carbon/nitrogen	29.5	16.4	37.0
Heavy metals (mg/kg):			
Zn	13.2	58.0	23.1
Mn	16.8	23.4	34.5
Cu	3.3	9.6	1.2
Co	5.7	9.9	4.5
Cd	0.03	0.07	0.01
Pb	1.2	1.5	1.7
Cr	1.2	4.1	0.4
Fe	90	480	720

* No Data Available.

3. Results and Discussion

There are four types of anaerobic digester configuration: covered lagoons, complete-mix, plug-flow, and fixed-film. The majority of commercially available digesters operate at mesophilic temperatures except for covered lagoons, which run at ambient temperatures.

For the purpose of this study, a complete-mix digester was chosen. It is the only system that can treat both cow and pig organic material in cold climate. Looking at Table 1, the total solid content (dry matter) of the swine manure was 25%, whereas that of the cow manure was around 17%. The design of the storage tank of the biogas plant took the volume of water that was to be added to the organic material into account. The manure was diluted up to 7.5–10% w/w as total solid concentration, which resulted in better performance in terms of biogas production as well as total solids and chemical oxygen demand degradation [27].

The estimated total amount of biogas was calculated as 13,231 m^3/day (4,829,163 m^3/year) from 279,501 kg/day of pig and cow manure.

A summary of the total capacity, heat and electricity consumption, and energy generated by the combined heat and power (CHP) unit is presented in Table 2.

Table 2. List of utilities involved in the biogas plant.

	Number of Units	Power, kW	Total Power, kW	Working Time, h/Day	Energy, kWh/Year
ELECTRIC ENERGY					
Feeding pump and screw charger	1	9.0	9.0	4.0	13,140
Pumps of storage tanks	4	3.1	12.2	4.0	17,844
Pumps of primary digesters	8	10.2	81.5	2.0	59,481
Agitators of primary digesters	4	25.0	100.0	12.0	438,000
Propellers of primary digesters	4	63.5	254.0	4.0	370,784
Pumps of post-digesters	4	13.4	53.8	2.0	39,258
Agitators of post-digesters	2	33.0	66.0	8.0	192,720
Propellers of post-digesters	2	63.5	127.0	4.0	185,392
Compressor for removal of H_2S	1	2.8	2.8	24.0	24,333
Compressor of engine	1	3.5	3.5	24.0	30,660
Screw separator + pump			11.0	8.0	32,120
Unlisted equipment			36.0	6.0	78,917
Total energy required					1,482,650
Electric energy from CHP engine			1202	24	10,526,802
Balance					9,044,152
HEAT					
Heat for primary digestion			406	24	3,555,300
Heat for post-digestion			406	24	3,555,300
Heat lost			189	24	1,654,748
Total heat required					8,765,348
Heat from CHP engine			1878	24	16,448,128
Excess of heat					7,682,780

The total installed power of the biogas plant was 757 kW, even though many of the devices did not work continuously during the day.

Utilities were not considered, as the only one used in the plant is heat produced by the plant itself. Regarding royalties, no patented processes ran in the plant.

In the calculation of total revenues, only the sale of electric power and solid fertilizers were taken into account, plus the saving from the avoided sewage disposal cost. Liquid fertilizer was supposed to be given free of charge to farmers in the plant neighborhood. The excess of heat was not considered, as it seemed unlikely that heat could be sold in the biogas plant area. The overall economic evaluation is shown in Table 4. Many European countries give carbon credits for renewable energy produced from biomass such as crop residues, mud, organic fraction of municipal wastes, and any other renewable organic feedstock. The German government, for instance, gives up to 0.21 EUR/kWh for 20 years and even a partial financial support for construction of plants. In 2005, Italy introduced a green-card system where each kWh from renewable energy could be sold for 0.1–0.15 EUR/kWh more than the current price of the energy, according to the condition of the energy market. In fact, energy companies must produce 2.7% of their total energy production from renewable sources. The total annual revenues are listed in Table 3.

Table 3. List of the annual revenues.

	(kg/Year)	(Tonn/Year)	EUR/Tonn	EUR/Year
Solid fertilizer *	8,955,178	8955	10	89,552
Liquid fertilizer	NOT estimated			
Dumping of WWTP sludge avoided (meat plant)				1716
		kWh/Year	EUR/kWh	
Net electric energy **		9,044,152	0.10	904,415
Excess of heat		7,682,780	0	0
Annual total revenues				995,683

* Prices of fertilizer based on information received from "Mineral Fertilizers", plant, Rossosh, Russian Federation (https://www.minudo.ru). ** Prices of energy derived from Ministry of Energy of Russian Federation (https://minenergo.gov.ru).

Table 4. Economic evaluation report of the biogas plant.

	Without Carbon Credit System	With Carbon Credit System
Direct fixed capital	EUR 3,800,160	EUR 3,800,160
Working capital	EUR 353,847	EUR 353,847
Startup cost	EUR 10,000	EUR 10,000
Up-front R&D	EUR 6500	EUR 6500
Up-front Royalties	EUR 0	EUR 0
Total investment	EUR 4,170,507	EUR 4,170,507
Revenues	EUR 995,683	EUR 2,126,202
Annual operating cost	EUR 654,999	EUR 654,999
Cross profit	EUR 340,684	EUR 1,471,203
Taxes	EUR 136,274	EUR 588,481
Net profit	EUR 558,258	EUR 1,236,569
Gross margin	34.2%	69.2%
Return on investment	13.4%	29.7%
Payback time	7.5 years	3.4 years

Considering a carbon credit system like the ones used in Germany, Italy, and other European countries, the electricity produced from renewables was supposed to be sold with a premium of 0.125 EUR/kWh over the current cost from fossil fuels. This meant that the total price was 0.225 EUR/kWh. In this case, the payback time is reduced by half.

4. Conclusions

The study demonstrates the financial viability of biogas plants in Russia, provided that the Russian Government adopted Western European-type incentive policies. The environmental and economic benefits of using anaerobic digestion processes to produce biogas from agricultural and livestock industry were evidentiated with calculations based on actual waste data from a cluster of farms and meat plants selected in a region of Central Russia.

In summary, the full potential production of biogas in Russia per year would be up to 72 billion m^3, which corresponds to the production of up to 172,500 GWh of electricity and up to 207,100 GWh of heat energy, per year [28].

Hence the expectation for the hasty advent of Western-type financial incentives for biogas technologies in Russia.

Author Contributions: Conceptualization, S.Z. and A.A.K.; methodology, N.M.I.; software, V.I.; validation, N.M.I. and A.A.K.; formal analysis, Y.V.L.; investigation, S.Z.; resources, Y.V.L.; data curation, A.A.K.; writing—original draft preparation, S.Z.; writing—review and editing, Y.V.L. and

S.Z.; visualization, V.I.; supervision, I.D.M.; project administration, I.D.M.; funding acquisition, I.D.M. All authors have read and agreed to the published version of the manuscript.

Funding: This research received no external funding.

Conflicts of Interest: The authors declare no conflict of interest.

References

1. Bielecki, A.; Ernst, S.; Skrodzka, W.; Wojnicki, I. The externalities of energy production in the context of development of clean energy generation. *Environ. Sci. Pollut. Res. Int.* **2020**, *27*, 11506–11530. [CrossRef]
2. IEA. Outlook for Biogas and Biomethane: Prospects for Organic Growth. Available online: https://www.iea.org/reports/outlook-for-biogas-and-biomethane-prospects-for-organic-growth (accessed on 15 March 2020).
3. Neugebauer, Z. Expensive Certificates: How the Ministry of Energy "Turns Green". *Gazeta.ru*, 29 February 2020. Available online: https://www.gazeta.ru/business/2020/02/28/12981601.shtml (accessed on 29 February 2020).
4. Davudova, A. Biofuel Is in the Air. *Kommersant*, 3 December 2018. Available online: https://www.kommersant.ru/doc/3819069 (accessed on 3 December 2018).
5. Abad, V.; Avila, R.; Vicent, T.; Font, X. Promoting circular economy in the surroundings of an organic fraction of municipal solid waste anaerobic digestion treatment plant: Biogas production impact and economic factors. *Bioresour. Technol.* **2019**, *283*, 10–17. [CrossRef] [PubMed]
6. Blades, L.; Morgan, K.; Douglas, R.; Glover, S.; De Rosa, M.; Cromie, T.; Smyth, B. Circular Biogas-Based Economy in a Rural Agricultural Setting. *Energy Procedia* **2017**, *123*, 89–96. [CrossRef]
7. Caposciutti, G.; Baccioli, A.; Ferrari, L.; Desideri, U. Biogas from Anaerobic Digestion: Power Generation or Biomethane Production. *Energies* **2020**, *13*, 743. [CrossRef]
8. Zupancic, G.D.; Uranjek-Ževart, N.; Roš, M. Full-scale anaerobic co-digestion of organic waste and municipal sludge. *Biomass Bioenergy* **2008**, *32*, 162–167. [CrossRef]
9. Fitamo, T.; Boldrin, A.; Boe, K.; Angelidaki, I.; Scheutz, C. Co-digestion of food and garden waste with mixed sludge from wastewater treatment in continuously stirred tank reactors. *Bioresour. Technol.* **2016**, *206*, 245–254. [CrossRef]
10. Cavinato, C.; Fatone, F.; Bolzonella, D.; Pavan, P. Thermophilic anaerobic co-digestion of cattle manure withagro-wastes and energy crops: Comparison of pilot and full scale experiences. *Bioresour. Technol.* **2010**, *101*, 545–550. [CrossRef]
11. Zhang, C.; Xiao, G.; Peng, L.; Su, H.; Tan, T. The anaerobic co-digestion of food waste and cattle manure. *Bioresour. Technol.* **2013**, *129*, 170–176. [CrossRef]
12. Spyridonidis, A.; Vasiliadou, I.A.; Akratos, C.S.; Stamatelatou, K. Performance of a Full-Scale Biogas Plant Operation in Greece and Its Impact on the Circular Economy. *Water* **2020**, *12*, 3074. [CrossRef]
13. Dashkovsky, I. Leaky ecology. *Agrotechnics and Technologies*, 22 March 2018. Available online: https://www.agroinvestor.ru/technologies/article/29525-dyryavaya-ekologiya/ (accessed on 22 March 2018).
14. Klepicova, S. Alternative disposal. *Agrotechnics and Technologies*, 2 December 2013. Available online: https://www.agroinvestor.ru/technologies/article/15078-alternativnaya-utilizatsiya/ (accessed on 2 September 2013).
15. Ishkov, A.G.; Pystina, N.B.; Akopova, G.S.; Yulkin, G.M. Research Associate, Role of Biogas in Modern Energy. *Energetika* **2014**, *5*, 130–137.
16. Meegoda, J.N.; Li, B.; Patel, K.; Lily, B.; Wang, L.B. A review of the processes, parameters, and optimization of anaerobic digestion. *Int. J. Environ. Res. Public Health* **2018**, *15*, 2224. [CrossRef]
17. Fu, Y.; Luo, T.; Mei, Z.; Li, J.; Qiu, K.; Ge, Y. Dry Anaerobic digestion technologies for agricultural straw and acceptability in China. *Sustainability* **2018**, *10*, 4588. [CrossRef]
18. Paolinia, V.; Petracchinia, F.; Segretoa, M.; Tomassettia, L.; Najab, N.; Cecinato, A. Environmental impact of biogas: A short review of current knowledge. *J. Environ. Sci. Health Part A* **2018**, *53*, 899–906. [CrossRef]
19. Pucker, J.; Jungmeier, G.; Siegl, S.; Pötsch, E.M. Anaerobic digestion of agricultural and other substrates—Implications for greenhouse gas emissions. *Animal* **2013**, *7* (Suppl. 2), 283–291. [CrossRef]
20. Florio, C.; Fiorentino, G.; Corcelli, F.; Ulgiati, S.; Dumontet, S.; Güsewell, J.; Eltrop, L. A Life Cycle Assessment of Biomethane Production from Waste Feedstock through Different Upgrading Technologies. *Energies* **2019**, *12*, 718. [CrossRef]
21. McCarty, P.L. The development of anaerobic treatment and its future. *Water Sci. Technol.* **2001**, *44*, 149–156. [CrossRef]
22. Ghangrekar, M.M.; Behera, M. Comprehensive Water Quality and Purification. *Wastewater Treat. Reuse* **2014**, *3*. [CrossRef]
23. Anukam, A.; Mohammadi, A.; Naqvi, M.; Granström, K. A Review of the Chemistry of Anaerobic Digestion: Methods of Accelerating and Optimizing Process Efficiency. *Processes* **2019**, *7*, 504. [CrossRef]
24. Nozhevnikova, A.N.; Russkova, Y.I.; Litti, Y.V.; Parshina, S.N.; Zhuravleva, E.A.; Nikitina, A.A. Syntrophy and Interspecies Electron Transfer in Methanogenic Microbial Communities. *Microbiology* **2020**, *89*, 129–147. [CrossRef]
25. Kallistova, A.Y.; Goel, G.; Nozhevnikova, A.N. Microbial diversity of methanogenic communities in the systems for anaerobic treatment of organic waste. *Microbiology* **2014**, *83*, 462–483. [CrossRef]
26. Environmental protection and environmental management. The procedure for calculating the economic efficiency of biogas complexes. *TKП 17*, 2 May 2011; 29.

27. Deepanraj, B.; Sivasubramanian, V.; Jayaraj, S. Solid Concentration Influence on Biogas Yield from Food Waste in an Anaerobic Batch Digester. In Proceedings of the International Conference and Utility Exhibition 2014 on Green Energy for Sustainable Development (ICUE 2014) Jomtien Palm Beach Hotel and Resort, Pattaya City, Thailand, 19–21 March 2014.
28. Chernin, S.Y.; Parubets, Y.S. Russian experience in the implementation of biogas technologies for the production of electricity and heat. *Heat Supply News Magazine*, 8 August 2011. Available online: www.rosteplo.ru/nt/132 (accessed on 15 August 2011).

 energies MDPI

Communication

Pilot-Scale Experiences with Aerobic Treatment and Chemical Processes of Industrial Wastewaters from Electronics and Semiconductor Industry

Valentina Innocenzi *, Svetlana B. Zueva, Francesco Vegliò and Ida De Michelis

Department of Industrial and Information Engineering and Economics, University of L'Aquila, Piazzale Ernesto Pontieri, Monteluco di Roio, 67100 L'Aquila, Italy; svetlanaborisovna.zueva@univaq.it (S.B.Z.); francesco.veglio@univaq.it (F.V.); ida.demichelis@univaq.it (I.D.M.)
* Correspondence: valentina.innocenzi1@univaq.it

Abstract: TMAH is quaternary ammonium salt, consists of a methylated nitrogen molecule, and is widely used in the electronics industry as a developer and silicon etching agent. This substance is toxic and fatal if ingested. It can also cause skin burns, eye damage, and organ damage. Moreover, TMAH exhibits long-lasting toxicity to aquatic systems. Despite this known toxicity, the authorities currently do not provide emission limits (i.e., discharge concentrations) for wastewater by EU regulation. The current scenario necessitates the study of the processes for industrial wastewater containing TMAH. This work aims to present a successful example of the treatment process for the degradation of TMAH waste solutions of the E&S industry. Research was conducted at the pilot scale, and the process feasibility (both technical and economic) and its environmental sustainability are demonstrated. This process, which treats three exhausted solutions with a high concentration of toxic substances, is considered to be innovative.

Keywords: tetramethylammonium hydroxide; industrial and organic wastewater; E&S industry; pilot scale activity; aerobic treatment; chemical precipitation; safeguard and groundwater quality

Citation: Innocenzi, V.; Zueva, S.B.; Vegliò, F.; De Michelis, I. Pilot-Scale Experiences with Aerobic Treatment and Chemical Processes of Industrial Wastewaters from Electronics and Semiconductor Industry. *Energies* **2021**, *14*, 5340. https://doi.org/10.3390/en14175340

Academic Editor: Antonio Zuorro

Received: 14 July 2021
Accepted: 20 August 2021
Published: 27 August 2021

Publisher's Note: MDPI stays neutral with regard to jurisdictional claims in published maps and institutional affiliations.

1. Introduction

The microelectronic industry produces a series of devices as memory and high-performance chips. These devices are produced on silicon wafers, which are integrated circuits for special applications, such as LEDs/OLEDs [1–4].

Different production processes involve the consumption of several chemicals, some of which are toxic and/or highly flammable. These substances should not be released into the environment without a specific treatment able to reduce the toxicity of wastewater.

The main substances contained in the residual effluents are organic ones, such as acetic acid, tetramethylammonium hydroxide (TMAH), and mineral acids, including nitric and hydrofluoric acid. All these compounds are hazardous materials and, with the exception of TMAH, can be removed by traditional processes [5]. TMAH is used as high-alkaline matter for the microelectronic devices. TMAH has toxic properties and can cause poisoning and, in some cases, death, even in diluted solutions. Moreover, TMAH is classified as ecotoxic, according to the ECHA [6]: "Hazardous to the aquatic life with ling lasting effects (H411). No observed Effect Concentration (NOEC) is 0.02 mg/L, max value to prevent chronic toxic effects on invertebrate organisms."

At the moment, EU regulation does not provide discharge limit values for TMAH. Despite this, TMAH is classified in the "Substances which have an unfavorable influence on the oxygen balance" and "Substances which contribute to eutrophication (in particular, nitrates and phosphates" in accordance with the Directive 2010/75/EU and EU Water Framework Directive 2000/60/EC (Annex VIII).

Regarding these aspects and NOEC values, the Italian National Institute of Health recommends 0.4 mg/L and 0.2 mg/L for TMAH concentrations discharged into sewage and water bodies, respectively [7].

The problem of TMAH wastewater treatment is of European importance, considering that several industries produce and use TMAH for different industrial applications.

Unlike other pollutants, the removal of TMAH remains an ongoing challenge, and researchers have only recently began to find a sustainable method to treat effluents containing TMAH. Both biological and chemical treatments can be adopted. Regarding biological treatments, anaerobic treatment has been further analyzed [8–10], achieving 95% of TMAH degradation and producing methane and carbon dioxide.

Alternately, chemical oxidation can be applied [11], which releases a poisonous gas containing NOx. Thus, it is necessary to insert a further operation to dissociate the NOx into nitrogen, carbon dioxide, and water [12]. Chemical oxidation can be combined with biological processes to improve the removal efficiency [13]. Other techniques have been studied, such as adsorption with activated carbon [14], zeolites [15], and ultrafiltration [16].

Considering the current scenario of TMAH classification and treatment, the Life Bitmaps project was proposed. The project studies a technical and economical approach to treat TMAH solutions from a microelectronic manufacturing plant, LFoundry Srl (Avezzano, Italy), to meet the Italian recommendation. The Life Bitmaps project addresses the current problem of TMAH treatment, and in this paper, the main results are described. After a laboratory-scale experiment [16–20], a pilot plant was designed and realized to validate the results obtained on a smaller scale. Finally, the pilot data were used to design an industrial plant for the treatment of TMAH effluent, and a study on the sustainability of the processes from an environmental and economic point of view was discussed. The main goal of the project was to test the performance and prove the feasibility of the E&S industrial wastewater from an integrated point of view (technical, economical, environment, social) while also considering the risk analysis of the processes and plants [21].

2. Materials and Methods

TMAH wastewater treatment was performed in the transportable mobile plant built in the ambit of the Life Bitmaps research project. Besides the TMAH degradation with aerobic bioreactions, the plant can also process chemical precipitation, as well as the other two selected solutions from microelectronic production, BOE and SEZ, which contain mainly fluorides and inorganic and organic acids at high concentrations.

After a laboratory-scale optimization, as reported in our recent paper [19], technical solutions were outlined, carrying out a mass and energy balance for the treatment of the three residual wastes.

The plant was realized in two containers located at the Lfoundry site, the industrial partner of the project (Figure 1).

Figure 1. Mobile pilot plant for the wastewater treatment.

The treatment of TMAH included a neutralization with sulfuric acid to achieve a neutral pH and three bioreactors in series (R101, R102, and R103) with a capacity of 1.1 m^3, which were fed by the activated sludge of the plant located on the industrial site (WWT). The equipment used for this treatment was located in the first container. The biological process operated in continuous (maximum flowrate: 25 L/h). The processes for BOE and SEZ included chemical precipitation by adding calcium hydroxide, a coagulant (aluminum sulfate), and a filtration. The second container contained the equipment for the process of BOE or SEZ (maximum treatable volume per batch: 180 L). This line worked in batch mode. At the end of the processes, all treated wastes were safety sent to industrial plant for the treatment of black waters. The block diagrams of the processes carried out in the Life Bitmaps portable plant are reported in Figure 2.

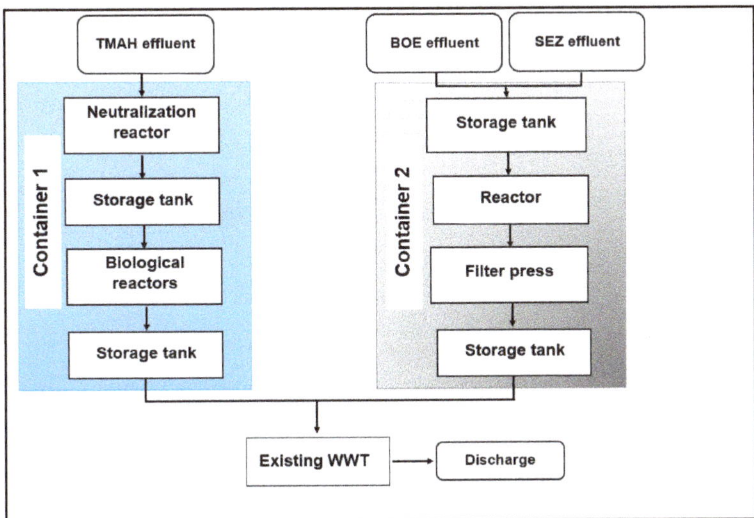

Figure 2. Block scheme of the processes performed in the Life Bitmaps portable plant (WWT = wastewater treatment).

2.1. Biological Processes for TMAH Degradation at the Pilot Scale

The experimental test for TMAH degradation began with the inoculation of the biological reactors by activated sludge coming from WWT. TMAH effluent was fed, changing the flowrate in a range equal to 5–20 L/h. The plant worked both in continuous mode and batch mode to stress and study the response of the microorganism and the efficiency of the degradation. The pilot research activity was carried out from June 2018 to December 2018. Over 50 liquid samples were collected, and around 800 chemical analyses were performed to monitor the Chemical Oxygen Demand (COD), pH, Total Suspended Solid (TSS), TMAH and its intermediates concentrations, dimethylamine (DMA), ammonium ions, and nitrates. Chemical oxygen demand (COD), N-NH_4, N-NO_3, TMAH, and dimethylamine (DMA) concentrations were regularly determined according to the APHA Standard Methods. Each sample was centrifuged at 4000 rpm in a 50 mL Falcon tube centrifuge (Rotofix 32A, Hettich, Westphalia, Germany). The supernatant was analyzed to determine the parameters indicated above, whereas the dried settled material (105 °C for 24) provided the TSS, the value proportional to the microbial mass.

TMAH and DMA were measured using the GC-MS technique (DX5000, Dionex, Sunnyvale, CA, USA).

COD, N-NH_4, and N-NO_3 were measured using the UV-Vis method with the spectrophotometer CADAS 200 (Hach Lange, Loveland, CO, USA) and Dr. Lange cuvette kits (LCI 400 for COD, LCK 303 for ammonium ion, and LCK 339/340 for nitrate ion) [22].

2.2. Chemical Processes for BOE and SEZ Effluent Treatment at the Pilot Scale

Chemical treatment for the BOE and SEZ effluents included lime precipitation in the presence of aluminum sulfate, as reported by the authors of [19]. $Ca(OH)_2$ was added at 16% concentrated solutions. Several experiments were performed to determine the optimal dosage of the reagents to remove the pollutants. At an optimal pH value, the system worked under constant stirring. After 2 h (reaction time), the suspensions were filtered. Samples of reactor suspension and filtrate were collected at the end of each test. Solid cake was dried at 105 °C. The liquors were analyzed to measure the fluorides, nitrates, and COD using Dr. Lange's kit, cuvette-test LCK 153 and LCK 114A. Other chemical analyses on the solutions were performed by atomic spectroscopy Agilent Synchronous Vertical Dual View (5100 ICP-OES). The recovered solids were analyzed by XRF spectrophotometer (Spectro XEPOS 2000) and infrared spectroscopy (FTIR, Impact 410 Nicolet spectrophotometer).

3. Results

The processes addressed to treat TMAH, BOE and SEZ waste were performed in the transportable pilot plant from the Life Bitmaps research project. The main results for the pilot scale experience are reported in the following subsections.

3.1. Pilot Scale Process for the Treatment of TMAH Effluent by Biological Degradation

TMAH effluent from the LFoundry industrial plant was collected into reactor N101. Then, sulfuric acid was added to modify the pH level from 12 to 7. Neutralized solution fed the reactor R101, which fed reactor R102, which fed reactor R103. A storage tank collected the solution coming from R103. Then, this suspension was fed WWT. The biological reactions transformed ammonium ions into nitrates, followed by denitrification reactions that produced nitrogen by denitrification [19]. Two series of tests were carried out: (1) the start-up of the plant with several interruptions due to technical problems, and (2) a series of experiments without technical interruptions. The results of the first series showed an abatement of TMAH over 90%. More interesting data were obtained in the second series, as described below. For the first period (until day 60), the flowrate of the TMAH effluent was 5 L/h. After that, there was a period of batch for 2 weeks. Then, the plant worked in a continuous mode, feeding 10 L/h. In the last days (from day 47), the plant was fed 20 L/h. Table 1 reports the TMAH effluent composition (before the treatment).

Table 1. Average composition of TMAH effluent.

Parameter	Value
pH	12
Tetramethylammonium hydroxide	2161–3200 mg/L
Ammonia nitrogen	6.05–8.6 mg/L
Nitrate nitrogen	<0.15 mg/L
Dimethylamine	<0.05 mg/L

The results of the second series are shown in Figures 3 and 4. The parameters TMAH, DMA, ammonium, and nitrates were analyzed because they are important to understand the functionality of the biological system. In fact, N-NH4 concentration is an index that degrades substances containing ammonia (in this case, TMAH).

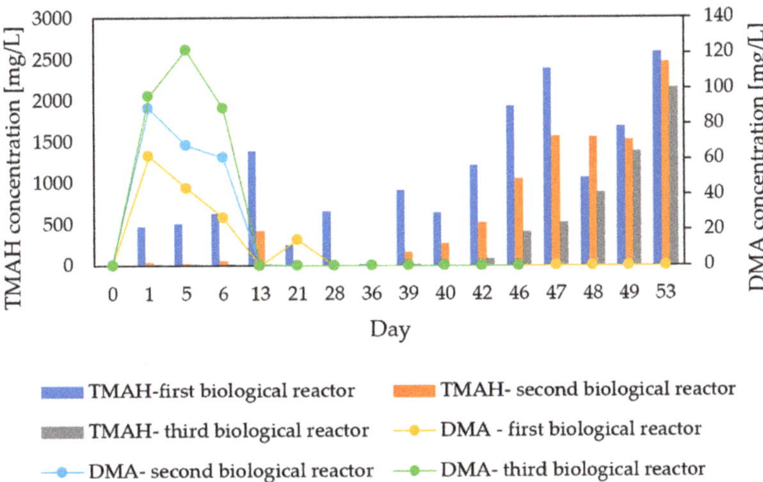

Figure 3. TMAH and DMA concentration (mg/L) trends as a function of the research pilot time.

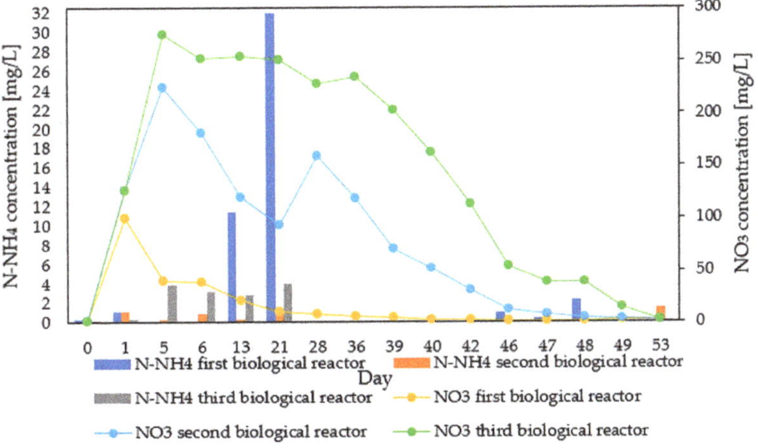

Figure 4. N-NH_4 and NO_3 concentration (mg/L) trends as a function of the research pilot time.

The TMAH concentration in the first biological reactor was variable, reaching close to 0 mg/L at the end of the batch period. When the 10 L/h flowrate was selected, the TMAH concentration started to increase, and this increment was also evident at the flowrate of 20 L/h. After 53 days of processing, the outlet TMAH concentration was close to 2.5 g/L (close to the inlet concentration). This result highlights the low biological degradation of TMAH that occurred when the flowrate was increased. As a direct consequence, the concentrations of ammonium nitrogen and nitrates decreased. The COD values were almost close to 135 mg/L, with a peak near 200 mg/L in the last period when the flowrate reached its maximum. The same situation was also recorded for the second and third biological reactors.

Obviously, the TMAH concentration was close to 0 mg/L in R102 and R103 until the end of batch operation. The N-NH_4^+ concentration was around 1 mg/L, and the NO_3^- concentration was higher with respect to the values in R101. These data indicate that the nitrification process of N-NH_4^+ happens in the last reactor. At the end of the test, the nitrates concentrations decreased, probably due to the changing conditions inside the reactors.

The system performance collapse was likely related to the phenomenon of washout. This situation was also confirmed by the trend of TSS. In fact, the TSS value decreased in the initial phase of adaptation (about 1600 mg/L). The TSS value remained constant until the end of the batch procedure, and then decreased during the subsequent continuous reactors feeding. When the process was stopped, the TSS value was close to 0.2 mg/L.

The experimental data were analyzed to define the TMAH degradation, which was 99%, in optimal conditions (5 L/h) with a process time of 192 h. We also performed an analysis to develop a potential full-scale plant schema.

3.2. Chemical Processes for BOE and SEZ Effluent Treatment at the Pilot Scale

SEZ and BOE come from the washing processes of wafers (thin slices of semiconductor) and are highly polluting effluents. SEZ contains fluorides (18,200 mg/L $\pm$ 1200), nitric acid (145 $\pm$ 15 g/L), and acetic acid (62 $\pm$ 30 g/L). Three experiments were replicated, which involved adding lime to increase the pH from 1.5 to 5. In all tests, a fluoride removal yield greater than 99% was obtained. After treatment, its concentration was about 7.5 mg/L $\pm$ 2.5. The removal yield for other acids was negligible, as these substances are nutrients for bacteria during subsequent biological treatment. The residue solid recovered after the filtration had around 60% humidity and mainly constituted calcium (59.2%), aluminum (3.4%), sodium (2.36%), phosphate (20.66%), and sulfates (7.2%).

The other effluent, BOE, contains fluorides (22.3 $\pm$ 2.3 g/L), phosphates (1.5 $\pm$ 0.6 g/L) ammonium (22.6 $\pm$ 1.5 g/L), and nitrates (12.4 $\pm$ 0.4 g/L). Three pilot experiments were performed for the BOE treatment, which involved adding lime to remove the fluorides and phosphates. The removal results were higher than 99% for both fluorides and phosphates. The final concentration of fluorides was near 9 mg/L, while the final concentration of phosphates was 0 mg/L. The residue solid recovered after filtration had around 50% humidity and mainly constituted calcium (91%), aluminum (2.2%), sodium (1.23%), and phosphate (5%).

Process analyses were carried out using the experimental data with the aim of proposing two process schemas.

Solid wastes characterization displayed that these residues are not dangerous for landfill disposal.

4. Discussion

The process analysis was useful to define the mass and energy balances able to treat the TMAH, BOE, and SEZ effluents for the full-scale plant. The details of the process analysis are not reported in this manuscript. Several process schemes were proposed and tested, fixing a continuous flowrate of 800 L/h for TMAH and assuming a batch mode for BOE and SEZ with a capacity of 435 t/y and 145 t/y, respectively. The mass balances, based on 100 kg of BOE and SEZ, showed that 3 kg of aluminum sulfates was necessary for both wastewater treatments. Lime consumption was 90 kg for BOE and 145 kg for SEZ. The residual solids were 17 kg and 22 kg for BOE and SEZ, respectively. Regarding the economic aspects, the total cost for the TMAH, BOE, and SEZ treatment was EUR 15/m^3, EUR 110/m^3, and EUR 207m^3. The main cost for the TMAH treatment was the energy used for the aeration of the biological reactors. The main cost for the BOE treatment, which was even greater for the SEZ treatment, was the disposal cost.

The environmental study performed by Gabi software (not reported here) showed that electricity utilization represented the principal contribution (larger than 90%) on the whole environmental load (TMAH treatment). In addition, lime consumption for precipitation caused about 70% of the impact, which was correlated to the BOE and SEZ treatment. The highest impact classes were climate change, ionizing radiation, and reserve use.

The process analysis showed that the proposed technologies were more sustainable, both economically and environmentally, when compared with the current disposal processes adopted by Lfoundry.

5. Conclusions

The E&S industry produces a large amount of toxic wastewater, and TMAH is among the most dangerous pollutants. At the moment, discharge limit values are not provided by the EU regulation (i.e., concentrations at discharge). However, the Directive 2010/75/EU and the EU Water Framework Directive 2000/60/ EC (Annex VIII) provide a list of contaminants that affect the water quality as "Substances which have an unfavorable influence on the oxygen balance (and can be measured using parameters such as BOD, COD, etc.)" and "Substances which contribute to eutrophication (in particular, nitrates and phosphates)", with TMAH belonging to both categories. Due the large use of TMAH in many industries in Europe, the problem of TMAH wastewater treatment assumes increasing importance. In fact, many industries synthetize and use TMAH for several industrial applications. The development of advanced treatment of this kind of industrial waste represents a crucial challenge for the European E&S industry that must be improved for the protection of surface and groundwater. Hence, in this study, an innovative wastewater treatment process was proposed and tested at the pilot scale within the Life Bitmaps research project. The pilot plant was able to treat the three effluents produced by microelectronic manufacturing, called TMAH, BOE, and SEZ, in an integrated way. For the first time, biological degradation with three biological reactors has been proposed, and BOE and SEZ, a chemical treatment has been tested. The results show that TMAH can be degraded with a yield of 99%. The same efficiency was also obtained for the removal of pollutants from the BOE and SEZ effluents.

Our economical and sustainability studies demonstrate that the proposed technologies are more sustainable, both economically and environmentally, when compared with the currently adopted disposal processes. The processes addressed in this study are under improvement. Our future work will address an optimization for the full-scale plant for TMAH and for the valorization of the solid residues from BOE and SEZ treatment.

6. Patents

Vegliò, F., Prisciandaro, M., Ferella, F., Innocenzi, V., Di Renzo, A., Saraullo, M., Zueva, S., De Michelis, I., Tortora, F., 2018. Process and plant for the treatment of a wastewater containing tmah, Patent number WO/2020/012240A1.

Author Contributions: Conceptualization, V.I., S.B.Z., I.D.M. and F.V.; validation, V.I., S.B.Z. and I.D.M.; formal analysis, V.I., I.D.M.; investigation, V.I., I.D.M. and F.V.; resources, F.V.; data curation, V.I., S.B.Z., I.D.M. and F.V.; Writing—review & editing, V.I., I.D.M.; visualization, I.D.M., F.V.; supervision, I.D.M., F.V.; project administration I.D.M., F.V.; funding acquisition, F.V. All authors have read and agreed to the published version of the manuscript.

Funding: This research was funded by the EU Life Program, LIFE15 ENV/IT/000332. https://www.lifebitmaps.eu/ (accessed on 24 August 2021).

Institutional Review Board Statement: Not applicable.

Informed Consent Statement: Not applicable.

Data Availability Statement: Not applicable.

Acknowledgments: The authors are thankful to all partners of the Project Life Bitmaps, Lfoundry Srl, Univaq, BME Biomaterials & Engineering S.R.L, and BFC Sistemi S.R.L. for their support to the research.

Conflicts of Interest: The authors declare no conflict of interest.

References

1. Tang, B.; Miao, J.; Liu, Y.; Wan, H.; Li, N.; Zhou, S.; Gui, C. Enhanced Light Extraction of Flip-Chip Mini LEDs with Prism-Structured Sidewall. *Nanomaterials* **2019**, *9*, 319. [CrossRef]
2. Hu, H.; Tang, B.; Wan, H.; Sun, H.; Zhou, S.; Dai, J.; Chen, C.; Liu, S.; Guo, L.J. Boosted ultraviolet electroluminescence of InGaN/AlGaN quantum structures grown on high-index contrast patterned sapphire with silica array. *Nano Energy* **2020**, *69*, 104427. [CrossRef]

3. Zhou, S.; Liu, X.; Yan, H.; Chen, Z.; Liu, Y.; Liu, S. Highly efficient GaN-based high-power flip-chip light-emitting diodes. *Opt. Express* **2019**, *27*, A669–A692. [CrossRef] [PubMed]
4. Chang, D.T.; Park, D.; Zhu, J.-J.; Fan, H.-J. Assessment of an MnCe-GAC treatment Process for Tetramethylammonium - Contaimined Wastewater from Optoelectronic Industries. *Appl. Sci.* **2019**, *9*, 4578. [CrossRef]
5. Huang, C.J.; Liu, J.C. Precipitate flotation of fluoride-containing wastewater from a semiconductor manufactures. *Wat. Res.* **1999**, *33*, 3403–3412. [CrossRef]
6. ECHA Site. Available online: https://echa.europa.eu/it/registration-dossier/-/registered-dossier/14295/2/1 (accessed on 9 July 2021).
7. Life Bitmaps Site. Available online: http://www.lifebitmaps.eu/pdf/presentazione-iuliano.pptx (accessed on 9 July 2021).
8. Lei, C.N.; Wang, L.M.; Chen, P.C. Biological treatment of thin-film transistor liquid crystal display (TFT-LCD) wastewater using aerobic and anoxic/oxic sequencing batch reactors. *Chemosphere* **2010**, *81*, 57–64. [CrossRef] [PubMed]
9. Asakawa, S.; Sauer, K.; Liesack, W.; Thauer, R.K. Tetramethylammonium: Coenzyme in methytrasferase system from *Methanococcoidess* sp. *Arch. Microbiol.* **1998**, *170*, 220–226.
10. Chang, K.F.; Yang, S.Y.; You, H.S.; Pan, J.R. Anaerobic treatment of tetramehylammonium hydroxide (TMAH) containing wastewater. *IEEE Trans. Semicond. Manuf.* **2008**, *21*, 486–491.
11. Karatza, D.; Prisciandaro, M.; Lancia, A.; Musmarra, D. Sulfide oxidation catalyzed by cobalt ions in flue gas desulfurization processes. *J. Air Waste Manage. Assoc.* **2010**, *60*, 675–680. [CrossRef]
12. Hirano, K.; Okamura, J.; Taira, T.; Sano, K.; Toyoda, A.; Ikeda, M. An efficient treatment tecnique for TMAH wastewater by catalytic oxidation. *IEEE Trans. Semicond. Manuf.* **2011**, *14*, 202–206. [CrossRef]
13. Den, W.; Ko, F.H.; Huang, T.Y. Treatment of organic wastewater discharged from semiconductor manufacturing process by ultraviolet/hydrogen peroxide and biodegradation. *IEEE Trans. Semicond. Manuf.* **2002**, *12*, 540–551. [CrossRef]
14. Prahas, D.; Liu, J.C.; Ismadji, S.; Wang, M.J. Adsorption of tetramethylammonium hydroxide on activated carbon. *J. Environ. Eng.* **2012**, *138*, 232–238. [CrossRef]
15. Nishihama, S.; Murakami, M.; Igarashi, N.Y.; Yamamoto, K.; Yoshizuka, K. Separation and recovery of tetramethylammonium hydroxide with mesoporous silica having ahexagonal structure (MCM-41). *Solvent Extr. Ion Exch.* **2012**, *30*, 724–734. [CrossRef]
16. Moretti, G.; Matteucci, F.; Saraullo, M.; Vegliò, F.; Del Gallo, M. Selection of a Very Active Microbial Community for the Coupled Treatment of Tetramethylammonium Hydroxide and Photoresist in Aqueous Solutions. *Int. J. Environ. Res. Public Health* **2017**, *15*, 41. [CrossRef]
17. Tortora, F.; Innocenzi, V.; Prisciandaro, M.; De Michelis, I.; Vegliò, F.; Mazziotti di Celso, G. Removal of tetramethyl ammonium hydroxide from synthetic liquid wastes of electronic industry through micellar enhanced ultrafiltration. *J. Dispers. Sci. Technol.* **2018**, *39*, 207–213. [CrossRef]
18. Innocenzi, V.; Prisciandaro, M.; Veglio, F. Effect of the hydrodynamic cavitation for the treatment of industrial wastewater. *Chem. Eng. Trans.* **2018**, *67*, 529–534.
19. Innocenzi, V.; Zueva, S.; Prisciandaro, M.; De Michelis, I.; Di Renzo, A.; Mazziotti di Celso, G.; Vegliò, F. Treatment of TMAH solutions from the microelectronics industry: A combined process scheme. *J. Water Process Eng.* **2019**, *31*, 100780. [CrossRef]
20. Ferella, F.; Innocenzi, V.; Zueva, S.; Corradini, V.; Ippolito, N.M.; Birloaga, I.B.; De Michelis, I.; Prisciandaro, M.; Vegliò, F. Aerobic Treatment of Waste Process Solutions from the Semiconductor Industry: From Lab to Pilot Scale. *Sustainability* **2019**, *11*, 3923. [CrossRef]
21. Innocenzi, V.; De Michelis, I.; Prisciandaro, M.; Iuliano, G.; Vegliò, F. Safety Analysis of Industrial Wastewater Pilot Plant for the Removal of Pollutants from Microelectronic Industry Effluents. *Chem. Eng. Trans.* **2020**, *82*, 325–330.
22. Ferella, F.; Innocenzi, V.; Moretti, G.; Zueva, S.; Pellegrini, M.; De Michelis, I.; Ippolito, N.M.; Del Gallo, M.; Prisciandaro, M.; Vegliò, F. Water reuse in a circular economy perspective in a microelectronics industry through biological effluents treatments. *J. Clean. Prod.* **2021**, 128820. [CrossRef]

MDPI

Article

Adsorptive Recovery of Cd(II) Ions with the Use of Post-Production Waste Generated in the Brewing Industry

Tomasz Kalak [1,*], Jakub Walczak [1] and Malgorzata Ulewicz [2]

[1] Department of Industrial Products and Packaging Quality, Institute of Quality Science, Poznań University of Economics and Business, 61-875 Poznań, Poland; kuba.jakub98@gmail.com

[2] Faculty of Civil Engineering, Częstochowa University of Technology, 42-200 Częstochowa, Poland; malgorzata.ulewicz@pcz.pl

* Correspondence: tomasz.kalak@ue.poznan.pl

Abstract: Post-production waste generated in the brewing industry was used to analyze the possibility of Cd(II) ion recovery in biosorption processes. Brewer's grains (BG), which are waste products from beer manufacturing processes, are a promising material that can be reused for biosorption. The biomass contains appropriate functional groups from fats, proteins, raw fibers, amino acids, carbohydrates and starch, showing a strong affinity for binding metal ions and their removal from wastewater. The biosorbent material was characterized by several research methods, such as particle size distribution, elemental composition and mapping using SEM-EDX analysis, specific surface area and pore volume (BET, BJH), thermogravimetry, electrokinetic zeta potential, SEM morphology and FT-IR spectrometry. Initial and equilibrium pH, adsorbent dosage, initial metal concentration and contact time were parameters examined in the research. The highest biosorption efficiency was obtained at a level of 93.9%. Kinetics analysis of the processes and sorption isotherms were also carried out. Based on the conducted experiments, it was found that this material has binding properties in relation to Cd(II) ions and can be used for wastewater treatment purposes, being a low-cost biosorbent. This research studies are in line with current global trends of circular and sustainable economies.

Keywords: water quality; brewing industry; residual brewer's grains; biosorption; Cd(II) ions; kinetics

Citation: Kalak, T.; Walczak, J.; Ulewicz, M. Adsorptive Recovery of Cd(II) Ions with the Use of Post-Production Waste Generated in the Brewing Industry. *Energies* **2021**, *14*, 5543. https://doi.org/10.3390/en14175543

Academic Editor: Gabriele Di Giacomo

Received: 30 June 2021
Accepted: 1 September 2021
Published: 5 September 2021

Publisher's Note: MDPI stays neutral with regard to jurisdictional claims in published maps and institutional affiliations.

1. Introduction

Cadmium (Cd) is one of the most toxic and harmful metals. It exists naturally in water, air, soils and foodstuffs. The most common forms of the metal are pure cadmium oxide (87.5% Cd), cadmium sulfide (77.6% Cd), otavite, cadmium carbonate (61.5% Cd) and greenockite (CdS). Cadmium is released into the environment as a result of many industrial processes, such as production of alloy, cadmium-nickel batteries, pesticides, fertilizers, plastics, pigments and dyes, smelting, electroplating, mining, refining and textile operations [1–3]. Cadmium has toxic properties in relation to living organisms and is a problem for the environment and food chain when it is present in excess amounts. The World Health Organisation recommends the maximum contaminant level (MCL) of Cd(II) in drinking water at up to 0.005 mg/L [4].

The brewing industry occupies a strategic economic position in the global economy and is open to technological development and scientific progress. In 2019, global beer production amounted to approximately 1.91 billion hectoliters [4]. The following countries are the leading producers in the world (in million liters in 2017): China (39.788), USA (21.775), Brazil (14.000), Mexico (11.000), Germany (9.301), Russia (7.440), Japan (5.247), Great Britain (4.405), Vietnam (4.375), Poland (4.050). Beer is ranked fifth in the ranking of the most consumed beverages in the world after tea, carbonated drinks, milk and coffee, with an average consumption of 23 L per person per year [5–7].

Increasing production in the brewing industry also brings with it millions of tons of waste, which is both an ecological and economic problem. Huge amounts of brewer's grains (BG) each year lead to environmental degradation, but also to the loss of valuable material that could be used as animal feed, fuel, raw material for methane production and many other purposes. Due to its valuable composition, one of the directions of reuse may be the extraction of ingredients such as amino acids, fats, proteins, polysaccharides, fiber, enzymes, vitamins, flavor compounds or phytochemicals, which can be reused as functional ingredients in food or pharmacological production [8]. Another alternative solution may be the use of brewer's grains (BG) to treat wastewater from heavy metals and other toxic pollutants through adsorption processes. Due to the presence of many functional groups (e.g., carboxyl, amide, amine, etc.), it is possible to bind metal ions in chemical and physical reactions. The biosorption process is one of the most promising and alternative techniques because of its low operating costs, mainly due to the use of cheap biowaste, the ease of obtaining materials, minimizing significant amounts of waste and high efficiency in removing metal ions from wastewater. The costs of water treatment by adsorption are difficult to determine as they depend on many factors. However, it can be assumed that they are in the range of about USD 5–200 per m^3 and are significantly lower compared to other commercial methods of metal removal from wastewater of about 10–USD 450 per m^3) [9]. In 2018, beer production in Poland amounted to approx. 40.93 million hectoliters [10]. Assuming that per 100 hectoliters of beer produced, there is an average of approx. 15–19 kg of the by-product in the form of brewer's grains (BG) with a dry matter content in the range of 35–40% [11], it can be estimated that in 2018, from 6.139 to 7776.7 tons of this waste were generated in Poland. Such amounts can be considered significant, which offers the possibility of their potential use in biosorptive processes for removing metals from wastewater.

The aim of this study was to determine the efficiency of the biosorption process of post-production waste in the form of brewer's grains in relation to Cd(II) ions. In addition, the purpose was also to characterize the physicochemical properties of the biosorbent using selected analytical methods as well as to study sorption kinetics, equilibrium and isotherms.

2. Experimental Procedure

2.1. Materials and Methods

2.1.1. Brewer's Grains Preparation

In this study, samples of the brewer's grains (BG) (Figure 1) were obtained as a result of a beer production process in the brewing industry in a Polish factory (Greater Poland voivodeship). The brewing effluents collected from the production process were wet biomass; therefore, they were dried in a laboratory dryer (BINDER Gmbh, Tuttlingen, Germany) at a temperature of 60 °C to constant weight and then placed in polyethylene containers and cooled in a desiccator. Then, the dried BG were ground and screened through a sieve in order to select grains with a diameter less than 0.212 mm. The samples prepared in this way were used for analyses. All chemical compounds used in the study were pure for analysis. Distilled water was used in the research.

2.1.2. Brewer's Grains Characterization

Particles of brewer's grains (BG) with a diameter less than 0.212 mm were used in these studies. At the beginning, the physicochemical properties of the biomass were analyzed with the use of selected analytical techniques such as particle size distribution, elemental composition and mapping using SEM-EDS analysis, electrokinetic zeta potential, specific surface area and pore diameter (BET adsorption isotherms), pore volume and pore volume distribution (BJH), thermogravimetry (TGA, DTG), SEM morphology, and FT-IR analysis. The detailed description of the methods are presented as supplementary material (SM ethods).

Figure 1. Sample of brewer's grains.

2.1.3. Biosorption Process of Cd(II) ions

The efficiency of the Cd(II) biosorption process on brewer's grains (BG) was determined by shaking at room temperature (23 ± 1 °C). A standard solution (Cd^{2+}, 1 g/L, Sigma-Aldrich) was used in the experiments. Biosorbent samples (1–50 g/L) and a Cd(II) stock solution at a concentration of 10 mg/L (volume of the solutions V = 20 mL) and at an initial pH 2–5 were placed in conical flasks and shaken at rotational speed 200 rpm for 60 min until equilibrium was reached. NaOH and HCl solutions (both 0.1 M) were used to adjust the pH of the Cd(II) stock solutions. After the biosorption processes, the solutions were centrifuged for 15 min (4000 rpm) for phase separation. The concentration of cadmium(II) ions in the solutions after biosorption was analyzed using an atomic absorption spectrophotometer SpectrAA 800 (F-AAS, at a wavelength λ = 228.8 nm for cadmium, Varian, Palo Alto, USA). The measurements were repeated three times at room temperature (23 ± 1 °C) and normal pressure and the results are presented as mean values of all measurements. Measurement errors were calculated, and confidence intervals are presented in tables and figures. The biosorption efficiency A (%) and the biosorption capacity q_e (mg/g) were calculated on the basis of Equations (1) and (2):

$$A = \left[\frac{C_0 - C_e}{C_0}\right] \times 100\% \quad (1)$$

$$q_e = \frac{(C_0 - C_e) \times V}{\mathrm{m}} \quad (2)$$

where: C_0 and C_e (mg/L) are initial and equilibrium metal ion concentrations, respectively, and V (L) is the volume of solution and m (g)—mass of BG.

Kinetics and isotherms were studied using pseudo-first-order and pseudo-second-order, Langmuir and Freundlich models according to Equations (3)–(6).

$$q_t = q_e\left(1 - e^{k_1 t}\right) \quad (3)$$

$$q_t = \frac{q_e^2 k_2 t}{1 + q_e k_2 t} \quad (4)$$

$$q_e = \frac{q_{max} K_L C_e}{1 + K_L C_e} \quad (5)$$

$$q_e = K_F C_e^{\frac{1}{n}} \quad (6)$$

where q_t (mg/g) is the amount of Cd(II) ions adsorbed at any time t (min.), q_e (mg/g) is the maximum amount of Cd(II) ions adsorbed per mass of BG at equilibrium, k_1 (1/min) is the rate constant of pseudo-first-order adsorption, k_2 (g/(mg·min.)) is the rate constant of pseudo-second-order adsorption, q_{max} (mg/g) is the maximum adsorption capacity, K_L is the Langmuir constant, C_e (mg/L) is the equilibrium concentration after the adsorption process, K_F is the Freundlich constant and 1/n is the intensity of adsorption.

3. Results and Discussion

3.1. The Process of Beer Production and Post-Production Brewer's Grains

Four main raw materials are used in the production of beer: water, barley malt, hops and yeast. Water is necessary primarily for the malting of barley, mashing of malt and production of malt wort, but also for cooling processes, in the boiler room and for washing equipment and packaging. The quality of water in brewing should be appropriate, hence it is subjected to treatment processes. Water quality requirements are specified in the Regulation of the Minister of Health on the quality of water intended for human consumption [12]. The general scheme of beer production is shown in Figure 2. Initially, wort is produced in many different processes. First, malt is ground mechanically to allow enzymes to break down insoluble substances into soluble compounds. The next and most important stage of wort production is mashing, in which insoluble malt components (cellulose, starch) at an optimal temperature and under influence of enzymes pass into the aqueous solution, creating a soluble extract in the form of minerals, proteins and sugars. Then, in the wort filtration process, solids are separated into so-called brewing waste or spent grains. During the boiling of wort, resins and essential oils are released from hops, protein-tannic compounds are precipitated, enzymes are decomposed and the wort becomes acidic and colored. Hops and hot sludge in the settling vat are removed from the hop wort followed by cooling and aeration. In the fermentation process, sugars are converted by yeast enzymes into C_2H_5OH and CO_2. In the last stage of fermentation, maturing and conditioning at low temperatures take place, in which yeast cells draw energy from the decomposition of spare substances still present in the wort. The yeast flocculates and sinks to the bottom of the tank. At this stage, the taste of the beer is finally achieved. At the end, beer is filtered to remove particles that could cause turbidity. The finished beer is poured into bottles or cans and sold [8].

The amount of brewer's grains produced is difficult to estimate. Producers keep their data secret because for the protection of intellectual property and the fear of inspection by organizations responsible for waste management. On the other hand, production processes are varied, and depend on technologies used and the size of brewery. Based on the literature, it is assumed that an average of 15–19 kg of by-product in the form of brewer's grains with a dry matter content of approx. 35–40% is produced per 100 hl of produced beer [11,13].

3.2. Characterization of the Biosorbent

Pursuant to the Act on Waste and the Regulation of the Polish Minister of Environment on waste recovery outside of installations and devices, the brewer's grains are classified as waste with the code 02 07 80 and are defined as bagasse, post fermentation must sediments (trub) and decoctions [14,15]. The brewing waste is brown or light brown in color and has a characteristic grain or bread aroma. Moist biomass contains approx. 18–23% of dry matter consisting mainly of barley, barley malt, wheat, rice, maize and hops. The most important nutrients are proteins, fats, crude fibers, crude ash, starch, carbohydrates, phosphorus, calcium, lysine, methionine, potassium, sodium, magnesium, B vitamins and others depending on the origin of the biomass. It is estimated that the total amount of nutrients is approx. 66% of dry matter, starch and sugars-approx. 14% of dry matter and protein substances-approx. 26% of dry matter. According to the Regulation of the Polish Minister of Environment on waste recovery outside of installations and devices, the method of recovering brewer's grains (BG) as waste is the R14 recovery process, stating that the waste can be used for 'feeding animals-in accordance with the rules of feeding individual species' [16]. Feeding livestock with BG waste has many benefits, such as elimination of costs related to storage and disposal, nutritional components having a good effect on health and condition of animals, as well as lower costs of feeding animals compared to the use of commercial feed. There are also other applications, such as use as a raw material for the production of various feeds, the use of modified BG for the production of edible mushrooms, the use as biomass for gasification processes in the production of methane in biogas plants and the use as an additive to fossil fuels in power plants as well as processing into solid fuel (briquettes, pellets) [13]. Another alternative application of post-production BG may be the removal of heavy metals from aqueous solutions in biosorption processes.

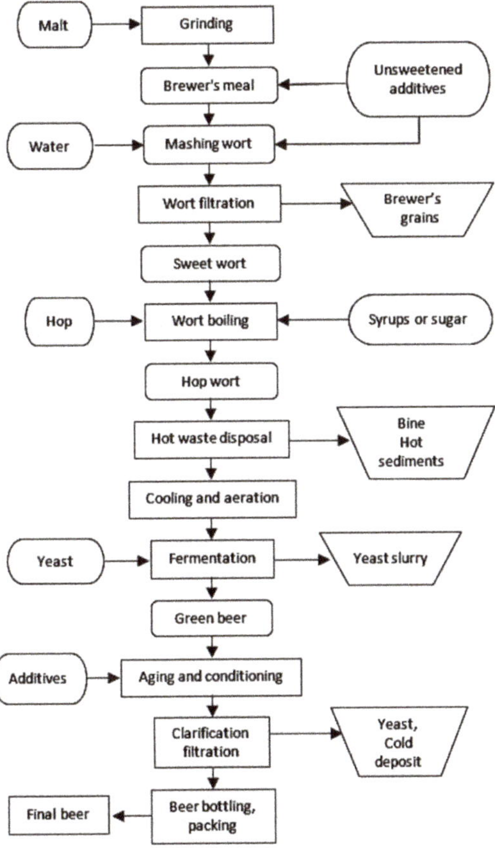

Figure 2. General scheme of beer production.

In preliminary experiments, BG waste was characterized in terms of physical and chemical properties using a number of methods. The analysis of particle size distribution using a laser diffraction method showed the presence of one peak with a particle size of 67.8 nm (Figure S1). The strength of the material and the rate of hydration are properties that depend on particle size distribution. Moreover, this parameter serves as an important indicator of particle quality and performance in many different processes. It should be noted that the laser diffraction method has limitations, in that only particles capable of being suspended in solution can be analyzed. Due to the complexity and nature of the BG mixture, the larger and heavier particles fall to the bottom of the solution. Smaller particles are able to dissolve faster in solution and form a more stable suspension compared to larger ones. Therefore, it was possible to measure only particles suspended in the solution. Based on the results, it can be stated that the BG particles are not homogeneous. In accordance with the literature data, as the biosorbent particle size decreased, the surface area of particles and the number of active sites on the sorbent increased, which resulted in an increase in cadmium ion removal efficiency [17,18]. Hence, it is suggested to use biosorbent fractions with the smallest particle size in experiments.

Elemental analysis was carried out using the SEM-EDS method and the results are presented in Figure 3 and Table 1. The biomass mainly contained elements, such as: O, C, P, Mg, Si, Ca, K, Na, S and oxides: CO_2, P_2O_5, MgO, SiO_2, CaO, K_2O, Na_2O, SO_3. The content of oxides was not the result of direct measurements, but the result of stoichiometric calculations based on the estimation in the EDS microanalyzer. The measurement in this method was carried out pointwise on the surface of the biomass sample. Therefore, the content of elements and their amounts may slightly differ in different places of the sample due to complexity of the BG mixture and its inhomogeneity.

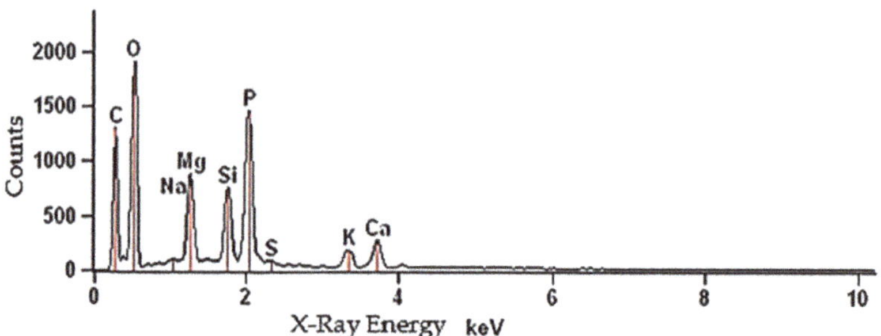

Figure 3. The EDS spectrum of brewer's grains.

Table 1. Elemental composition of BG waste (EDS microanalyser).

Elements	C	O	Na	Mg	Si	P	S	K	Ca
Content, weight (%)	26.44 ± 0.4	48.46 ± 0.5	0.36 ± 0.02	5.81 ± 0.2	4.27 ± 0.1	9.65 ± 0.3	0.24 ± 0.02	1.91 ± 0.03	2.86 ± 0.09
Content, atomic (%)	36.23 ± 0.3	49.85 ± 0.4	0.26 ± 0.01	3.93 ± 0.1	2.5 ± 0.2	5.13 ± 0.4	0.12 ± 0.01	0.8 ± 0.02	1.17 ± 0.3
Oxides	CO_2		Na_2O	MgO	SiO_2	P_2O_5	SO_3	K_2O	CaO
Content in oxides (%)	26.44 ± 0.2	48.46 ± 0.5	0.36 ± 0.02	5.81 ± 0.03	4.27 ± 0.03	9.65 ± 0.04	0.24 ± 0.01	1.91 ± 0.02	2.86 ± 0.03

As a result of BET analysis, specific surface area (4.218 m^2/g), pore volume (0.0046 cm^3/g) and pore diameter (4.386 nm) were determined (Figures 4–6). The BET adsorption isotherm, which depends on the pore size and the intensity of adsorbent-adsorbate interaction, slightly resembles the type III isotherm in that it is slightly convex towards the pressure axis. The isotherm indicates cooperative adsorption, which indicates that previously adsorbed molecules can contribute to increasing the adsorption efficiency of other free molecules. On the other hand, low adsorption efficiency may occur in the case of low relative pressure and weak interaction at the adsorbent-adsorbate interface. In any case, the isotherm becomes convex into the pressure axis in a situation where, after a single adsorption, the interaction between adsorbent and adsorbate promotes adsorption of further metal ions [19].

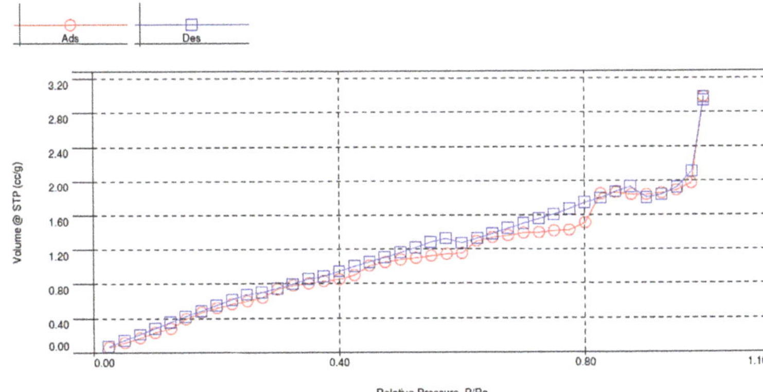

Figure 4. Linear form of the BET adsorption and desorption isotherm of brewer's grains.

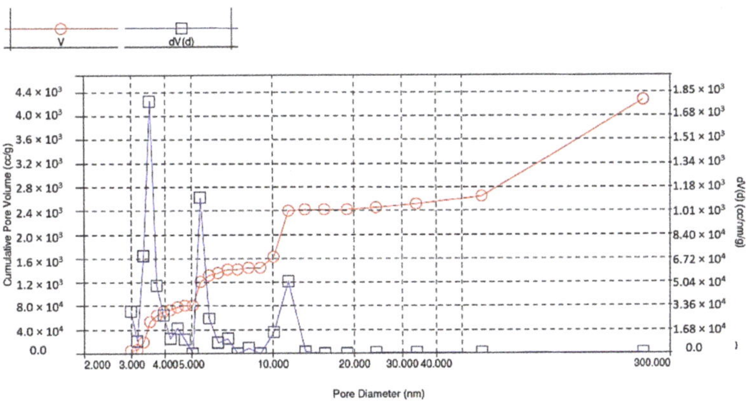

Figure 5. Pore size distribution in brewer's grains during adsorption.

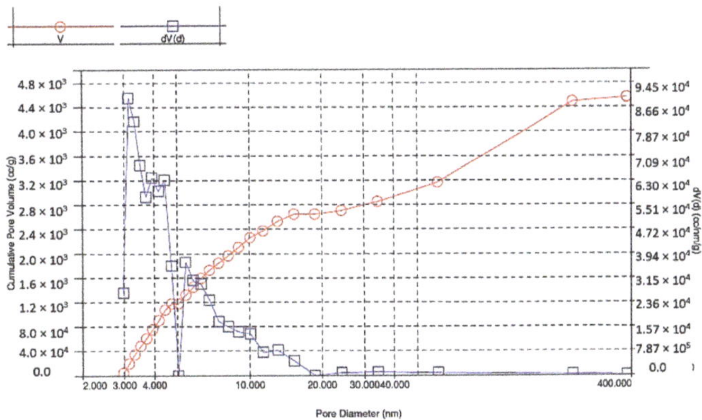

Figure 6. Pore size distribution in brewer's grains during desorption.

Thermogravimetric analysis was carried out at a temperature ranging from 29 to 600 °C (Figure 7). As a result of these measurements, a gradual loss of weight of the biomass material with increasing temperature was observed. The first phase of decomposition of the BG pomace occurred in the temperature range from 30 to 120 °C, and the second in the range of about 180–500 °C, as shown by the TGA and DTG curves. In the first stage there was a slight decrease in weight equal to approx. 1.8%, which may have resulted from evaporation of adsorbed water from the pomace sample. In the next phase, greater weight loss (about 38%) was observed, which was a consequence of the pyrolysis process and the evaporation of most of the volatile substances (e.g., CO_2). At a temperature of 355 °C, the spectrum shows a very intense peak (DTG) that relates to the breakdown of lipids, carbohydrates and proteins. Similar research results have been published in the literature [20,21].

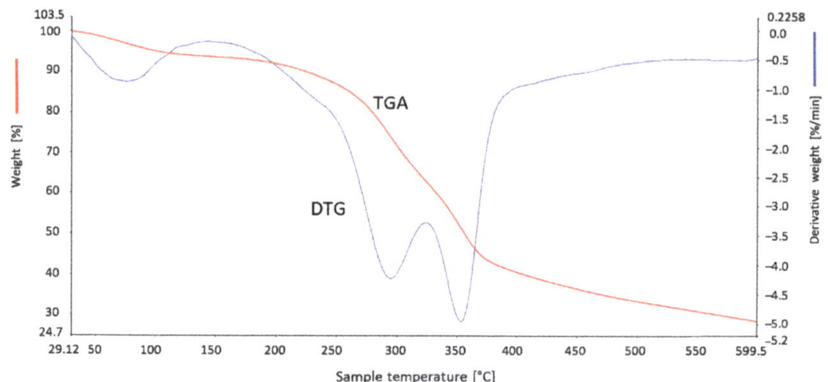

Figure 7. Thermogravimetric curves of brewer's grains.

The analysis of electrokinetic zeta potential was carried out by combining electrophoresis and measuring the velocity of particles in liquid with red laser (λ = 633 nm) based on the Doppler effect (LDV). Determining this parameter is important for many industrial activities, including wastewater treatment in terms of environmental protection. It makes it possible to determine the influence of surface charge on the metal ion sorption process and to estimate its efficiency [22]. According to the literature, aqueous slurry is more stable at higher zeta potential values, as electrostatically charged particles repel each other, overcoming the tendency to aggregate and agglomerate [23,24]. These studies showed that pH values had an influence on the surface charge, which gradually decreased from −3.01 mV (pH 2.01) to −32.1 mV (pH 6.93) (Figure 8). The isoelectric point (IEP) is set at the zero point (no electrical charge) and represents the equilibrium balance between negative and positive ions (the system is the least stable). At the IEP point, the particles of the suspension are characterized by the lowest solubility, viscosity and osmotic pressure. In this analysis, the electrostatic charge did not reach the isoelectric point (IEP) and took only negative values below the IEP. This means that there is an advantage of positively charged particles over negatively charged ones on the biomass surface in the pH range from 2.0 to 7.0.

Brewer's grains were analyzed using SEM morphology and images are shown in Figure 9A,B. Generally speaking, the brewer's grains consist of particles of varying size, including larger, elongated particles and smaller spherical particles placed between them. Shape irregularities occur with both smaller and larger particles. The structure is not homogeneous, and developed flat surfaces are visible. Some particles have a fibrous structure, suggesting they are derived from the same barley grains or embryonic tissue. The remaining particles are covered with small spherical projections that represent the characteristic formations on the outer cells of the seed coat. The visible spherical material is less cohesive, has a looser structure, and the particles are not densely packed. Such particles have uneven surfaces and edges and can chip off easily. A similar structure to this biomass was described by other authors [25–30].

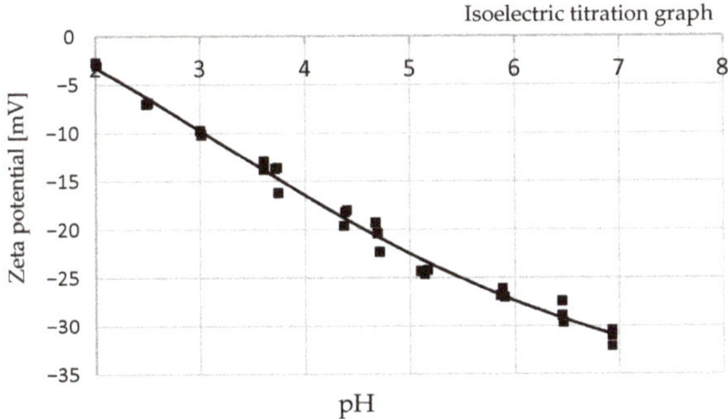

Figure 8. The change in zeta potential with equilibrium pH.

Figure 9. SEM images of BG residues: (**A**) magn.: ×10,000, scale bar: 2 µm, (**B**) ×10,000, 2 µm.

3.3. FT-IR Analysis

FT-IR measurements of brewer's grain residues before and after the sorption process were carried out and Figure 10 shows the spectra. The following experimental conditions were used to examine the samples: BG dose 25 g/L, initial concentration of Cd(II) ions 10 mg/L, initial pH 4, T = 23 ± 1 °C, contact time 60 min. The description of the FT-IR peaks is presented in Table 2. The spectra were analysed before and after the biosorption process in terms of differences in shape, intensity of bands, frequency and possible interaction of functional groups with Cd(II) ions. Figure 10 shows that after the sorption process, the intensity of the bands shifted towards lower transmittance values, and their location remained at the same wavelengths or was slightly shifted. These changes can be identified as follows: 3284.83 (shift to 3289.81 cm^{-1}), 2923.72 (shift to 2919.92 cm^{-1}), 2852.02 (shift to 2851.84 cm^{-1}), 1643.84 (shift to 1633.69 cm^{-1}), 1515.88 (shift to 1539.91 cm^{-1}), 1428.13 (shift to 1427.82 cm^{-1}), 1032.92 (shift to 1032.7 cm^{-1}). The occurring phenomena of peak shifts can be explained by the interaction and bonding of Cd(II) ions with functional groups of compounds present in BG residues, probably due to the process of ion exchange. This shifting could occur due to the formation of a bond between Cd^{2+} with oxygen and nitrogen.

During the biosorption process with BG sorbent, the mechanism of ion exchange between Cd(II) ions and various metal cations (e.g., Mg^{2+}, Ca^{2+}) probably took place. The presence of various metal cations was determined by the SEM-EDS method (Table 2). It is assumed that ion exchange between protons or ionizable cations on the surface of the biosorbent participated in binding of Cd(II) ions at initial pH 4.0.

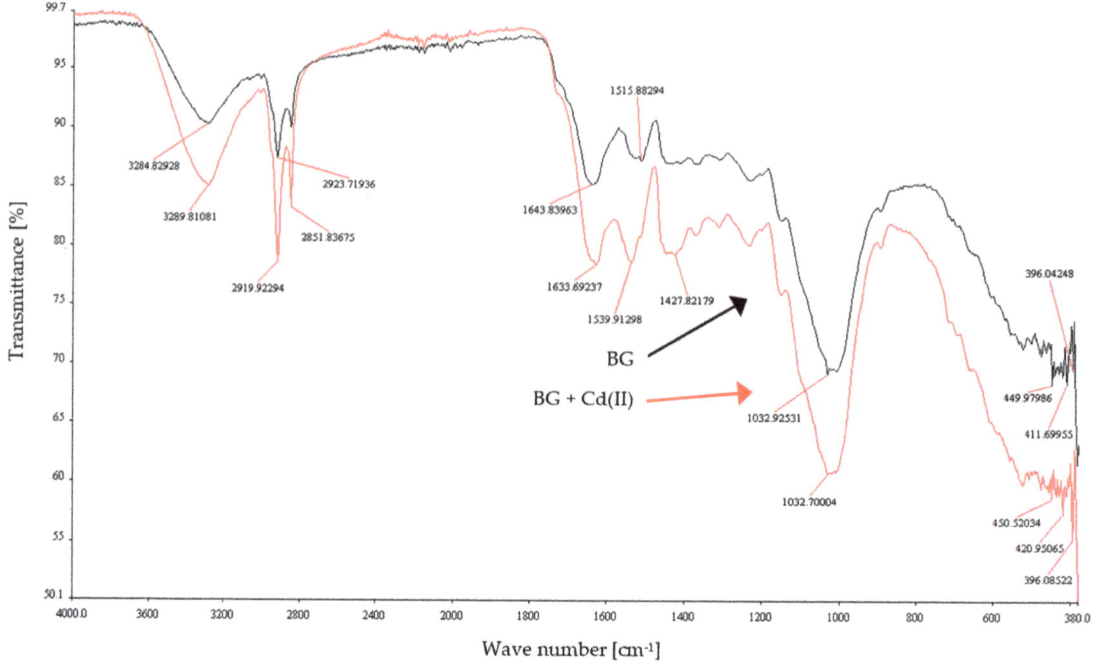

Figure 10. FT-IR spectrum of BG residues before and after Cd(II) ions adsorption.

Zeta potential studies showed that BG particles contained a lot of negative charges on their surface (Figure 8). It is therefore possible to assume that electrostatic attraction could also occur during the biosorption of Cd(II) ions. In these studies, at initial pH 4.0, the BG sorbent had the highest sorption capacity, where zeta potential ranging from −16 to −17 mV was observed. This phenomenon could have been a consequence of the presence of a greater number of localized electrons, which could attract more Cd(II) ions on the BG surface [31].

According to this research, the probable mechanism of Cd(II) biosorption can be explained by ion exchange reactions based on the Equations (7)–(9):

$$Cd^{2+} + BG_m - Ca \rightarrow BG_m - Cd + Ca^{2+} \tag{7}$$

$$Cd^{2+} + BG_m - Mg \rightarrow BG_m - Cd + Mg^{2+} \tag{8}$$

$$Cd^{2+} + BG_m - H + H_2O \rightarrow BG_m - Cd^{+} + H_3O^{+} \tag{9}$$

where: BG_m is the brewer's grains residues matrix.

It should be mentioned that high sorption efficiency of Cd(II) ions at the level of at least 90% was observed in this study under various experimental conditions. Chemical properties of cadmium, which facilitate the attraction of the sorbent to active sites, promote satisfactory results. Additionally, the electronegativity constant (Pauling's) of Cd(II) is 1.69, which is responsible for electrostatic attraction [32]. High sorption capacity of Cd(II) may be also influenced by such factors as first hydrolysis constant pK_H 10.1 (negative logarithm of hydrolysis constant), ionic radius of Cd^{2+} (1.54 Å) and ionization energy (207 kcal/g/mol). These parameters favor easier sorption on the brewer's grains surface [33].

Table 2. FT-IR peaks of BG residues.

FT-IR band (cm^{-1})	Assignment (Vibrations, Species)
3284.8, 3289.8	O–H stretching vibrations (carbohydrates and amino groups in proteins)
2919.9, 2923.7, 2851.8	O–H stretching vibrations (methyl and methylene groups in phospholipids) [34,35]
1643.84, 1633.69,	C = O stretching vibrations of protein bonds (amide band I)
1539.91, 1515.88	C = O stretching vibrations of protein bonds (amide band II)
1427.82	stretching vibrations of carboxyl groups (amide band III) [29]
1032.7, 1032.92	nucleic acid vibrations of phospholipids, phosphodiester groups and carbohydrates [34,35]

3.4. Biosorption Process of Cd(II) Ions

3.4.1. Impact of Adsorbent Dosage

The impact of BG dosage on biosorption efficiency of Cd(II) ions was investigated and Figure 11 shows the results. The ffollowing conditions were used during the research: initial concentration of Cd(II) ions 10.0 mg/L, pH 2–5, agitation speed 200 rpm, T = 23 ± 1 °C, contact time 60 min. In general, increasing sorbent dosage up to 50 g/L increased the biosorption efficiency. The results clearly show that maximum sorption was reported at a dose of 10 g/L (93.9%, initial pH 4) and the lowest efficiency was reported at the initial pH 2. Further increasing the BG dose was not necessary as no further improvement was observed, and optimization of the process was achieved. Besides, biosorption capacity decreased from 8.4 mg/g (dose 1 g/L, pH 5) to 0.04 mg/g (dose 50 g/L, pH 2) (Figure 11B). The parameter was the highest at the smallest doses (1 g/L), while Cd(II) removal (%) was the lowest, which means that the saturation value was reached. This phenomenon can be attributed to the increased surface area of the BG biosorbent, availability of more sorption sites, and it may result from completed interactions of some factors. The decrease in biosorption capacity in the subsequent reactions can be attributed to aggregation or overlapping of active sites, resulting in a reduction in the total surface area of the biosorbent. The results obtained in this way can be explained by the fact that biosorption capacity depends on both the concentration of adsorbed Cd(II) ions and biomass dosage [36,37].

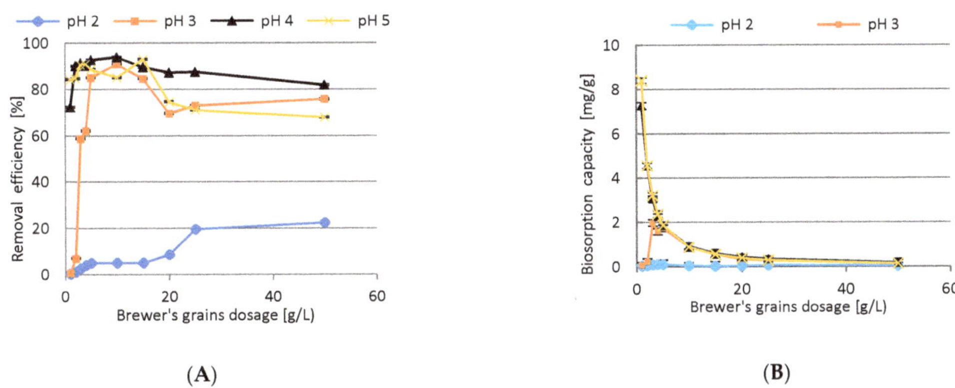

Figure 11. The impact of BG dosage on biosorption efficiency (**A**) and biosorption capacity (**B**) of Cd(II) ions.

3.4.2. Impact of Initial Concentration of Cd(II)

Biosorption behavior under different initial concentration of Cd(II) ions was studied and the results are presented in Figure 12. After analyzing previous research results, it was decided to apply the following experimental conditions: initial concentration of Cd(II) ions (2.5–50 mg/L), adsorbent dosages 1–50 g/L, initial pH 4, contact time 60 min, rotation speed 200 rpm, T = 23 ± 1 °C. An increase in adsorption efficiency was observed in the case of initial concentrations of 1 mg/L and 50 mg/L. In other cases, there was no upward or downward trend, and the results remained stable. The highest removal equal to 93.43% was achieved for the biomass dosage of 10 g/L and initial concentration of 10 mg/L.

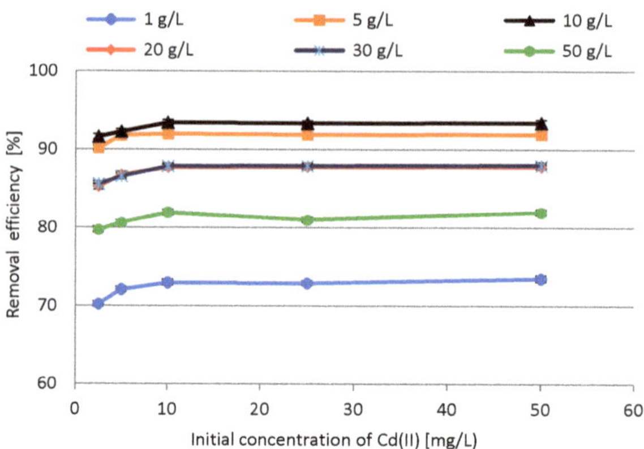

Figure 12. The impact of initial concentration of Cd(II) ions on BG residues removal efficiency.

3.4.3. Analysis of pH Profile

The influence of initial pH on the efficiency of the process was investigated and the results are presented in Figure 13. The applied conditions of the experiments were as follows: initial concentration of Cd(II) ions 10.0 mg/L, BG dosage 1.0–50 g/L, pH range of 2–5, contact time 60 min, rotation speed 200 rpm, T = 23 ± 1 °C. After analysis of the results, it was found that the best performance was obtained at pH 4.0. Maximum sorption was observed for the following biosorbent doses: 1 g/L (84.2%, initial pH 5), 2 g/L (89.99%, pH 4), 3 g/L (91.35%, pH 4), 4 g/L (90.86%, pH 5), 5 g/L (92.62%, pH 4), 10 g/L (93.87%, pH 4), 15 g/L (93%, pH 5), 20 g/L (87.2%, pH 4), 25 g/L (87.47%, pH 4), 50 g/L (81.82%, pH 4). In all cases an increase at pH 2 and 3, and a decrease in adsorption at pH 5 were demonstrated. At pH 2 the formation of cadmium chloride probably occurred, which may make adsorption more difficult. The adsorbent dosage of 1.0 g/L revealed the lowest efficiency. As seen in Figure 13B, the highest biosorption capacity was observed as follows: 8.4 mg/g (1 g/L sorbent dosage, pH 5), 4.56 mg/g (2 g/L, pH 5), 3.22 mg/g (3 g/L, pH 5), 2.37 mg/g (4 g/L, pH 5), 1.87 mg/g (5 g/L, pH 5), 0.93 mg/g (10 g/L, pH 4), 0.59 mg/g (15 g/L, pH 4), 0.43 mg/g (20 g/L, pH 4), 0.35 mg/g (25 g/L, pH 4), 0.16 mg/g (50 g/L, pH 4). On the basis of the obtained results, the relationship between equilibrium pH and the highest biosorption efficiency obtained at initial pH 4.0 was investigated. It was shown that equilibrium pH increased to pH 6.4–7.3. Hence, a change in pH after the sorption processes in the examined systems was noticeable. To conclude, the highest adsorption efficiency was observed with the biosorbent dose of 10 g/L at initial pH 4.

The cation exchange mechanism took place during binding of Cd^{2+} ions in an aqueous solution. The surface of brewer's grains and functional groups could be protonated with a large number of H^+ ions. As a consequence, the number of negatively charged active centers decreased, and the number of positively charged centers increased. In an aqueous solution, Cd^{2+} ions compete with H^+ ions which, through electrostatic repulsion, interfere with adsorption of positively charged Cd^{2+} ions. As the pH increased to 3–4, the surface of the brewer's grains became more negatively electrostatically charged. There was a deprotonation of acid groups and initiation of ion exchange by increasing electrostatic affinity, which influenced the binding of more Cd^{2+} ions. Based on the literature, in the pH range of 2–5, Cd exists in the ionic form, while at a higher alkaline pH, Cd precipitates as hydroxides. Therefore, in these studies, the maximum sorption efficiency was obtained at pH 4. At pH 5, a decrease in sorption was observed, which could result from the competition of hydroxyl ions in active sites. Other forms of cadmium, such as $Cd(OH)^+$ and

$Cd(OH)_2$, can interfere and slow down the sorption process. The influence of many factors (including pH) on Cd(II) sorption mechanism has been presented in the literature [38].

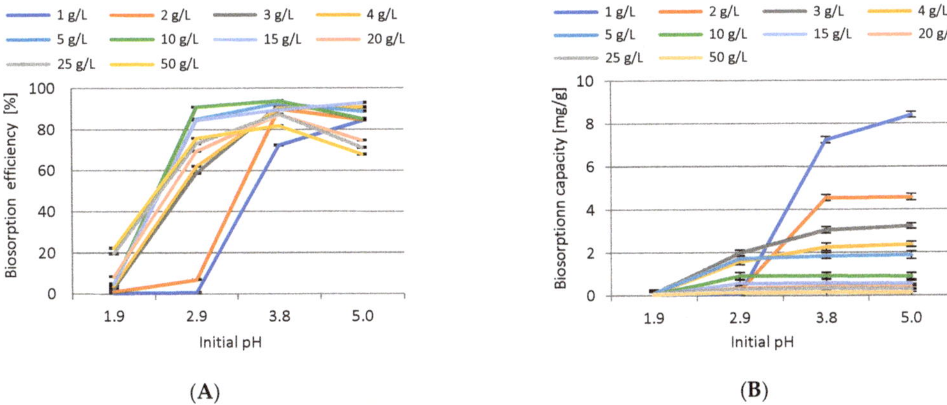

(A) **(B)**

Figure 13. The impact of initial pH on removal efficiency (**A**) and biosorption capacity (**B**) of Cd(II) ions.

3.4.4. Studies of Contact Time

The influence of contact time on biosorption was investigated, and the results are presented in Figure 14. Our previous research helped to establish the following optimal experimental conditions: initial concentration of Cd(II) 10 mg/L, initial pH 4.0, BG dosage 10 g/L, rotation speed 200 rpm, T = 23 ± 1 °C. Based on the results it was shown that the best biosorption efficiency was noticed within the first 5–10 min, and no changes up to 60 min were observed. Thus, there was no need to carry out further experiments on longer times of the process. After 10 min. the sorption efficiency was 93.34% and remained at this level.

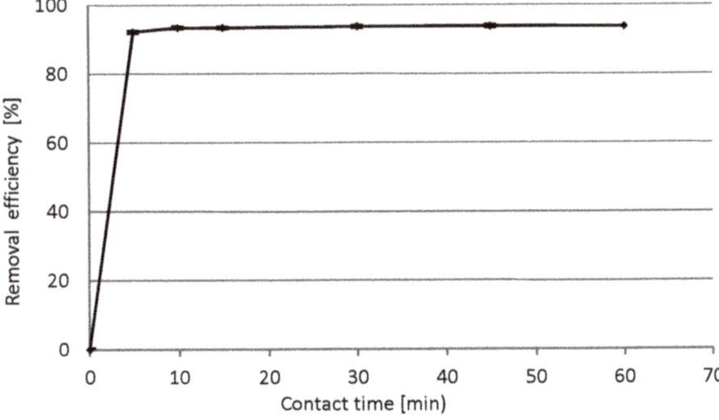

Figure 14. The impact of contact time on the Cd(II) ions biosorption efficiency.

3.4.5. Pseudo-First-Order and Pseudo-Second-Order Kinetic Models

The kinetics analysis of the biosorption of Cd(II) ions onto BG residues was performed. Pseudo-first-order and pseudo-second order models were used for calculations. The model parameters were estimated and the results are shown in Table 3 and Figures S2 and S3 (Supplementary Material). The biosorption process was very quick, and equilibrium was achieved after 5–10 min. In the case of the pseudo-second reaction model the correlation coefficient was R^2 = 0.999, which means that the biosorption process better fits the kinetic model indicating possible ion exchange and sharing of electrons between cadmium ions and BG surface [39]. It is assumed that biosorption may be the rate-limiting step in the process. Such parameters as contact time, initial and equilibrium concentration of Cd(II) were experimental variables. As brewer's grain particle sizes increased, K_1 values decreased. During biosorption diffusion probably occurred, and ion exchange appeared in the process. Cd(II) ions could be attracted by the active sites of the BG waste surface, which was related to an increase in coordination number with the surface. The effect was the phenomenon of adhesion to the biosorbent surface and the formation of chemical bonds [40].

Table 3. Parameters of pseudo-first-order and the pseudo-second-order rate equations.

Metal Ion	Adsorbent Dosage (g/L)	Pseudo-First-Order Kinetic Model			Pseudo-Second-Order Kinetic Model		
		k_{ad} (min^{-1})	q_e (mg/g)	R^2	k (g/mg min)	q_e (mg/g)	R^2
Cd(II)	10	0.0249	0.010	0.982	1270.45	0.0614	0.999

3.4.6. Isothermal Studies

The analysis of Langmuir and Freundlich isotherms were performed in relation to biosorption of Cd(II) using the BG biosorbent. The isotherm parameters were calculated and are presented in Table 4 and Figures S4–S13 (supplementary material). Based on the data, it can be assumed that the Freundlich model better describes the biosorption process. Other investigators, such as Khajavian et al. [41], Kaparapu and Prasad [42], and Naeem et al. [43], reported that the biosorption processes fits the Freundlich model better than the Langmuir model. The R^2 coefficient, related to dimension separation factor or the equilibrium parameter, determines whether the biosorption system is favorable or unfavorable. The R_L parameter determines the shape of isotherms in the following way: irreversible (R_L = 0), favorable (0 < R_L < 1), linear (R_L = 1), unfavorable (R_L > 1). In the present study, the obtained data were in the range of 0 < R_L < 1, which translates into favorable biosorption for the removal of Cd(II) ions [40]. The Langmuir equation contains the K_L parameter, which is the energy constant related to biosorbent and the binding energy of a solute as well as biosorption heat. It defines the interaction between adsorbate and sorbent surface, and the higher the values of the K_L parameter, the stronger interaction between them [44]. The Freundlich isotherm is associated with the relationship between the concentration of metal ions dissolved in a liquid at equilibrium (C_e) and the concentration of dissolved ions on the surface of a biosorbent (q_e). The determined isotherm parameters (K_f-biomass sorption capacity index, n-sorption intensity index) indicate the ease of separation of Cd(II) ions in the solution and biosorption on BG. In this research, the highest biosorption efficiency was 93.9% (initial pH 4.0, biosorbent dosage 10 g/L, initial concentration of Cd(II) 10 mg/L, contact time 60 min, agitation speed 200 rpm, T = 23 ± 1 °C). A comparison of biosorption capacity of BG for the removal of Cd(II) ions with different adsorbents is presented in Table 5, where q_{max} of BG is comparable or even greater than that of various adsorbents published in the literature.

Table 4. Langmuir and Freundlich isotherm parameters for biosorption of Cd(II) using BG residues.

Metal Ion	Adsorbent Dosage (g/L)	Langmuir Isotherm			Freundlich Isotherm		
		Calculated q_m (mg/g)	K_L (L/mg)	R^2	K_f (mg/g) $(L/mg)^{(1/n)}$	n	R^2
Cd(II)	1	53.491	0.152	0.946	6.201	1.086	0.982
	5	57.982	0.086	0.997	4.630	1.021	0.999
	10	57.885	0.048	0.992	2.757	0.977	0.997
	30	74.764	0.006	0.997	0.469	0.976	0.999
	50	85.848	0.002	0.998	0.174	0.991	0.999

Table 5. A comparison of biosorption capacity of BG for the removal of Cd(II) ions with different adsorbents in the literature.

Adsorbents	q_{max} (mg/g)	References
Brewer's grains (BG)	85.85	these studies
Cystoseira baccata	77.56	[45]
Water hyacinth	70.30	[46]
Scenedesmus obliquus	68.60	[47]
Dairy manure	54.60	[48]
Eichornia Crassipes	49.84	[49]
Wheat straw	45.00	[50]
Grape husk	29.20	[50]
Rice straw	34.13	[51]
Lemna aequinoctialis	32.98	[52]
Peanut shells	32.00	[53]
Bamboo	24.95	[54]
KOH activated Cassava stem	24.88	[55]
Cassava stem	10.46	[55]
Loquat leaves	29.24	[56]
Loquat ash	21.32	[56]
Gelidium	18.00	[57]
Chelating resin	17.7	[58]
Rice straw	13.89	[59]
Buffalo weed	11.63	[60]
Diatomaceous earth	10.49	[61]
Activated carbon	8.93	[62]
Lentinus edodes	6.45	[63]
Kaolinite	5.32	[64]
Pistia stratiotes	4.16	[65]
Lemna minor	3.71	[65]
Bentonite	4.13	[66]
Modified lignin	3.52	[67]

4. Conclusions

Brewer's grains obtained during processing in the brewing industry were examined for the removal of Cd(II) ions from aqueous solutions. The physicochemical properties of the biosorbent material were determined using several analytical methods. Biosorbent dosage, initial concentration of Cd(II) ions, initial pH and contact time were factors examined in terms of their influence on biosorption. The results revealed that the maximum biosorption efficiency was equal to 93.9% under the following conditions: initial pH 4.0, biosorbent dosage 10 g/L, initial concentration of Cd(II) 10 mg/L, contact time 60 min, rotation speed 200 rpm, T = 23 $\pm$ 1 °C. In general, the average process efficiency was above 80% under different experimental conditions. The possible biosorption mechanism may involve ion exchange and electrostatic attraction. The FT-IR analysis revealed slight shifts of peaks and changes in the intensity of bands, which can indicate Cd(II) ion binding by various functional groups. The kinetics analysis showed that the pseudo-second order kinetic model and the Freundlich model fit the experimental data better.

In conclusion, it is assumed that post-production brewer's grains are capable of removing Cd(II) ions with high efficiency. These results may be an opportunity for industrial use of brewer's grains to remove heavy metals as well as improve water quality, which is in line with current global trends in sustainable development and the circular economy.

Supplementary Materials: The following are available online at https://www.mdpi.com/article/10.3390/en14175543/s1. Figure S1: Particle size distribution of brewer's grains, Figure S2: Pseudo-first-order isotherm for adsorption of Cd(II) ions with brewer's grains, Figure S3: Pseudo-second-order isotherm for adsorption of Cd(II) ions with brewer's grains, Figure S4: Langmuir kinetic isotherm for the biosorption of Cd(II) with brewer's grains (BG dosage 1 g/L), Figure S5: Langmuir kinetic isotherm for the biosorption of Cd(II) with brewer's grains (BG dosage 5 g/L), Figure S6: Langmuir kinetic isotherm for the biosorption of Cd(II) with brewer's grains (BG dosage 10 g/L), Figure S7: Langmuir kinetic isotherm for the biosorption of Cd(II) with brewer's grains (BG dosage 30 g/L), Figure S8: Langmuir kinetic isotherm for the biosorption of Cd(II) with brewer's grains (BG dosage 50 g/L), Figure S9: Freundlich kinetic isotherm for the biosorption of Cd(II) with brewer's grains (BG dosage 1 g/L), Figure S10: Freundlich kinetic isotherm for the biosorption of Cd(II) with brewer's grains (BG dosage 5 g/L), Figure S11: Freundlich kinetic isotherm for the biosorption of Cd(II) with brewer's grains (BG dosage 10 g/L), Figure S12: Freundlich kinetic isotherm for the biosorption of Cd(II) with brewer's grains (BG dosage 30 g/L), Figure S13: Freundlich kinetic isotherm for the biosorption of Cd(II) with brewer's grains (BG dosage 50 g/L).

Author Contributions: Conceptualization, T.K.; methodology, T.K.; validation, T.K.; formal analysis, T.K.; investigation, T.K. and J.W.; resources, T.K. and M.U.; data curation, T.K.; writing—original draft preparation, T.K.; writing—review & editing, T.K.; visualization, T.K.; supervision, T.K.; project administration, T.K.; funding acquisition, M.U. All authors have read and agreed to the published version of the manuscript.

Funding: This research did not receive a specific grant from any a funding agency in the public, commercial or not-for-profit sectors.

Institutional Review Board Statement: Not applicable.

Informed Consent Statement: Not applicable.

Data Availability Statement: Data is contained within the article.

Conflicts of Interest: The authors declare no conflict of interest.

References

1. Rao, K.S.; Mohapatra, M.; Anand, S.; Venkateswarlu, P. Review on cadmium removal from aqueous solutions. *Int. J. Eng. Sci. Technol.* **2010**, *2*, 81–103. [CrossRef]
2. Cheung, C.W.; Porter, J.F.; McKay, G. Elovich equation and modified second- order equation for adsorption of cadmium ions onto bone char. *J. Chem. Technol. Biotechnol.* **2000**, *75*, 963–970. [CrossRef]
3. Wu, J.; Lu, J.; Chen, T.H.; He, Z.; Su, Y.; Yao, X.Y. In situ biotreatment of acidic mine drainage using straw as sole substrate. *Environ. Earth Sci.* **2010**, *60*, 421–429. [CrossRef]
4. Conway, J. Beer Production Worldwide from 1998 to 2019. Available online: https://www.statista.com/statistics/270275/worldwide-beer-production/ (accessed on 20 June 2021).
5. Levinson, J. Malting-brewing: A changing sector. *BIOS Int.* **2002**, *5*, 12–15.
6. Ciancia, S. Micro-brewing: A new challenge for beer. *BIOS Int.* **2000**, *2*, 4–10.
7. de Oliveira Dias, M. Heineken brewing industry in Brazil. *Int. J. Manage. Technol. Eng.* **2018**, *8*, 1304–1310.
8. Fărcaş, A.C.; Socaci, S.A.; Mudura, E.; Dulf, F.V.; Vodnar, D.C.; Tofană, M.; Salanță, L.C. *Exploitation of Brewing Industry Wastes to Produce Functional Ingredients, Brewing Technology*; Chapter 7; IntechOpen: London, UK, 2017; pp. 137–156.
9. Gupta, V.K.; Ali, I. *Environmental Water Advances in Treatment, Remediation and Recycling*; Elsevier: New Delhi, India, 2013; pp. 1–232.
10. Portal, S. Available online: https://www.portalspozywczy.pl/alkohole-uzywki/wiadomosci/produkcja-piwa-wzrosla-w-2018-roku-do-40-93-mln-hektolitrow,167388.html (accessed on 20 June 2021).
11. Związek Pracodawców Przemysłu Piwowarskiego w Polsce. *Najlepsze Dostępne Techniki (BAT), Wytyczne dla Przemysłu Piwowarskiego*; Ministerstwo Środowiska: Warszawa, Poland, 2005.
12. *Regulation of the Polish Minister of Health of 7 December 2017 Amending the Regulation on the Quality of Water Intended for Human Consumption, Item 2294*; Polish J. Laws: Warsaw, Poland, 2017.
13. Czekała, W.; Pawlisiak, A. Produkcja i wykorzystanie wysłodzin browarnianych. *Tech. Rol. Ogrod. Leśna* **2017**, *5*, 23–25.
14. *The Act of 14 December 2012 on Waste, Item. 21, as Amended*; Pol. J. Laws: Warsaw, Poland, 2013.
15. *Regulation of the Minister of Environment of 11 May 2015 on Waste Recovery Outside of Installations and Devices*; No. 796; Pol. J. Laws: Warsaw, Poland, 2015.
16. *Regulation of the Minister of Environment of 21 March 2006 on Waste Recovery Outside of Installations and Devices*; No. 49, item 356; Pol. J. Laws: Warsaw, Poland, 2006.
17. Alquzweeni, S.S.; Alkizwini, R.S. Removal of Cadmium from Contaminated Water Using Coated Chicken Bones with Double-Layer Hydroxide (Mg/Fe-LDH). *Water* **2020**, *12*, 2303. [CrossRef]
18. Nnaji, C.H.; Emefu, S.C.H. Effect of Particle Size on the Sorption of Lead from Water by Different Species of Sawdust: Equilibrium and Kinetic Study. *Bioresources* **2017**, *12*, 4123–4145. [CrossRef]
19. Alinnor, I.J. Adsorption of heavy metal ions from aqueous solution by fly ash. *Fuel* **2007**, *86*, 853–857. [CrossRef]

20. Cui, F.M.; Zhang, X.Y.; Shang, L.M. Thermogravimetric Analysis of Biomass Pyrolysis under Different Atmospheres. *Appl. Mech. Mater.* **2013**, *448–453*, 1616–1619. [CrossRef]
21. Kalak, T.; Dudczak-Halabuda, J.; Tachibana, Y.; Cierpiszewski, R. Residual biomass of gooseberry (*Ribes uva-crispa* L.) for the bioremoval process of Fe(III) ions. *Desalin. Water Treat.* **2020**, *202*, 345–354. [CrossRef]
22. Erdemoglu, M.; Sarikaya, M. Effects of heavy metals and oxalate on the zeta potential of magnetite. *J. Colloid Interface Sci.* **2006**, *300*, 795–804. [CrossRef] [PubMed]
23. Demirbas, Ö.; Alkan, M.; Doğan, M.; Turhan, Y.; Namli, H.; Turan, P. Electrokinetic and adsorption properties of sepiolite modified by 3-aminopropyltriethoxysilane. *J. Hazard. Mater.* **2007**, *149*, 650–656. [CrossRef]
24. Kalak, T.; Kłopotek, A.; Cierpiszewski, R. Effective adsorption of lead ions using fly ash obtained in the novel circulating fluidized bed combustion technology. *Microchem. J.* **2019**, *145*, 1011–1025.
25. Zdanowska, P.; Florczak, I.; Słoma, J.; Tucki, K.; Orynycz, O.; Wasiak, A.; Świć, A. An Evaluation of the Quality and Microstructure of Biodegradable Composites as Contribution towards Better Management of Food Industry Wastes. *Sustainability* **2019**, *11*, 1504. [CrossRef]
26. Olkku, J.; Kotaviita, E.; Salmenkallio-Marttila, M.; Sweins, H.; Home, S. Connection between structure and quality of barley husk. *J. Am. Soc. Brew. Chem.* **2005**, *63*, 17–22. [CrossRef]
27. Boateng, A.A.; Cooke, P.H.; Hicks, K.B. Microstructure development of chars derived from high-temperature pyrolysis of barley (*Hordeum vulgare* L.) hulls. *Fuel* **2007**, *86*, 735–742. [CrossRef]
28. Ferraz, E.; Coroado, J.; Gamelas, J.; Silva, J.; Rocha, F.; Velosa, A. Spent Brewery Grains for Improvement of Thermal Insulation of Ceramic Bricks. *J. Mater. Civ. Eng.* **2013**, *25*, 1638–1646. [CrossRef]
29. Rubio, F.T.V.; Maciel, G.M.; Silva, M.V.; Correa, V.G.; Peralta, R.M.; Haminiuk, C.W.I. Enrichment of waste yeast with bioactive compounds from grape pomace as an innovative and emerging technology: Kinetics, isotherms and bioaccessibility. *Innov. Food Sci. Emerg. Technol.* **2018**, *45*, 18–28. [CrossRef]
30. Tao, Y.; Han, Y.; Liu, W.; Peng, L.; Wang, Y.; Kadam, S.; Show, P.L.; Ye, X. Parametric and phenomenological studies about ultrasound-enhanced biosorption of phenolics from fruit pomace extract by waste yeast. *Ultrason. Sonochem.* **2019**, *52*, 193–204. [CrossRef] [PubMed]
31. Nasution, A.N.; Amrina, Y.; Zein, R.; Aziz, H.; Munaf, E. Biosorption characteristics of Cd(II) ions using herbal plant of mahkota dewa (*Phaleria macrocarpa*). *J. Chem. Pharm. Res.* **2015**, *7*, 189–196.
32. Shaheen, S.M.; Derbalah, A.S.; Moghanm, F.S. Removal of Heavy Metals from Aqueous Solution by Zeolite in Competitive Sorption System. *Int. J. Environ. Sci. Develop.* **2012**, *3*, 362–367. [CrossRef]
33. Sulaymon, A.H.; Mohammed, A.A.; Al-Musawi, T.J. Competitive biosorption of lead, cadmium, copper, and arsenic ions using algae. *Environ. Sci. Pollut. Res.* **2013**, *20*, 3011–3023. [CrossRef] [PubMed]
34. Stafussa, A.P.; Maciel, G.M.; Anthero, A.G.S.; Silva, M.V.; Zielinski, A.A.F.; Haminiuk, C.W.I. Biosorption of anthocyanins from grape pomace extracts by waste yeast: Kinetic and isotherm studies. *J. Food Eng.* **2016**, *169*, 53–60. [CrossRef]
35. Kim, T.-Y.; Lee, J.-W.; Cho, S.-Y. Application of residual brewery yeast for adsorption removal of Reactive Orange 16 from aqueous solution. *Adv. Powder Technol.* **2015**, *26*, 267–274. [CrossRef]
36. Kalak, T.; Dudczak, J.; Cierpiszewski, R. Adsorption behaviour of copper ions on elderberry, gooseberry and paprika waste from aqueous solutions. In Proceedings of the 12th International Interdisciplinary Meeting on Bioanalysis (CECE), Brno, Czech Republic, 21–23 September 2015; pp. 123–127.
37. Kalak, T.; Dudczak-Hałabuda, J.; Tachibana, Y.; Cierpiszewski, R. Effective use of elderberry (Sambucus nigra) pomace in biosorption processes of Fe(III) ions. *Chemosphere* **2020**, *246*, 125744. [CrossRef]
38. Ahmad, I.; Akhtar, M.J.; Jadoon, I.B.K.; Imran, M.; Ali, S. Equilibrium modeling of cadmium biosorption from aqueous solution by compost. *Environ. Sci. Pollut. Res.* **2017**, *24*, 5277–5284. [CrossRef] [PubMed]
39. Usman, A.; Sallam, A.; Zhang, M.; Vithanage, M.; Ahmad, M.; Al-Farraj, A.; Ok, Y.S.; Abduljabbar, A.; Al-Wabel, M. Sorption process of date palm biochar for aqueous Cd (II) removal: Efficiency and mechanisms. *Water Air Soil Pollut.* **2016**, *227*, 1–16. [CrossRef]
40. Ho, Y.S.; Ng, J.C.Y.; McKay, G. Kinetics of pollutant sorption by biosorbents: Review. *Sep. Purif. Rev.* **2000**, *29*, 189–232. [CrossRef]
41. Khajavian, M.; Wood, D.A.; Hallajsani, A.; Nasrollah, M. Simultaneous biosorption of nickel and cadmium by the brown algae *Cystoseria indica* characterized by isotherm and kinetic models. *Appl. Biol. Chem.* **2019**, *62*, 69. [CrossRef]
42. Kaparapu, J.; Prasad, M.K. Equilibrium, kinetics and thermodynamic studies of cadmium(II) biosorption on Nannochloropsis oculata. *Appl. Water Sci.* **2018**, *8*, 1–9. [CrossRef]
43. Naeem, M.A.; Imran, M.; Amjad, M.; Abbas, G.; Tahir, M.; Murtaza, B.; Zakir, A.; Shahid, M.; Bulgariu, L.; Ahmad, I. Batch and Column Scale Removal of Cadmium from Water Using Raw and Acid Activated Wheat Straw Biochar. *Water* **2019**, *11*, 1438. [CrossRef]
44. Asuquo, E.; Martin, A.; Nzerem, P.; Siperstein, F.; Fan, X. Adsorption of Cd(II) and Pb(II) ions from aqueous solutions using mesoporous activated carbon adsorbent: Equilibrium, kinetics and characterisation studies. *J. Environ. Chem. Eng.* **2017**, *5*, 679–698. [CrossRef]
45. Lodeiro, P.; Barriada, J.L.; Herrero, R.; Vicente, M.E.S. The marine macroalga Cystoseira baccata as biosorbent for cadmium(II) and lead(II) removal: Kinetic and equilibrium studies. *Environ. Pollut.* **2006**, *142*, 264–273. [CrossRef]

46. Zhang, F.; Wang, X.; Yin, D.; Peng, B.; Tan, C.; Liu, Y.; Tan, X.; Wu, S. Efficiency and mechanisms of Cd removal from aqueous solution by biochar derived from water hyacinth (*Eichornia crassipes*). *J. Environ. Manag.* **2015**, *153*, 68–73. [CrossRef]
47. Chen, C.Y.; Chang, H.W.; Kao, P.C.; Pan, J.L.; Chang, J.S. Biosorption of cadmium by CO2-fixing microalga Scenedesmus obliquus CNW-N. *Bioresour. Technol.* **2012**, *105*, 74–80. [CrossRef] [PubMed]
48. Xu, X.; Cao, X.; Zhao, L.; Wang, H.; Yu, H.; Gao, B. Removal of Cu, Zn, and Cd from aqueous solutions by the dairy manure-derived biochar. *Environ. Sci. Pollut. Res.* **2013**, *20*, 358–368. [CrossRef]
49. Li, F.; Shen, K.; Long, X.; Wen, J.; Xie, X.; Zeng, X.; Liang, Y.; Wei, Y.; Lin, Z.; Huang, W.; et al. Preparation and Characterization of Biochars from Eichornia crassipes for Cadmium Removal in Aqueous Solutions. *PLoS ONE* **2016**, *11*, 0148132. [CrossRef] [PubMed]
50. Trakal, L.; Bingöl, D.; Poho ˇrelý, M.; Hruška, M.; Komárek, M. Geochemical and spectroscopic investigations of Cd and Pb sorption mechanisms on contrasting biochars: Engineering implications. *Bioresour. Technol.* **2014**, *171*, 442–451. [CrossRef] [PubMed]
51. Han, X.; Liang, C.-F.; Li, T.-Q.; Wang, K.; Huang, H.-G.; Yang, X.-E. Simultaneous removal of cadmium and sulfamethoxazole from aqueous solution by rice straw biochar. *J. Zhejiang Univ. Sci. B* **2013**, *14*, 640–649. [CrossRef]
52. Chen, L.; Fang, Y.; Jin, Y.; Chen, Q.; Zhao, Y.; Xiao, Y.; Zhao, H. Biosorption of Cd^{2+} by untreated dried powder of duckweed *Lemna aequinoctialis*. *Des. Water Treat.* **2015**, *53*, 183–194. [CrossRef]
53. Ahmad, M.; Lee, S.S.; Dou, X.; Mohan, D.; Sung, J.-K.; Yang, J.E.; Ok, Y.S. Effects of pyrolysis temperature on soybean stover-and peanut shell-derived biochar properties and TCE adsorption in water. *Bioresour. Technol.* **2012**, *118*, 536–544. [CrossRef] [PubMed]
54. Zhang, S.; Zhang, H.; Cai, J.; Zhang, X.; Zhang, J.; Shao, J. Evaluation and Prediction of Cadmium Removal from Aqueous Solution by Phosphate-Modified Activated Bamboo Biochar. *Energy Fuels* **2017**, *32*, 4469–4477. [CrossRef]
55. Prapagdee, S.; Piyatiratitivorakul, S.; Petsom, A. Activation of Cassava Stem Biochar by Physico-Chemical Method for Stimulating Cadmium Removal Efficiency from Aqueous Solution. *Environ. Asia* **2014**, *7*, 60–69.
56. Al-Dujaili, A.H.; Awwad, A.M.; Salem, N.M. Biosorption of cadmium (II) onto loquat leaves (*Eriobotrya japonica*) and their ash from aqueous solution, equilibrium, kinetics, and thermodynamic studies. *Int. J. Ind. Chem.* **2012**, *3*, 1–7. [CrossRef]
57. Vilar, V.J.P.; Botelho, C.M.S.; Boaventura, R.A.R. Equilibrium and kinetic modelling of Cd(II) biosorption by algae Gelidium and agar extraction algal waste. *Water Res.* **2006**, *40*, 291–302. [CrossRef] [PubMed]
58. Baraka, A.; Hall, P.J.; Heslop, M.J. Preparation and characterization of melamine-formaldehyde-DTPA chelating resin and its use as an adsorbent for heavy metal removal from wastewater. *React. Funct. Polym.* **2007**, *67*, 585–600. [CrossRef]
59. Ding, Y.; Jing, D.B.; Gong, H.L.; Zhou, L.B.; Yang, X.S. Biosorption of aquatic cadmium(II) by unmodified rice straw. *Bioresour. Technol.* **2012**, *114*, 20–25. [CrossRef] [PubMed]
60. Yakkala, K.; Yu, M.-R.; Roh, H.; Yang, J.-K.; Chang, Y.-Y. Buffalo weed (*Ambrosia trifida L. var. trifida*) biochar for cadmium (II) and lead (II) adsorption in single and mixed system. *Des. Water Treat* **2013**, *51*, 7732–7745. [CrossRef]
61. Al-karam, U.F.; Danil de Namor, A.F.; Derwish, G.A.W.; Al-Dujaili, A.H. Removal of chromium, copper, cadmium and lead ions from aqueous solutions by diatomaceous earth. *Environ. Eng. Manag. J.* **2012**, *11*, 811–819.
62. Marzal, P.; Seco, A.; Gabaldon, C.; Ferrer, J. Cadmium and zinc adsorption onto activated carbon: Influence of temperature, pH and metal/carbon ratio. *J. Chem. Technol. Biotechnol.* **1996**, *66*, 279–285. [CrossRef]
63. Zhang, D.; Zeng, X.D.; Ma, P.; He, H.J. The sorption of Cd(II) from aqueous solutions by fixed Lentinus edodes mushroom flesh particles. *Desalin. Water Treat.* **2012**, *46*, 21–31. [CrossRef]
64. Bhattacharyya, K.; Gupta, S. Kaolinite, montmorillonite, and their modified derivatives as adsorbents for removal of Cd(II) from aqueous solution. *Sep. Purif. Technol.* **2006**, *50*, 388–397. [CrossRef]
65. Miretzky, P.; Saralegui, A.; Cirelli, A.F. Simultaneous heavy metal removal mechanism by dead macrophytes. *Chemosphere* **2006**, *62*, 247–254. [CrossRef] [PubMed]
66. Lacin, O.; Bayrak, B.; Korkut, O.; Sayan, E. Modeling of adsorption and ultrasonic desorption of cadmium(II) and zinc(II) on local bentonite. *J. Colloid Interface Sci.* **2005**, *292*, 330–335. [CrossRef] [PubMed]
67. Demirbas, A. Adsorption of lead and cadmium ions in aqueous solutions onto modified lignin from alkali glycerol delignication. *J. Hazard. Mater.* **2004**, *B109*, 221–226. [CrossRef] [PubMed]

 *energies*

MDPI

Review

Odour Nuisance at Municipal Waste Biogas Plants and the Effect of Feedstock Modification on the Circular Economy—A Review

Marta Wiśniewska *, Andrzej Kulig and Krystyna Lelicińska-Serafin

Faculty of Building Services, Hydro and Environmental Engineering, Warsaw University of Technology, 20 Nowowiejska Street, 00-653 Warsaw, Poland; Andrzej.Kulig@pw.edu.pl (A.K.); Krystyna.Lelicinska@pw.edu.pl (K.L.-S.)
* Correspondence: Marta.Wisniewska@pw.edu.pl

Abstract: The increase in the amount of municipal solid waste (MSW) generated, among other places, in households is a result of the growing population, economic development, as well as the urbanisation of areas with accompanying insufficiently effective measures to minimise waste generation. There are many methods for treating municipal waste, with the common goal of minimising environmental degradation and maximising resource recovery. Biodegradable waste, including selectively collected biowaste (BW), also plays an essential role in the concept of the circular economy (CE), which maximises the proportion of waste that can be returned to the system through organic recycling and energy recovery. Methane fermentation is a waste treatment process that is an excellent fit for the CE, both technically, economically, and environmentally. This study aims to analyse and evaluate the problem of odour nuisance in municipal waste biogas plants (MWBPs) and the impact of the feedstock (organic fraction of MSW-OFMSW and BW) on this nuisance in the context of CE assumptions. A literature review on the subject was carried out, including the results of our own studies, showing the odour nuisance and emissions from MWBPs processing both mixed MSW and selectively collected BW. The odour nuisance of MWBPs varies greatly. Odour problems should be considered regarding particular stages of the technological line. They are especially seen at the stages of waste storage, fermentation preparation, and digestate dewatering. At examined Polish MWBPs c_{od} ranged from 4 to 78 ou/m^3 for fermentation preparation and from 8 to 448 ou/m^3 for digestate dewatering. The conclusions drawn from the literature review indicate both the difficulties and benefits that can be expected with the change in the operation of MWBPs because of the implementation of CE principles.

Keywords: anaerobic digestion; biogas plant; biowaste; circular economy; feedstock; municipal waste; odorants; odour nuisance; organic fraction from municipal solid waste

Citation: Wiśniewska, M.; Kulig, A.; Lelicińska-Serafin, K. Odour Nuisance at Municipal Waste Biogas Plants and the Effect of Feedstock Modification on the Circular Economy—A Review. *Energies* **2021**, *14*, 6470. https://doi.org/10.3390/en14206470

Academic Editor: Gabriele Di Giacomo

Received: 10 September 2021
Accepted: 8 October 2021
Published: 10 October 2021

1. Introduction

The generation of municipal solid waste (MSW), mainly from households and catering and other places where the waste of a similar morphology is generated, seems to be an inherent element of intensive urban development. Minimising waste generation is the first and most important element of the waste hierarchy and an element of the CE. The CE concept is based on a "take-use-reuse" approach that consists of closing the cycle of the extended product life cycle and treating waste as valuable recyclable materials. The CE involves minimising the negative impact of the production line on the environment [1,2]. The CE has many definitions, but it is most often defined according to Ellen MacArthur Foundation [3]. The CE is a systemic approach to economic development designed to benefit businesses, society, and the environment. In contrast to the "take-make-waste" linear model, a circular economy is regenerative by design and aims to decouple growth from the consumption of finite resources gradually. An increase in the amount of waste requiring further management results from the growing population, economic development, and the urbanisation of areas with accompanying insufficiently effective waste minimisation activities [4,5]. The problems associated with the MSW economy, however widely varied, affect all countries in the world. Improperly managed waste management is reflected in the form of soil degradation, water bodies (including the organisms living in them), atmospheric pollution, and negative impacts on human health [6,7]. Legislation at the European level [8,9] indicates numerous requirements necessary to reduce the amount of waste deposited in landfills, to maximise the use of generated waste as raw materials, but above all to minimise its generation, especially that of food waste, which is in line with the Agenda for Sustainable Development 2030 adopted by the United Nations (UN) General Assembly [10]. Waste prevention is the most crucial part of the waste hierarchy [11].

There are many methods of municipal waste treatment, which can generally be divided into mechanical, biological, and thermal methods [12–14]. All these processes have one common goal: to minimise environmental degradation and maximise resource recovery [15–19]. They should be used according to the waste hierarchy [7]. An example of a biological method is organic recycling, which is limited to selectively collected BW. The main advantage of this process is the possibility of producing non-waste organic fertiliser. Mechanical methods are applied primarily to sort waste materials and prepare them for raw material recycling. Mechanical–biological methods are used mainly for mixed MSW as a disposal method. After this process, waste (so-called stabilised waste) remains and requires further processing (landfilling or thermal treatment). Waste incineration should only be used for non-recyclable, combustible waste in the form of energy recovery (waste combustion without energy recovery is against the principle of the CE) [4,11–13,20]. In the strategy for handling biodegradable waste, biological processes play a dominant role [21].

There is excellent potential in wastes undergoing biological decomposition (both under aerobic and anaerobic conditions), including biowaste, for example, food waste [9]. Studies by Das et al. [22] and Slorach et al. [23] show that a lower content of biodegradable fraction characterises municipal waste generated in highly developed countries than in medium- and low-developed countries.

There has been a growing interest in biodegradable waste due to obtaining energy from it in recent years. One of the main reasons for this increase is the change in the European Union (EU) energy policy (as of 2023, separate collection of biowaste will be obligatory for EU member states), which is dictated by the increasing demand for energy but also by the growing greenhouse effect due to the emission of greenhouse gases into the atmosphere (including methane and carbon dioxide), which is caused by, among other reasons, the use of fossil fuels, on which about 88% of the produced energy is based [24,25]. Biodegradable waste, including BW, also plays an essential role in the CE concept, which involves maximising the proportion of waste that can be returned to the system after organic recycling and energy recovery [26–30]. The origin of BW determines its composition. The amount of it is increasing every year, causing problems in regard to its disposal and management [31–37]. In the case of MSW, the organic fraction may be separated mechanically from the mixed waste stream (using a system of separators and screens connected by conveyors) or selectively collected at the source of waste generation [38–40]. Previous analyses in various countries show that the organic fraction contained in municipal solid waste (OFMSW) constitutes about 30–75% of the total content [41–45]. This content is mainly determined by factors such as the geographical location of the region, the degree of industrialisation, the socio-economic situation, lifestyle, education, or aspects of families (number and age of family members) [46,47]. Waste-to-energy technologies (WETs) are the basis for managing organic waste and converting it into valuable fuels, fertilisers, and electricity [48,49]. According to the 2030 Agenda for Sustainable Development with its Sustainable Development Goals (SDGs) [10] there is a big concern for the steady supply of affordable, renewable and clean energy sources, so solid waste is great hope among them. WETs are a crucial issue of a waste management system. Incineration dominates the WET market all over the world, and specifically in developed countries. After thermal processes, anaerobic digestion is the emerging technology in clean energy production. Reduction in greenhouse gas emissions and the generation of alternatives to fossil fuels are major goals of WETs. Among the research trends, there are also studies that focus on environmental impacts, energy technology innovations, improved energy recovery efficiency, and climate change impacts. They are supposed to contribute to the development of a low-carbon society. According to CE ideas and bioeconomy concepts, countries with the most advanced WETs should always encourage recycling and stricter policies for waste reduction [50,51]. Reports have shown that Poland is among the top countries in Europe when it comes to bioeconomies. Among the Visegrad group, Poland is leading the way with bio-based fuel, bioenergy, biomass processing and conversion, and other bio industries such as biorefinery, biochemicals, and biopharmaceuticals, whereas others have made some progress in the agro-food sector [52,53].

There are many different WETs for biodegradable fractions: thermochemical methods such as incineration, gasification, and pyrolysis, biochemical methods such as anaerobic digestion, and chemical methods such as esterification [54,55]. Aerobic methods of biological treatment also play an important role in both composting (for selectively collected waste) and aerobic stabilisation (for the organic fraction separated mechanically from the stream of mixed waste) [56–59]. Factors that determine the suitability of different types of waste for particular processes include moisture content, organic matter content, C/N ratio, calorific value (CV), and the content of non-flammable fractions [60]. When analysing the rate of organic matter degradation, factors such as the initial microbial community, oxygen availability, the physical availability to degrade, temperature, and the chemical composition of organic matter play an important role [58,61].

The most promising processing technology for OFMSW and BW is methane fermentation [62–64]. The first biogas stations were built in wastewater treatment plants in order to stabilise the sewage sludge. They were equipped with open digestate storage tanks with free digestate surfaces, without covers. In that case, limited mixing of digestate was recommended to allow a solid crust to form on the surface to limit odour emissions [65]. Many rural biogas plants in the world, especially operated by small and medium farmers, run under psychrophilic conditions—so-called psychrophilic rural digesters [66]. However, according to Environmental Protection Agency (EPA) guidelines, conventional AD biogas systems are commonly designed to operate in either the mesophilic or thermophilic temperature range [67]. Some biogas plants use open digester chambers, which is a phenomenon on the European scale. They are mainly used in sewage treatment plants. The feedstock in sewage sludge is kept in the chamber for up to 6 months under psychrophilic conditions (<20 °C), which contributes to the high mineralisation of the material. The disadvantage of this solution is the emission of gases, including odorants [68]. Thi et al. [69], in a comparative analysis of different BW processing technologies, indicated that AD is a suitable solution for developing countries with temperate climates. Among the possibilities of biogas (the main product of this process) utilisation are, e.g., heat and electricity production, use as vehicle fuel, and injection into the gas grid (after upgrading) [70]. Research conducted by Swedish scientists indicates that 1 Mg of food waste can be generated into 1200 kWh of energy, which is enough to drive 1900 km in a gas-powered car. In turn, the energy obtained from the digestion of food waste generated by 3000 households is enough to cover the annual fuel requirements of one gas-powered bus [71]. The method that fits perfectly into CE assumptions is biogas-to-biomethane upgrading. To obtain high-quality biomethane via upgrading biogas from waste anaerobic digestion there are such techniques as membrane separation, water scrubbing, chemical absorption with amine solvent, and pressure swing adsorption [72–75]. In Polish municipal waste biogas plants, biogas is most often used in cogeneration systems to produce electricity and heat, often for their own needs. In the case of surplus or insufficient parameters, biogas is burnt in flares. EPA guidelines require flares to be enclosed and operated at a minimum temperature of 1000 °C with a 0.3 s retention time to ensure adequate destruction and minimisation of emissions [67]. Other compounds, such as aliphatic and aromatic hydrocarbons, siloxanes, volatile organosulfur and organohalogen compounds, and oxygenated organic compounds, are present in small amounts in biogas [76,77]. A modern method of biogas purification is the adsorptive packed column system. Piechota presented the results of studies on the effectiveness of this method on biogas from a wastewater treatment plant [78]. The author proved the removal efficiency of 99.76% for total non-silica impurities and 100% for siloxanes. This study aims to analyse and evaluate the problem of odour nuisance of MWBPs and the impact of the input on this nuisance in the context of CE assumptions. A literature review on the subject was carried out, including the results of our own research, showing the odour nuisance and emissions from MWBPs processing both mixed MSW and selectively collected BW. The conclusions drawn from the literature review indicate both the difficulties and benefits that can be expected with the change in the operation of MWBPs due to the implementation of CE principles.

2. Municipal Waste Biogas Plants (MWBPs)

2.1. Anaerobic Digestion Process

Anaerobic stabilisation is the anaerobic decomposition of organic matter, a two-step biological process involving the conversion and stabilisation of waste [79–81]. Methane fermentation, which plays a significant role in the anaerobic stabilisation of waste, occurs with the participation of microorganisms, whose main gaseous products of decomposition of organic compounds are methane and carbon dioxide. The fermentation process consists of four steps: hydrolysis, acidogenesis, acetogenesis, and methanogenesis [34].

The course of the fermentation process is influenced by many factors, including the type of waste processed, which has a very significant impact on the final post-fermentation product, and many other parameters, including environmental parameters. One of them is the temperature at which the process is carried out (psychrophilic, <20 °C; mesophilic, 25 ÷ 40 °C; and thermophilic, 45 ÷ 60 °C) [82,83]. This parameter mainly affects the physicochemical properties of the treated wastes, which are essential for thermodynamic reactions and kinetic biological processes [84]. A higher process temperature has a beneficial effect on the hydrolysis of soluble substances, making them more accessible to microorganisms, increasing the kinetics of chemical and biological processes, which in turn causes a reduction in the hydraulic retention time (HRT) of the digester reactor, and improves the physicochemical properties of soluble substances as well as diffusivity [85,86]. Another benefit of using a high process temperature is the elimination of pathogenic bacteria in the fermented material [87–89].

Another critical environmental parameter for fermentation is pH, which depends on the activity of bacteria in particular stages of fermentation. In the case of methanogenic bacteria, the optimum pH level is within the range of 6.5 ÷ 7.2. A drop in pH below 6.5 rapidly inhibits the process, which stops the removal of acids from the treated raw material. Other types of bacteria are less demanding and show their activity in a broader pH range: 4.0 ÷ 8.5 [89,90]. The pH level also determines the end products of the process at low pH; acetic acid (CH_3COOH) and butyric acid ($CH_3(CH_2)_2COOH$) are formed, while at a higher pH propionic acid (CH_3CH_2COOH) is formed [91]. Digester failures are often the result of acid gathering when too many volatile solids are fed into the process (per unit reactor volume). Volatile fatty acids (VFAs) generated during anaerobic digestion of organic waste are considered a promising substrate for microbial oil production [92] and the production of renewable green chemicals. Due to this, anaerobic digestion supports the implementation of the waste management hierarchy as it enables the production of renewable green products [93]. VFAs have various applications; they are used in the production of biodiesel fuel, the synthesis of complex biopolymers, and the generation of electricity through microbial fuel cells [94]. On the other hand, a pH level above 8.0 is toxic to most anaerobic organisms, causing the inhibition of their vital functions. This increase in pH may be due to intensive methanogenesis, resulting in higher ammonia concentration, which hinders acidogenesis [95].

Another critical parameter for the fermentation process is the nutrients necessary for microorganisms' proper growth and functionality. One of the essential components is nitrogen, necessary for synthesising amino acids, which can be converted into ammonia, acting as a buffer to neutralise the acidification process of the fermented material. As reported by Rajeshwari et al. [96], the authors in their work indicate that the fermented feedstock should contain carbon, nitrogen, and phosphorus in the ratio (C:N:P) 100:3:1, which is optimal for process efficiency and a high methane yield. An imbalance in these proportions can result in a deficiency in the buffering capacity of the material or insufficient nutrients for life functions and microbial growth. Other nutrients needed by microorganisms for optimal functioning are phosphorus, sulphur, potassium, sodium, magnesium, calcium, and microelements: iron, molybdenum, manganese, copper, zinc, cobalt, nickel, selenium, or tungsten. The corresponding C/N/P/S ratio was determined by Weiland [97] to be 600/15/5/3.

Other parameters affecting the AD process include moisture content [98], particle size [99], organic loading rate (OLR) [100], solid retention time [101], sulphate reduction [102–104], denitrification [105], and ammonium concentration [106–109]. All the parameters mentioned above can play an important role in modifying the reaction rate of individual phases of the fermentation process [110–113]. The products of the fermentation process are biogas and digestate. Biogas is produced and captured during the process, and its dominant component is methane, which is the raw material required for electricity and heat generation [114]. The methane fermentation process carried out in biogas installations contributes to preventing uncontrolled methane emission into the atmosphere (e.g., during waste disposal in landfills) and increases the potential of renewable energy sources [115,116]. Besides, the encapsulation of the methane fermentation process contributes to preventing the emission of other compounds such as ammonia, NH_3, hydrogen sulphide, H_2S, or volatile organic compounds, VOCs, characterised by an unpleasant odour [117–120].

Among the methods of digestate processing, the following solutions are used: dewatering, aerobic stabilisation (in closed or open conditions), sieving and other unit operations of post-treatment, such as compost or landfilling (in the case of compost, not complying with the requirements for plant improvement products) [11,13,19].

2.2. Odour Nuisance of MWBPs

The AD process, due to the lack of air supply, is a hermetic process and therefore odourless. However, processes such as feedstock preparation and aerobic stabilisation of the digestate are associated with odour and odorant emissions [121,122].

The level of odour and odorant concentrations at MWBPs varies widely, and this variation is mainly due to the type of waste processed, technological factors and processes, air temperature, humidity, and microclimatic conditions [120,123,124].

The study by Fang et al. [125] identified 60 different compounds belonging to nine chemical groups. The compounds determined were sulphides, terpenes, ketones, alcohols, alkenes, aromatic hydrocarbons, acids, and esters. Terpenes and sulphur-containing compounds are the leading cause of odours [126]. On the other hand, in their works, Komilis et al. [127] and Scaglia et al. [128] also demonstrated the presence of BTEX compounds—benzene, toluene, ethylbenzene, and xylene—which in addition to an unpleasant odour are also characterised by harmfulness to human health. Benzene shows carcinogenic effects [129], while long-term exposure to toluene, ethylbenzene, and xylene adversely affects the respiratory system, causing asthma or asthmatic symptoms such as dyspnea, coughing, wheezing, chest tightness, and difficulty in breathing, as well as the central nervous system, causing symptoms such as headache and dizziness, nausea, fatigue, agitation, and disorientation [130–134].

Byliński et al. [135], in their work, focused on the analysis of odorant emissions (for example, dimethyl sulphide-2.43–18.67 ppb, methanethiol-2.91–12.43 ppb, benzene-0.93–10.48 ppb, toluene-0.92–26.35 ppb, and xylene-1.72–18.18 ppb) at biogas plants processing sewage sludge, which is waste from the municipal sector. The results of this work also confirmed the release of odorants from the digestate. Costa et al. [136], also investigating odorants at biogas plants but processing a different type of feedstock (microalgae), indicated a method with which to regulate the concentrations of emitted compounds by controlling the fermentation process in such a way as to maximise the transition of these compounds to the volatile fraction (biogas), which should then be treated before further use.

An analysis of the rules at the global level in terms of odour regulation shows an extensive variation in both odour and odorant concentrations as well as types of selected compounds. In countries without these regulations, the odour limit can be determined by specific and relatively easy-to-determine chemicals such as hydrogen sulphide and ammonia. Table 1 summarises odorous compounds found in the Guidelines for Air Quality produced by the WHO [137].

Table 1. Odorous compounds and average ambient air concentrations [137].

Chemical Compound	Average Ambient Air Concentration (μg/m^3)
Acetone	0.5–125
Acrolein	15
Carbon disulphide	10–1500
Hydrogen sulphide	0–15
Isophorone	no data
Styrene	1.0–20
Tetrachloroethylene	1–5
Toluene	5–150

Wiśniewska et al. [138], in their work, defined five primary categories of odorant sources at MWBPs: waste storage, preRDF storage, mechanical waste treatment and fermentation preparation, digestate dewatering, and oxygen stabilisation of digestate. The research conducted at two Polish MWBPs shows the mean VOC, and NH_3 concentrations vary depending on the stage of the technological line and are in the following ranges: 0–38.64 ppm (0–0.169 mg/m^3) and 0–100 ppm (0–69.653 mg/m^3), respectively, while according to the best available technique (BAT) conclusions, for waste treatment channelled [139] VOC and NH_3 emissions to air from biological waste treatment should not exceed values of 40 mg/m^3 and 20 mg/m^3 [120]. Pilot studies carried out at six Polish MWBPs have shown that the most significant odour nuisance and odour emissions are caused by such elements of the process line as fermentation preparation, digestate dewatering, waste storage, etc., at a technological wastewater pumping station [5,124]. Tables 2 and 3 summarise the odour and odorant concentrations and odorant emissions at MWBPs in Poland for various elements of the process line.

On the other hand, studies presented in paper [5], conducted in 2020 at three Polish MWBPs, indicate that the highest odour nuisance and the highest VOC and NH_3 concentrations concern the oxygen stabilisation of digestate and technological wastewater pumping stations. In the VOC mixture emitted, the dominance of toluene is very clear, followed by phenol and styrene.

Table 2. Odour and odourant concentrations at MWBPs in Poland [124,138].

Location/Odour Source	Fermentation Preparation				Digestate Dewatering			
	c_{od} (ou/m^3)	C_{VOC} (ppm)	C_{H2S} (ppm)	C_{DMS} (ppm)	c_{od} (ou/m^3)	C_{VOC} (ppm)	C_{H2S} (ppm)	C_{DMS} (ppm)
Jarocin	16 ÷ 78	0.20 ÷ 0.53	0.279	0.360	142 ÷ 448	0.82 ÷ 1.30	0.406	1.317
Tychy	4 ÷ 22	1.37 ÷ 1.94	0.860	<0.001	31 ÷ 42	1.20 ÷ 1.83	0.114	<0.001
Promnik	5 ÷ 11	1.06 ÷ 5.71	0.007	0.624	8 ÷ 42	2.65 ÷ 6.41	0.267	0.997
Stalowa Wola	16 ÷ 31	2.35 ÷ 2.38	<0.001	0.009	16 ÷ 31	0.20 ÷ 2.13	<0.001	0.022
Wólka Rokicka	22 ÷ 42	1.30 ÷ 1.40	<0.001	0.026	-	-	-	-
Biała Podlaska	4 ÷ 5	0.50 ÷ 0.63	<0.001	0.002	-	-	-	-

–a technological process that does not occur at a biogas plant. c_{od}—odour concentration. C_{VOC}—volatile organic compounds concentration. C_{H2S}—hydrogen sulphide concentration. C_{DMS}—dimethyl sulphide concentration.

Table 3. Odourant concentrations and emissions from mixed MWBPs in Poland [123].

Waste Storage		Emission from the Hall of Waste Storage			Digestate Dewatering		Emission from the Hall of Digestate Dewatering		Oxygen Stabilisation of Digestate	
C_{NH3} (ppm)	C_{VOC} (ppm)	E_{NH3} (kg/h)	E_{VOC} (kg/h)	E_{H2S} (kg/h)	C_{NH3} (ppm)	C_{VOC} (ppm)	E_{NH3} (kg/h)	E_{VOC} (kg/h)	C_{NH3} (ppm)	C_{VOC} (ppm)
2–8	4.42–19.79	0.23–0.44	0.15–0.42	0.02–0.25	1–12	2.07–6.27	0.004–0.04	0.03–0.17	0–7	0.08–2.47

C_{NH3}—ammonia concentration. C_{VOC}—volatile organic compounds concentration. E_{NH3}—ammonia emission. E_{VOC}—volatile organic compounds emission. E_{H2S}–hydrogen sulphide emission.

2.3. MWBPs in the CE

There are many indicators of sustainable development in the literature, which are essential from the point of view of investments in CE assumptions [56,113,140–148]. In general, they can be divided into three main categories: economic, environmental, and social. From the point of view of this work, the environmental indicators presented in Table 4 seem to be the most relevant since the literature review shows that "odour", analysed in this paper, is most commonly classified as an environmental indicator. The indicators presented in this table are selected based on the literature review as those most frequently identified by the authors.

Table 4. The sustainable environmental indicators [56,110,140–148].

No	Environmental Indicators
1	Land use
2	Water use
3	Pollutant generation
4	Life cycle of CO_2 emission
5	Overall emissions
6	SO_x emissions
7	NO_x emissions
8	Particulate matters
9	Ash
10	Noise
11	Dust
12	Odour
13	Litter

When considering MWBPs in terms of sustainable development and the CE, the mentioned indicators supporting the investment assessment are essential for the analysis [149–152]. The vision of the CE should also support the implementation of the SDGs [153]. Important indicators characterising biogas plants are, first, minimising the emission of greenhouse gases to the atmosphere (CO_2, SO_x, and NO_x) [154–156], but also reducing the emission of odours, which in turn positively affects the well-being of residents living near the analysed investments [157–159]. Social indicators such as quality of life, health, and well-being are also important in terms of the nuisance associated with the operation of waste treatment facilities [160,161]. The concept of the CE should contribute to the well-being of individuals and communities, but many authors note that the CE focuses on the economic value of products while neglecting the social dimension. In the specific case presented, the emission of odours is important from both an environmental and a social viewpoint.

2.4. Feedstock for MWBPs

The methane fermentation process of waste can be qualified in different ways from the hierarchy of waste treatment methods defined in the Waste Directive [9]. This is determined primarily by the type of input in the process and, consequently, the type of products produced. The methane fermentation of waste may be implemented as a recovery process (organic recycling), where the input in the process is separately collected post-consumer waste and the product (besides biogas) is compost, or as a disposal process, where the input is mixed MSW and the mechanically sorted white-water fraction (OFMSW) is directed to the biological process. In this case, the aim of the process is primarily to reduce the volume of waste and reduce the activity of microorganisms, including pathogens dangerous to human and animal organisms. After this process, there remains (besides biogas) stabilised waste (stabilised digestate), which, when deposited in a landfill, has a much lower biogas production [162–165]. During AD, not all biodegradable substances are broken down [166,167]. However, if properly managed, the processed organic feedstock can achieve a high degree of stabilisation [168,169]—whether the product is a non-waste organic fertiliser (in the case of organic recycling) or stabilised waste (in the case of disposal).

Regarding the differentiation resulting from the hierarchy of waste handling methods, the basic types of raw materials intended for MWBPs may be the waste separated mechanically from the stream of mixed MSW (the undersized fraction after passing through sieves sized 50–100 mm [162,170]), as well as BW collected selectively (mainly kitchen waste) [171–173]. An unquestionable benefit of the BW fermentation process for BW collected selectively, and an added value from the CE's point of view, is the possibility to use not only the biogas produced in the process but also the digestate as an organic fertiliser, which used correctly, increases soil fertility [174–176]. On the other hand, O-Thong et al. [177] state that co-digestion, i.e., the joint processing of several types of raw materials, is most used at biogas plants in the world. The advantages of co-digestion are mainly: an increase in the amount of rapidly biodegradable matter [178,179], improvement in the buffer capacity of the material, which in turn helps to maintain an adequate pH level necessary for methanogenesis [180,181], an increase in the carbon-to-nitrogen ratio (C:N—optimum range $20 \div 30$ [182–187]) [187,188], a decrease in the effect of inhibitors on the process by their dilution [187–189], an increase in the volumetric production of methane [190–192], and others. However, this applies to agricultural biogas plants. According to the report Anaerobic Digestion of the Organic Fraction of Municipal Solid Waste in Europe, other products to the organic fraction of MSW used to be rather the exception, mainly due to the adjustment of the pre-treatment at biogas plants to the municipal waste stream. The use of co-digestion at MWBPs in Europe concerns about 13% of the installed capacity, where the number of co-products is at the level of about 10% to 15% [193]. However, under Polish conditions, at MWBPs, one specific type of waste is usually directed to digesters [40,124].

The benefit of using selectively collected BW as feedstock is its higher biogas potential in terms of the amount of biogas produced per unit weight of waste and the methane content of the biogas (biogas yield and stability), which is also crucial in the context of the CE. The literature analysis indicates an approximately 10% increase in the amount of biogas produced when it is produced from BW collected selectively compared to OFMSW separated mechanically from the stream of mixed MSW [40,62,194–197]. The MWBPs operating in Poland are based mainly on the fermentation of OFMSW, mechanically separated from the stream of mixed MSW, characterised by a biogas yield of 105 m^3/Mg of charge and methane concentration in biogas of 51 ÷ 53% [40,124]. Due to the changes introduced to the country in the system of selective collection of MSW (mainly because of the introduction and development of the selective collection of BW), biogas plants equipped with two digestion lines allocate one of them to BW. This fraction is characterised by a higher biogas yield—about 111 m^3/Mg charge—and methane concentration in biogas on the level of 58 ÷ 60% [40].

Between 2018 and 2020, studies have been conducted at six MWBPs in Poland, published in [5,120,123,124,138]. This study primarily considered odorant and odour emissions as well as technological aspects (including the type of feedstock) accompanying the waste treatment process. Table 5 presents a summary of the capacity and feedstock of MWBPs in Poland, while Figure 1a,b presents simplified schemes of typical process lines for mixed MSW (1a) and selectively collected BW (1b).

Table 5. The feedstock and capacity of MWBPs in Poland (own elaboration).

Location	Feedstock	Annual Biogas Plant Capacity (Mg/a)	Fermentation	Digestate Stabilisation
Jarocin	OFMSW	15,000	Horizontal reactor with one paddle agitator; thermophilic, semi-dry dynamic fermentation	Digestate dewatering followed by two-step oxygen stabilisation
Tychy	OFMSW	30,000	Two separate horizontal reactors, each with four agitators; mesophilic, dry dynamic fermentation	
Promnik	OFMSW	30,000		
Stalowa Wola	OFMSW	15,000	Horizontal reactor with four agitators; mesophilic, dry dynamic fermentation	
Wólka Rokicka	OFMSW	20,000	Seven open-feed reactors; thermophilic, dry static (garage) fermentation	One-step oxygen stabilisation
Biała Podlaska	BW	20,000	Two separate horizontal reactors, each with one paddle agitator; thermophilic, semi-dry dynamic fermentation	

(a)

(b)

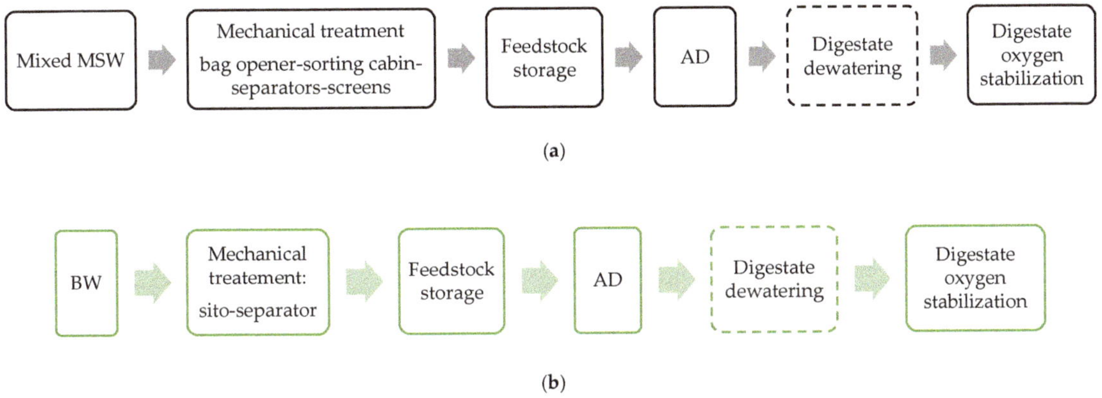

Figure 1. Typical process flowcharts of mixed MSW (**a**) and BW (**b**) at Polish MWBPs.

According to [191], by 2014 about 55% of the installed capacity in Europe (of MWBPs) was destined to treat BW.

3. The Feedstock Modification to Reduce Odour Nuisance at MWBPs–Analysis, Discussion, Recommendations

At MWBPs, the input in the process may be mixed MSW, from which OFMSW or selectively collected BW is then mechanically separated. Different morphological compositions characterise both these raw materials. Studies conducted by Seruga et al. [40] in one of the Polish biogas plants on the group composition of BW fractions from selective collection intended for the fermentation process indicate that the biofraction content is 68.1% ± 5.2% on average. The remaining input materials are wood—8.1% ± 0.5%, paper—2.4% ± 0.7%, plastics—1.1% ± 0.4%, glass—0.8% ± 0.4%, inert waste—1.4% ± 0.9%, textiles—0.1% ± 0.4%, metals—0.1% ± 0.1%, hazardous—0.1% ± 0.1%, tetra pack—0.3% ± 0.1%, others—0.4% ± 0.1%, and fine fraction 0 ÷ 15 mm—17.1% ± 2.3%. At the same time, the OFMSW mechanically sorted from the mixed MSW at this biogas plant has a food and green waste content of only 48.3% ± 2.7%. However, municipal biowaste is not always characterised by high "purity" and homogeneity. This is particularly the case in countries where the separate collection system has recently been introduced and is undergoing implementation and development. Unpublished morphological studies of BW diverted to one of the biogas plants studied between September 2017 and October 2020 show that the food waste content is highly variable throughout the year, ranging from 22.6% to 62.2%, with a mean value of 36.8% and a standard deviation of 7.6%. The rest consists of glass and stone (min. 3.7%, max. 39.3%, avg. 19.4%, standard deviation 7.6%), plastics and aluminium (min. 1.2%, max. 22.4%, avg. 6.7%, standard deviation 4.8%), and paper and textiles (min. 15.0%, max. 57.3%, avg. 37.1%, standard deviation 9.2%). The presented BW quality results from the fact that the selective collection system for biowaste is currently at the initial stage of development in Poland. Many hard fractions in the form of glass and stones, which are impurities of BW, can cause the failure of digesters through their accumulation in the lower part of reactors, resulting in the blockage of mixers. Each such failure is not only a technological problem (the necessity of emptying the chambers) but also an environmental problem (increased emission of odorants, in particular, ammonia), which was proved in the paper [123]. The above comparison also shows that the effectiveness of selective collection systems for MSW, particularly BW, varies considerably in different countries where this system has only just been introduced and is at the implementation stage. During this transition period, differences in feedstocks (OFMSW from the mixed MSW/BW stream) may be small and less significant, but this significance and variation will increase over time.

Our own research conducted in Poland in the period 2018–2020, the results of which have been published in a series of articles [5,120,123,124,138], allow us to state that the lower emissions of odorous compounds from biogas plants processing BW have been observed in comparison to installations where the input material for the fermentation process is OFMSW mechanically separated from the stream of mixed MSW, especially in some stages of the technological sequence—namely mechanical treatment and fermentation preparation. A comparison of odour and odorant concentrations at these process stages for BW and OFMSW at several biogas plants is presented in Figures 2 and 3.

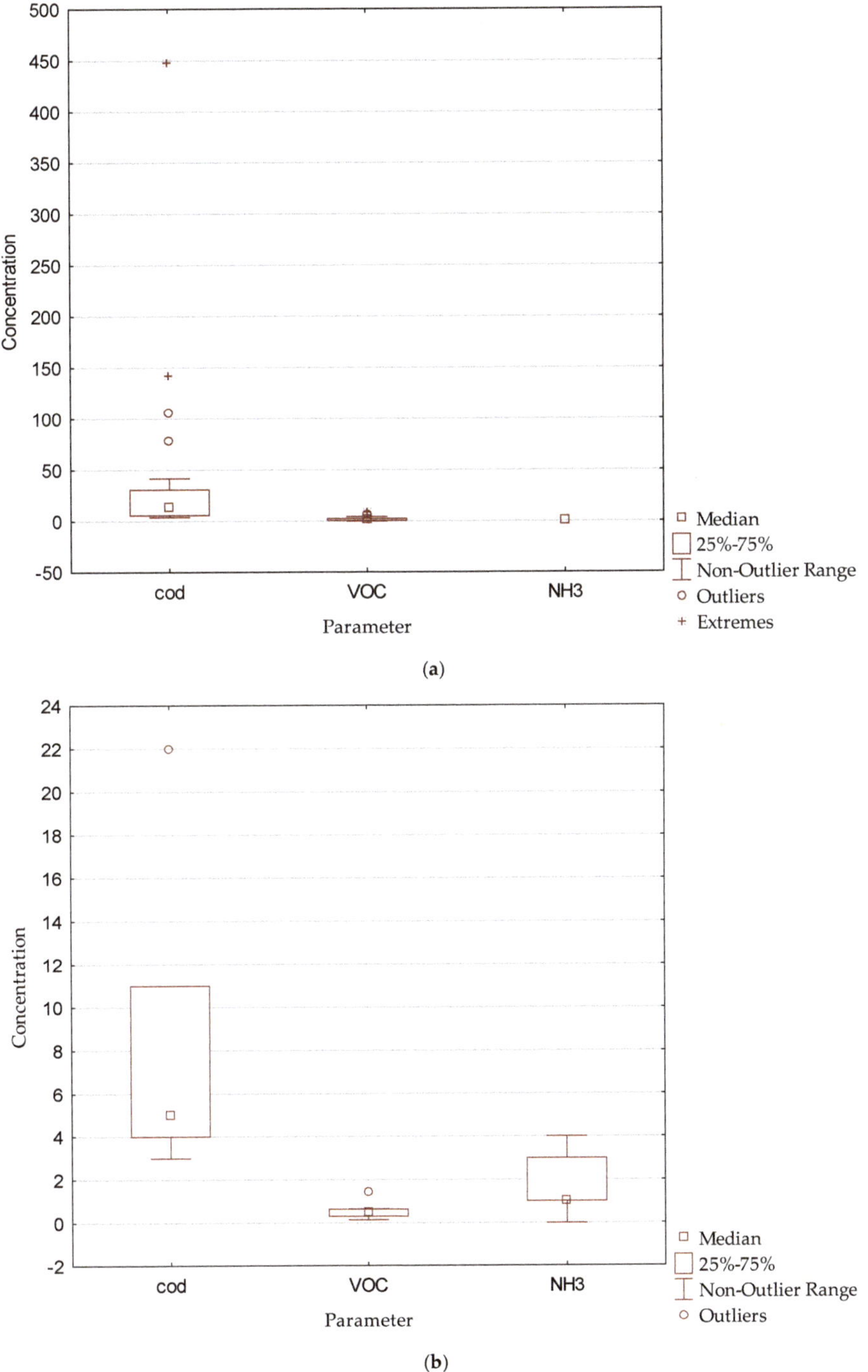

Figure 2. The range of odour and odorant concentration for OFMSW (**a**) and BW (**b**) at Polish MWBPs (own elaboration based on [122,136]).

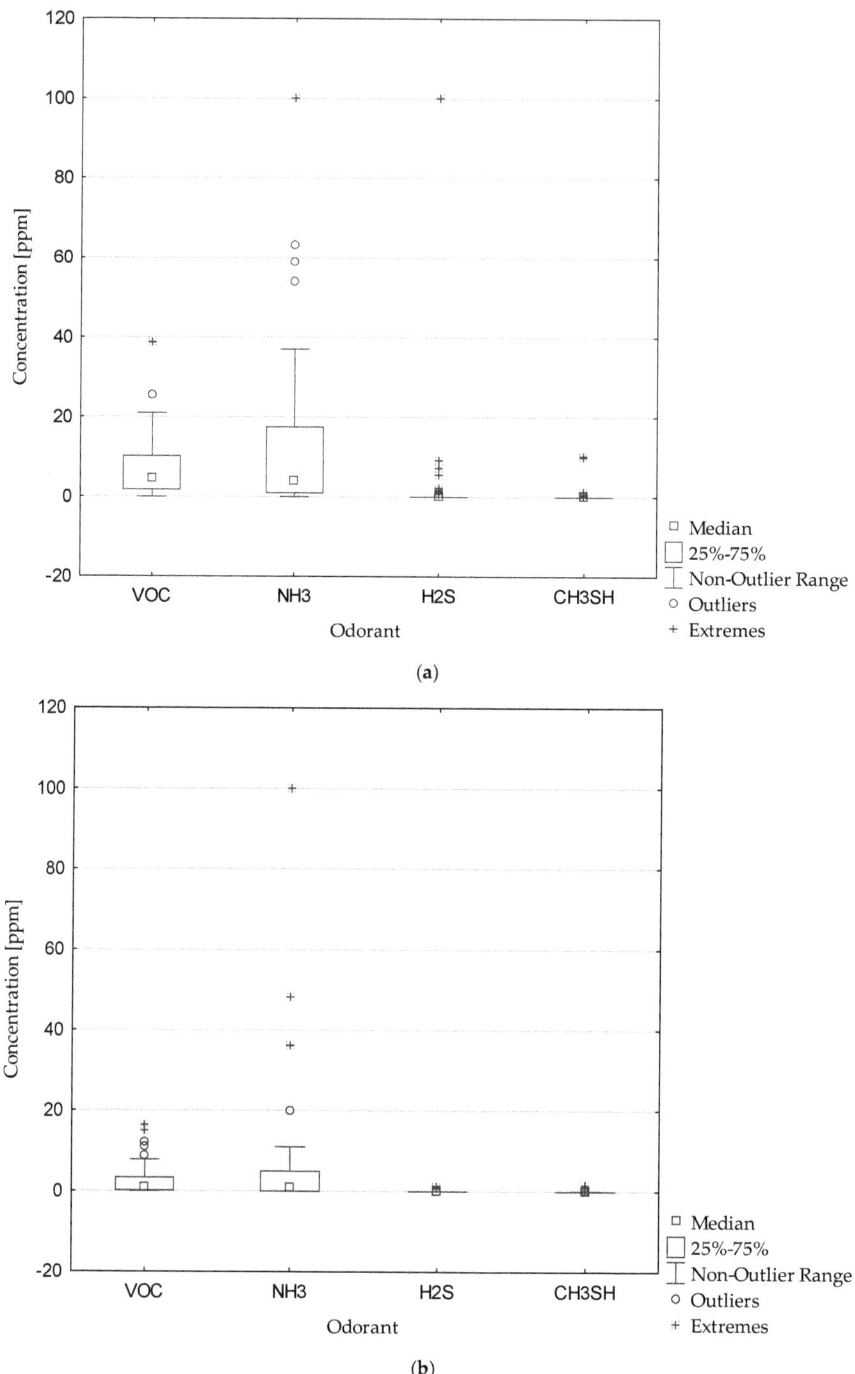

Figure 3. The range of odorant concentration for OFMSW (**a**) and BW (**b**) at Polish MWBPs (own elaboration based on [118]).

The odour concentration for the methane digestion feedstock preparation in a biogas plant processing selectively collected BW is significantly lower than biogas plants processing mixed MSW. Pre-treatment of mixed MSW is accompanied by varying concentrations of ash—the ranges of this parameter in the six biogas plants studied are presented in Table 1. Additionally, the concentration of odorants when BW is prepared for fermentation is much lower. Even about 6-fold lower CH_3SH concentrations and 1.5- to 9-fold lower VOC concentrations were observed compared to pre-treatment of mixed MSW. When selectively collected BW was processed, either no H_2S emissions were recorded or they were 100 times lower. However, no differences in NH_3 concentrations were evident. Similar relationships were not observed for the oxygen stabilisation of digestate, regardless of whether the digestate was from BW (1–4 ppm) or OFMSW (1–3 ppm) [120,124]. There are similar results in papers [198–201], where it was stated that waste treatment odours depend on the type of raw material.

The relationships presented are highly relevant to the CE indicators (Table 1). In attempting to determine the reason for these differences, one should first note the significantly shorter pre-treatment process sequence in the case of selectively collected BW, comprising of particular screening and sometimes pre-treatment with the use of a single separator in most cases (Figure 1b) in comparison to mixed MSW, for which bag tearing, manual segregation, as well as a system of screens and many different separators connected by conveyors are used (Figure 1a). Longer process lines mean longer waste processing times, resulting in longer odour and odorant impacts and more locations where malodorous emissions can occur. Additionally, in research conducted in a biogas plant processing BW, a short storage time for this fraction was observed (waste was collected from the inhabitants more frequently and in smaller amounts, and immediately directed to the fermentation on the same-day process). This is also a significant reason for lower odour nuisance, even despite temporary high concentrations of odours and odorants.

Odour nuisance of biogas plants is also connected with the production of process effluents, especially in the process of wet and semi-dry fermentation requiring appropriate management due to the high pollutant loads, mainly organic and nitrogenous. This wastewater, in Polish conditions, is treated and then directed to a wastewater treatment plant, treated on-site, or directly directed to the treatment plant by pipeline or using a slurry fleet [202–205]. Studies conducted so far in the analysed biogas plants indicate that the process wastewater from BW processing is characterised by a lower pollutant load and is a source of lower odour and odourant emissions (c_{od} = 4 ou/m^3; C_{VOC} = 0 ÷ 1.53 ppm) in comparison to wastewater from the treatment of OFMSW (undersized) mechanically separated from mixed MSW (c_{od} = 142 ÷ 394 ou/m^3; C_{VOC} = 1.11 ÷ 25.41 ppm) [204–206].

The guidelines and recommendations of waste treatment plant odours are indicated in the previously mentioned BAT conclusions for waste treatment [139]. This document lists, among other things, the recommended techniques to reduce emissions of organic compounds into the air (e.g., biological filter, fabric filter, thermal oxidation, and wet scrubbing) and the emission levels associated with the best available techniques for organised emissions. These levels follow for an odour concentration of 200–1000 ou/m^3, ammonia of 0.3–20 mg/Nm^3, and total VOC 5–40 mg/Nm^3 (the values indicated are the average over the sampling period).

4. Summary and Future Research Work

Environmental and social aspects are an essential part of the CE concept and monitoring progress in its implementation and delivery. In CE strategies, social indicators such as quality of life, health, and well-being of inhabitants, and environmental indicators such as pollutant generation, overall emissions, and odour are mentioned. Methane fermentation as a waste treatment process is an excellent fit for the CE, both technically, economically, and environmentally. Co-fermentation, i.e., using at least two complementary substrates in one digester, can be an exciting and promising way to achieve CE goals. Biogas installations have multiple functions in the CE, and feedstock modification at MWBPs and effective selective collection of BW is essential from the point of view of higher biogas yield, renewable energy production, minimising greenhouse gas emissions, and improving the efficiency of soil fertility (using the digestate as an organic fertiliser). In addition, it is shown that the modification of feedstock towards BW contributes to a reduction in odour and odorant emissions, which are the cause of odour nuisance among residents living in the vicinity of biogas plants, and thus additionally fits into the CE environmental and social objectives.

With the change of feedstock in municipal waste biogas plants due to the implementation of CE rules, both benefits and difficulties in the operation of these plants can be expected, especially during the initial period of change. During the implementation of a separate collection system, the differences in input (in the form of OFMSW mechanically separated from mixed MSW and BW) are smaller and of less importance. However, this importance will increase with time as the efficiency of separate collection improves.

Nevertheless, lower odour and odorant emissions at biogas plants processing BW compared to mixed MSW are observed especially due to the shorter processing time of BW. The time of waste processing is significant, especially at the stage of storage and pre-treatment. For mixed municipal waste, it is much longer (due to the greater complexity of the process). The situation is different for BW, for which the pre-treatment is simplified, and its time is shorter, thus reducing the inconvenience for the user. BW processing usually involves a less extensive pre-treatment process line and, therefore, fewer locations from which odour sources can be emitted.

This study has limitations that point to future works and research avenues due to changing requirements regarding waste collection systems and the implementation of new technological solutions. It would be interesting to carry out a study comparing two kinds of feedstock in one biogas plant. Moreover, survey research can be recommended to assess the impact of such installations on residents. Further studies should include an assessment of the biogas plant influence and proposals for measures to improve deodorisation and minimise odour nuisance. Similar studies can also be carried out in other countries to find the best technological and technic solutions, which are the nearest to the CE.

5. Conclusions

Odour nuisance of MWBPs varies greatly, and this variation is caused by technological factors and processes, air temperature, humidity, microclimatic conditions, and the type of waste processed. The problems of odour nuisance should be considered regarding particular phases of the MWBP technological line. They are especially seen at the stage of waste storage, fermentation preparation, and digestate dewatering. Research results indicate lower odour and odorant emissions at biogas plants processing BW compared to mixed MSW. This is particularly evident at the stage of mechanical treatment, fermentation preparation and technological wastewater management. In countries where a separate BW collection system is at an early stage of development BW biogas plants may initially be characterised by a similar odour nuisance as mixed MSW plants, because of the heterogeneity and low degree of BW "purity". However, the beneficial influence of BW feedstock on the odour effect will increase with time, as the efficiency of separate collection improves. Further research on odour nuisance of MSW treatment plants should be conducted for the sake of the health, comfort, and well-being of residents living in their vicinity—elements that are significant CE indicators.

Author Contributions: Conceptualisation, investigation, writing—original draft preparation, and visualisation, M.W.; supervision, A.K.; conceptualisation, writing—review and editing, and validation, K.L.-S. All authors have read and agreed to the published version of the manuscript.

Funding: This research was funded by the Warsaw University of Technology.

Institutional Review Board Statement: Not applicable.

Informed Consent Statement: Not applicable.

Data Availability Statement: Not applicable.

Conflicts of Interest: The authors declare no conflict of interest.

Abbreviations

AD	Anaerobic digestion
BAT	Best available techniques
BW	Selectively collected biowaste
CE	Circular economy
c_{od}	Odour concentration (ou/m^3)
MSW	Municipal solid waste
MWBP	Municipal waste biogas plant
OFMSW	Organic fraction of municipal solid waste
SDGs	Sustainable development goals
VFAs	Volatile fatty acids
VOCs	Volatile organic compounds
WETs	Waste-to-energy technologies

References

1. Commission of European Communities. *Closing the Loop—An EU Action Plan for the Circular Economy*; Communication No. 614 (COM (2015), 614); Commission of European Communities: Brussels, Belgium, 2015. Available online: https://eur-lex.euopa.eu/resource.html?uri=cellar:8a8ef5e8-99a0-11e5-b3b7-01aa75ed71a1.0012.02/DOC_1&format=PDF (accessed on 20 November 2020).
2. Commission of European Communities. *Towards a Circular Economy: A Zero Waste Programme for Europe*; Communication No. 398 (COM (2014), 398); Commission of European Communities: Brussels, Belgium, 2014. Available online: http://ec.europa.eu/environment/circular-economy/pdf/circular-economy-communication.pdf (accessed on 20 November 2020).
3. Ellen MacArthur Foundation. Available online: https://ellenmacarthurfoundation.org/ (accessed on 2 October 2021).
4. Kaza, S.; Yao, L.C.; Bhada-Tata, P.; Van Woerden, F. *What a Waste 2.0*; World Bank: Washington, DC, USA, 2018. [CrossRef]
5. Wiśniewska, M.; Kulig, A.; Lelicińska-Serafin, K. The Use of Chemical Sensors to Monitor Odour Emissions at Municipal Waste Biogas Plants. *Appl. Sci.* **2021**, *11*, 3916. [CrossRef]
6. Iqbal, A.; Liua, X.; Chen, G.-H. Municipal solid waste: Review of best practices in application of life cycle assessment and sustainable management techniques. *Sci. Total. Environ.* **2020**, *729*, 138622. [CrossRef]
7. Zulkifli, A.A.; Yusoff, M.Z.M.; Abd Manaf, L.; Zakaria, M.R.; Roslan, A.M.; Ariffin, H.; Hassan, M.A. Assessment of Municipal Solid Waste Generation in Universiti Putra Malaysia and Its Potential for Green Energy Production. *Sustainability* **2019**, *11*, 3909. [CrossRef]
8. The European Parliament and the Council of the European Union. Directive 2008/98/EC of the European Parliament and of the Council of 19 November 2008 on waste and repealing certain Directives. *Off. J. Eur. Union* **2008**, *312*, 3–30.
9. The European Parliament and the Council of the European Union. Directive (EU) 2018/851 of the European Parliament and of the Council of 30 May 2018 amending Directive 2008/98/EC on waste. *Off. J. Eur. Union* **2018**, *150*, 109–140.
10. United Nations. *Resolution Adopted by the General Assembly on 25 September 2015, Transforming Our World: The 2030 Agenda for Sustainable Development*; United Nations: Geneva, Switzerland, 2015.
11. Kledyński, Z.; Bogdan, A.; Jackiewicz-Rek, W.; Lelicińska-Serafin, K.; Machowska, A.; Manczarski, P.; Masłowska, D.; Rolewicz-Kalińska, A.; Rucińska, J.; Szczygielski, T.; et al. Condition of Circular Economy in Poland. *Arch. Civ. Eng.* **2020**, *66*, 37–80. [CrossRef]
12. Kuo, W.-C.; Lasek, J.; Słowik, K.; Głód, K.; Jagustyn, B.; Li, Y.H.; Cygan, A. Low-temperature pre-treatment of municipal solid waste for efficient application in combustion systems. *Energy Convers. Manag.* **2019**, *196*, 525–535. [CrossRef]
13. Wiśniewska, M.; Lelicińska-Serafin, K. The Role and Effectiveness of the MBT Installation in Poland Based on Selected Examples. *Civ. Environ. Eng. Rep.* **2019**, *29*, 1–12. [CrossRef]
14. Wienchol, P.; Szlęk, A.; Ditaranto, M. Waste-to-energy technology integrated with carbon capture—Challenges and opportunities. *Energy* **2020**, *198*, 117352. [CrossRef]
15. Huttunen, S.; Manninen, K.; Leskinen, P. Combining biogas LCA reviews with stakeholder interviews to analyse life cycle impacts at a practical level. *J. Clean. Prod.* **2014**, *80*, 5–16. [CrossRef]
16. Komakech, A.J.; Sundberg, C.; Jonsson, H.; Vinnerås, B. Life cycle assessment of biodegradable waste treatment systems for sub-Saharan African cities. *Resour. Conserv. Recycl.* **2015**, *99*, 100–110. [CrossRef]
17. Jensen, M.B.; Møller, J.; Scheutz, C. Comparison of the organic waste management systems in the Danish–German border region using life cycle assessment (LCA). *Waste Manag.* **2016**, *49*, 491–504. [CrossRef]
18. Bhatia, S.K.; Joo, H.S.; Yang, Y.H. Biowaste-to-bioenergy using biological methods—A mini-review. *Energy Convers. Manag.* **2018**, *177*, 640–660. [CrossRef]

19. Margallo, M.; Ziegler-Rodriguez, K.; V'azquez-Rowe, I.; Aldaco, R.; Irabien, A.; Kahhat, R. Enhancing waste management strategies in Latin America under a holistic environmental assessment perspective: A review for policy support. *Sci. Total Environ.* **2019**, *689*, 1255–1275. [CrossRef]
20. Brunner, P.H.; Rechberger, H. Waste to energy—Key element for sustainable waste management. *Waste Manag.* **2015**, *37*, 3–12. [CrossRef] [PubMed]
21. Saveyn, H.; Eder, P. *End-of-Waste Criteria for Biodegradable Waste Subjected to Biological Treatment (Compost & Digestate): Technical Proposals, JRC Scientific and Policy Reports*; Publications Office of the European Union: Luxembourg, 2014.
22. Das, S.; Lee, S.H.; Kumar, P.; Kim, K.H.; Lee, S.S.; Bhattacharya, S.S. Solid waste management: Scope and the challenge of sustainability. *J. Clean. Prod.* **2019**, *228*, 658–678. [CrossRef]
23. Slorach, P.C.; Jeswani, H.K.; Cuéllar-Franca, R.; Azapagic, A. Environmental and economic implications of recovering resources from food waste in a circular economy. *Sci. Total Environ.* **2019**, *693*, 133516. [CrossRef]
24. International Energy Agency. *World Energy Outlook Special Report 2015: Energy and Climate Change*; Final Report; OECD/IEA: Paris, France, 2015.
25. United Nations Environment Programme. *The Emissions Gap Report 2014: A UNEP Synthesis Report*; Final Report; UNEP: Nairobi, Kenya, 2014.
26. Rolewicz-Kalińska, A.; Lelicińska-Serafin, K.; Manczarski, P. The Circular Economy and Organic Fraction of Municipal Solid Waste Recycling Strategies. *Energies* **2020**, *13*, 4366. [CrossRef]
27. Diamantis, V.; Eftaxias, A.; Stamatelatou, K.; Noutsopoulos, C.; Vlachokostas, C.; Aivasidis, A. Bioenergy in the era of circular economy: Anaerobic digestion technological solutions to produce biogas from lipid-rich wastes. *Renew. Energy* **2021**, *168*, 438–447. [CrossRef]
28. Tisserant, A.; Pauliuk, S.; Merciai, S.; Schmidt, J.; Fry, J.; Wood, R.; Tukker, A. Solid Waste and the Circular Economy: A Global Analysis of Waste Treatment and Waste Footprints. *J. Ind. Ecol.* **2017**, *21*, 628–640. [CrossRef]
29. Abad, V.; Avila, R.; Vicent, T.; Font, X. Promoting circular economy in the surroundings of an organic fraction of municipal solid waste anaerobic digestion treatment plant: Biogas production impact and economic factors. *Bioresour. Technol.* **2019**, *283*, 10–17. [CrossRef] [PubMed]
30. Holm-Nielsen, J.B.; Al Seadi, T.; Oleskowicz-Popiel, P. The future of anaerobic digestion and biogas utilization. *Bioresour. Technol.* **2009**, *100*, 5478–5484. [CrossRef] [PubMed]
31. Jędrczak, A. *Biologiczne Przetwarzanie Odpadów (Biological Waste Treatment)*; PWN: Warszawa, Poland, 2008. (In Polish)
32. Nizami, A.S.; Rehan, M.; Waqas, M.; Naqvi, M.; Ouda, O.K.M.; Shahzad, K.; Miandada, R.; Khand, M.Z.; Syamsiroe, M.; Ismaila, I.M.I.; et al. Waste biorefineries: Enabling circular economies in developing countries. *Bioresour. Technol.* **2017**, *241*, 1101–1117. [CrossRef]
33. Tyagi, V.K.; Fdez-Güelfo, L.A.; Zhou, Y.; Álvarez-Gallego, C.J.; Garcia, L.I.R.; Ng, W.J. Anaerobic co-digestion of organic fraction of municipal solid waste (OFMSW): Progress and challenges. *Renew. Sustain. Energy Rev.* **2018**, *93*, 380–399. [CrossRef]
34. Cardoen, D.; Joshi, P.; Diels, L.; Sarma, P.M.; Pant, D. Agriculture biomass in India: Part 1. Estimation and characterisation. *Resour. Conserv. Recycl.* **2015**, *102*, 39–48. [CrossRef]
35. Cardoen, D.; Joshi, P.; Diels, L.; Sarma, P.M.; Pant, D. Agriculture biomass in India: Part 2. Post-harvest losses, cost and environmental impacts. *Resour. Conserv. Recycl.* **2015**, *101*, 143–153. [CrossRef]
36. Farmanbordar, S.; Karimi, K.; Amiri, H. Municipal solid waste as a suitable substrate for butanol production as an advanced biofuel. *Energy Convers. Manag.* **2018**, *157*, 396–408. [CrossRef]
37. Ravindran, R.; Jaiswal, A.K. Exploitation of food industry waste for high-value products. *Trends Biotechnol.* **2016**, *34*, 58–69. [CrossRef]
38. Banks, C.J.; Chesshire, M.; Heaven, S.; Arnold, R. Anaerobic Digestion of Source-Segregated Domestic Food Waste: Performance Assessment by Mass and Energy Balance. *Bioresour. Technol.* **2011**, *102*, 612–620. [CrossRef]
39. Di Maria, F.; Micale, C.; Morettini, E. Impact of the Pre-Collection Phase at Different Intensities of Source Segregation of Bio-Waste: An Italian Case Study. *Waste Manag.* **2016**, *53*, 12–21. [CrossRef]
40. Seruga, P.; Krzywonos, M.; Seruga, A.; Niedźwiecki, Ł.; Pawlak-Kruczek, H.; Urbanowska, A. Anaerobic Digestion Performance: Separate Collected vs. Mechanical Segregated Organic Fractions of Municipal Solid Waste as Feedstock. *Energies* **2020**, *13*, 3768. [CrossRef]
41. Lim, L.S.; Lee, L.H.; Wu, T.Y. Sustainability of Using Composting and Vermicomposting Technologies for Organic Solid Waste Biotransformation: Recent Overview, Greenhouse Gases Emissions and Economic Analysis. *J. Clean. Prod.* **2016**, *111*, 262–278. [CrossRef]
42. Bölükba¸s, A.; Akinci, G. Solid Waste Composition and the Properties of Biodegradable Fractions in Izmir City, Turkey: An Investigation on the Influencing Factors. *J. Environ. Health Sci. Eng.* **2018**, *16*, 299–311. [CrossRef]
43. Jędrczak, A. Composting and Fermentation of Biowaste—Advantages and Disadvantages of Processes. *Civ. Environ. Eng. Rep.* **2019**, *28*, 71–87. [CrossRef]
44. Edjabou, M.E.; Jensen, M.B.; Götze, R.; Pivnenko, K.; Petersen, C.; Scheutz, C.; Astrup, T.F. Municipal solid waste composition: Sampling methodology, statistical analyses, and case study evaluation. *Waste Manag.* **2015**, *36*, 12–23. [CrossRef] [PubMed]
45. Karak, T.; Bhagat, R.M.; Bhattacharyya, P. Municipal solid waste generation, composition, and management: The world scenario. *Crit. Rev. Environ. Sci. Technol.* **2012**, *42*, 1509–1630. [CrossRef]

46. Chen, Y.-C. Effects of urbanisation on municipal solid waste composition. *Waste Manag.* **2018**, *79*, 828–836. [CrossRef] [PubMed]
47. Khair, H.; Putri, C.N.; Dalimunthe, R.A.; Matsumoto, T. Examining of solid waste generation and community awareness between city center and suburban area in Medan City, Indonesia. *IOP Conf. Ser. Mater. Sci. Eng.* **2018**, *309*, 012050. [CrossRef]
48. Gasification Technologies Council (GTC). *Gasification: The Waste and Energy Solution*; Gasification Technologies Council (GTC): Penwell, TX, USA, 2011.
49. World Energy Council (WEC). *World Energy Resources: Waste to Energy*; World Energy Council: London, UK, 2013. Available online: https://www.worldenergy.org/wp-content/uploads/2013/10/WER_2013_7b_Waste_to_Energy.pdf (accessed on 2 December 2019).
50. Mubeen, I.; Buekens, A. Chapter 14—Energy from Waste: Future Prospects Toward Sustainable Development. In *Current Developments in Biotechnology and Bioengineering Waste Treatment Processes for Energy Generation*; Elsevier: Amsterdam, The Netherlands, 2019; pp. 283–305. [CrossRef]
51. Sun, L.; Liu, W.; Fujii, M.; Li, Z.; Ren, J.; Dou, Y. Chapter 1—An overview of waste-to-energy: Feedstocks, technologies and implementations. In *Waste-to-Energy, Multi-Criteria Decision Analysis for Sustainability Assessment and Ranking*; Academic Press: Cambridge, MA, USA, 2020; pp. 1–22. [CrossRef]
52. Haarich, S. *Bioeconomy Development in EU Regions: Mapping of EU Member States'/Regions' Research and Innovation Plans & Strategies for Smart Specialisation (RIS3) on Bioeconomy*; Publications Office of the European Union: Luxembourg, 2017.
53. Zaleski, P.; Chawla, Y. Circular Economy in Poland: Profitability Analysis for Two Methods of Waste Processing in Small Municipalities. *Energies* **2020**, *13*, 5166. [CrossRef]
54. WEC. World Energy Resources Waste to Energ. 2016. Available online: https://www.worldenergy.org/wp-content/uploads/2017/03/WEResources_Waste_to_Energy_2016.pdf (accessed on 30 September 2021).
55. Khan, I.; Zobaidul, K. Waste-to-energy generation technologies and the developing economies: A multi-criteria analysis for sustainability assessment. *Renew. Energy* **2020**, *150*, 320–333. [CrossRef]
56. Di Giacomo, G.; Gallifuoco, A.; Taglieri, L. Hydrothermal Carbonization of Mixed Biomass: Experimental Investigation for an Optimal Valorisation of Agrofood Wastes. In Proceedings of the 24th European Biomass Conference and Exibition (24th EUBCE), Amsterdam, The Netherlands, 6–9 June 2016; pp. 1252–1255. [CrossRef]
57. Panigrahi, S.; Dubey, B.K. A Critical Review on Operating Parameters and Strategies to Improve the Biogas Yield From Anaerobic Digestion of Organic Fraction of Municipal Solid Waste. *Renew. Energy* **2019**, *143*, 779–797. [CrossRef]
58. Carmo-Calado, L.; Hermoso-Orzáez, M.J.; Mota-Panizio, R.; Guilherme-Garcia, B.; Brito, P. Co-Combustion of Waste Tires and Plastic-Rubber Wastes with Biomass Technical and Environmental Analysis. *Sustainability* **2020**, *12*, 1036. [CrossRef]
59. Kluska, J.; Ochnio, M.; Karda's, D. Carbonization of Corncobs for the Preparation of Barbecue Charcoal and Combustion Characteristics of Corncob Char. *Waste Manag.* **2020**, *105*, 560–565. [CrossRef] [PubMed]
60. Ruggeri, B.; Tommasi, T.; Sanfilippo, S. *BioH_2 & BioCH_4 through Anaerobic Digestion: From Research to Full-Scale Applications*; Springer: London, UK, 2015. [CrossRef]
61. Agostini, F.; Sundberg, C.; Navia, R. Is biodegradable waste a porous environment? A review. *Waste Manag. Res.* **2012**, *30*, 1001–1015. [CrossRef]
62. Achinas, S.; Achinas, V.; Euverink, G.J.W. A Technological Overview of Biogas Production from Biowaste. *Engineering* **2017**, *3*, 299–307. [CrossRef]
63. Mirmohamadsadeghi, S.; Karimi, K.; Tabatabaei, M.; Aghbashlo, M. Biogas production from food wastes: A review on recent developments and future perspectives. *Bioresour. Technol. Rep.* **2019**, *7*, 100202. [CrossRef]
64. Sharma, U.C. Energy from Municipal Solid Waste. In *Environmental Science and Engineering Vol. 5 Solid Waste Management*; Studium Press LLC: Delhi, India, 2019.
65. Vanek, M.; Mitterpach, J.; Zacharova, A. Odour control in biogas plant. In Proceedings of the 15th International Multidisciplinary Scientific GeoConferences SGEM2015: Renewable Energy Sources and Clean Technologies, Albena Resort, Bulgaria, 16–25 June 2015; pp. 353–360.
66. Jaimes-Estévez, J.; Zafra, G.; Martí-Herrero, J.; Pelaz, G.; Morán, A.; Puentes, A.; Gomez, C.; Castro, L.D.P.; Escalante Hernández, H. Psychrophilic Full Scale Tubular Digester Operating over Eight Years: Complete Performance Evaluation and Microbiological Population. *Energies* **2021**, *14*, 151. [CrossRef]
67. United States Environmental Protection Agency. Available online: https://www.epa.gov/sites/default/files/2020-11/documents/agstar-operator-guidebook.pdf (accessed on 30 November 2020).
68. Pilarski, G.; Kyncl, M.; Stegenta, S.; Piechota, G. Emission of Biogas from Sewage Sludge in Psychrophilic Conditions. *Waste Biomass Valorization* **2020**, *11*, 3579–3592. [CrossRef]
69. Thi, N.B.D.; Kumor, G.; Liu, C.-Y. An overview of food waste management in developing countries: Current status and future perspectives. *J. Environ. Manag.* **2015**, *157*, 220–229. [CrossRef]
70. Fagerström, A.; Al Seadi, T.; Rasi, S.; Briseid, T. *The Role of Anaerobic Digestion and Biogas in the Circular Economy*; Murphy, J.D., Ed.; IEA Bioenergy Task 37; IEA Bioenergy: Didcot, UK, 2018; Volume 8.
71. Anderson, S. *Samhällsnyttan Med Biogas—En Studie i Jönköpings Län*; Energikontor Norra Småland: Stockholm, Sweden, 2016.
72. Ardolino, F.; Parrillo, F.; Arena, U. Biowaste-to-biomethane or biowaste-to-energy? An LCA study on anaerobic digestion of organic waste. *J. Clean. Prod.* **2018**, *174*, 462–476. [CrossRef]

73. Ardolino, F.; Arena, U. Biowaste-to-Biomethane: An LCA study on biogas and syngas roads. *Waste Manag.* **2019**, *87*, 441–453. [CrossRef]
74. Barbera, E.; Menegon, S.; Banzato, D.; D'Alpaos, C.; Bertucco, A. From biogas to biomethane: A process simulation-based techno-economic comparison of different upgrading technologies in the Italian context. *Renew. Energy* **2019**, *135*, 663–673. [CrossRef]
75. Ardolino, F.; Cardamone, G.F.; Parrillo, F.; Arena, U. Biogas-to-biomethane upgrading: A comparative review and assessment in a life cycle perspective. *Renew. Sustain. Energy Rev.* **2021**, *139*, 110588. [CrossRef]
76. Rasi, S.; Veijanen, A.; Rintala, J. Trace compounds of biogas from different biogas production plants. *Energy* **2007**, *32*, 1375–1380. [CrossRef]
77. Słupek, E.; Makoś, P.; Gębicki, J. Deodorization of model biogas by means of novel non-ionic deep eutectic solvent. *Arch. Environ. Prot.* **2020**, *46*, 41–46. [CrossRef]
78. Piechota, G. Multi-step biogas quality improving by adsorptive packed column system as application to biomethane upgrading. *J. Environ. Chem. Eng.* **2021**, *9*, 105944. [CrossRef]
79. McCarty, P.L. Anaerobic waste treatment fundamentals. *Public Work* **1964**, *95*, 9–12.
80. Taricska, J.R.; Long, D.A.; Chen, J.P.; Hung, Y.T.; Zou, S.W. Anaerobic Digestion. In *Biosolids Treatment Processes*; Wang, L.K., Shammas, N.K., Hung, Y.T., Eds.; Handbook of Environmental Engineering; Humana Press: Totowa, NJ, USA, 2007; Volume 6. [CrossRef]
81. Mihai, F.C.; Ingrao, C. Assessment of biowaste losses through unsound waste management practices in rural areas and the role of home composting. *J. Clean. Prod.* **2018**, *172*, 1631–1638. [CrossRef]
82. Khalid, A.; Arshad, M.; Anjum, M.; Mahmood, T.; Dawson, L. The anaerobic digestion of solid organic waste. *Waste Manag.* **2011**, *31*, 1737–1744. [CrossRef]
83. Mathew, A.K.; Bhui, I.; Banerjee, S.N.; Goswami, R.; Shome, A.; Chakraborty, A.K.; Balachandran, S.; Chaudhury, S. Biogas production from locally available aquatic weeds of Santiniketan through anaerobic digestion. *Clean Technol. Environ. Policy* **2015**, *17*, 1681–1688. [CrossRef]
84. Lohani, S.P.; Havukainenm, J. Anaerobic Digestion: Factors Affecting Anaerobic Digestion Process. In *Waste Bioremediation, Energy, Environment and Sustainability*; Varjani, S.J., Gnansounou, E., Baskar, G., Pant, D., Zakaria, Z.A., Eds.; Springer Nature Singapore Pte Ltd.: Singapore, 2018; pp. 343–360. [CrossRef]
85. Van Lier, J.B.; Sanz Martin, J.L.; Lettinga, G. Effect of temperature on the anaerobic thermophilic conversion of volatile fatty acids by dispersed and granular sludge. *Water Res.* **1996**, *30*, 199–207. [CrossRef]
86. Abdelgadir, A.; Chen, X.; Liu, J.; Xie, X.; Zhang, J.; Zhang, K.; Wang, H.; Liu, N. Characteristics, process parameters, and inner components of anaerobic bioreactors. *BioMed Res. Int.* **2014**, *2014*, 841573. [CrossRef]
87. Bendixen, H.J. Safeguards against pathogens in Danish biogas plants. *Water Sci. Technol.* **1994**, *30*, 171–180. [CrossRef]
88. Smith, S.R.; Lang, N.L.; Cheung, K.H.M.; Spanoudaki, K. Factors controlling pathogen destruction during anaerobic digestion of biowastes. *Waste Manag.* **2005**, *25*, 417–425. [CrossRef]
89. Appels, L.; Baeyens, J.; Degrève, J.; Dewil, R. Principles and potential of the anaerobic digestion of waste-activated sludge. *Prog. Energy Combust. Sci.* **2008**, *34*, 755–781. [CrossRef]
90. Van Lier, J.B.; Mohmoud, N.; Zeeman, G. Anaerobic wastewater treatment. In *Biological Wastewater Treatment, Principles, Modelling and Design*; Henze, M., van Loosdrecht, M., Ekama, G.A., Brdjanovic, D., Eds.; IWA Publishing: London, UK, 2008; pp. 415–457.
91. Hwang, M.H.; Jang, N.J.; Hyun, S.H.; Kim, I.S. Anaerobic bio-hydrogen production from ethanol fermentation: The role of pH. *J. Biotechnol.* **2004**, *111*, 297–309. [CrossRef] [PubMed]
92. Patel, A.; Mahboubi, A.; Horváth, I.S.; Taherzadeh, M.J.; Rova, U.; Christakopoulos, P.; Matsakas, L. Volatile Fatty Acids (VFAs) Generated by Anaerobic Digestion Serve as Feedstock for Freshwater and Marine Oleaginous Microorganisms to Produce Biodiesel and Added-Value Compounds. *Front. Microbiol.* **2021**, *12*, 614612. [CrossRef]
93. Tampio, E.A.; Blasco, L.; Vainio, M.M.; Kahala, M.M.; Rasi, S.E. Volatile fatty acids (VFAs) and methane from food waste and cow slurry: Comparison of biogas and VFA fermentation processes. *GCB Bioenergy* **2019**, *11*, 72–84. [CrossRef]
94. Swiatkiewicz, J.; Slezak, R.; Krzystek, L.; Ledakowicz, S. Production of Volatile Fatty Acids in a Semi-Continuous Dark Fermentation of Kitchen Waste: Impact of Organic Loading Rate and Hydraulic Retention Time. *Energies* **2021**, *14*, 2993. [CrossRef]
95. Lusk, P. Latest progress in anaerobic digestion. *Biocycle* **1999**, *40*, 52–54.
96. Rajeshwari, K.V.; Balakrishnan, M.; Kansal, A.; Lata, K.; Kishore, V.V.N. State-of-the-art of anaerobic digestion technology for industrial wastewater treatment. *Renew. Sustain. Energy Rev.* **2000**, *4*, 135–156. [CrossRef]
97. Weiland, P. Stand und Perspektiven der Biogasnutzung und -erzeugung in Deutschland. In *Gülzower Fachgespräche: Band 15: Energetische Nutzung von Biogas: Stand der Technik und Optimierungspotenzial;* Fachagentur Nachwachsende Rohstoffe e. V. (Hrsg.): Gülzow, Germany, 2000.
98. Fujishima, S.; Miyahara, T.; Noike, T. Effect of moisture content on anaerobic digestion of dewatered sludge: Ammonia inhibition to carbohydrate removal and methane production. *Water Sci. Technol.* **2000**, *41*, 119–127. [CrossRef]
99. Kim, S.I.; Kim, H.D.; Hyun, H.S. Effect of particle size and sodium ion concentration on anaerobic thermophilic food waste digestion. *Water Sci. Technol.* **2000**, *41*, 67–73. [CrossRef] [PubMed]
100. Liu, Y.; Tay, J. State of the art of biogranulation technology for wastewater treatment. *Biotechnol. Adv.* **2004**, *22*, 533–563. [CrossRef] [PubMed]

101. Halalsheh, M.; Koppes, J.; Den Elzen, J.; Zeeman, G.; Fayyad, M.; Lettinga, G. Effect of SRT and temperature on biological conversions and the related scum-forming potential. *Water Res.* **2005**, *39*, 2475–2482. [CrossRef] [PubMed]
102. Batstone, D.J.; Keller, J.; Angelidaki, R.I.; Kalyuzhnyi, S.V.; Pavlostathis, S.G.; Rozzi, A.; Sanders, W.T.M.; Siegrist, H.; Vavilin, V.A. *Anaerobic Digestion Model No. 1 (ADM1)*; IWA Publishing: London, UK, 2002.
103. Speece, R.E. *Anaerobic Biotechnology for Industrial Wastewaters*; Archae Press: Nashville, TN, USA, 1996.
104. Arceivala, S.J.; Asolekar, S.R. *Wastewater Treatment for Pollution Control and Reuse*; Tata McGraw-Hill: New Delhi, India, 2007.
105. Foxon, K.M.; Brouckaert, C.J.; Remigi, E.; Buckley, C.A. The anaerobic baffled reactor: An appropriate technology for onsite sanitation. *Water SA* **2006**, *30*, 44–50. [CrossRef]
106. Fidaleo, M.; Laveccio, R. Kinetic study of enzymatic urea hydrolysis in the pH range of 4–9. *Chem. Biochem. Eng.* **2003**, *17*, 311–318.
107. Fotidis, I.A.; Karakashev, D.; Kotsopoulos, T.A.; Martzopoulos, G.G.; Angelidaki, I. Effect of ammonium and acetate on methanogenic pathwayand methanogenic community composition. *FEMS Microbiol. Ecol.* **2013**, *83*, 38–48. [CrossRef] [PubMed]
108. Kadam, P.C.; Boone, D.R. Influence of pH on ammonia accumulation and toxicity in halophilic, Methylotrophic Methanogens. *Appl. Environ. Microbiol.* **1996**, *62*, 4486–4492. [CrossRef] [PubMed]
109. Shanmugam, P.; Horan, N.J. Simple and rapid methods to evaluate methane potential and biomass yield for a range of mixed solid wastes. *Bioresour. Technol.* **2008**, *100*, 471–474. [CrossRef]
110. Chen, Y.; Cheng, J.J.; Creamer, K.S. Inhibition of anaerobic digestion process: A review. *Bioresour. Technol.* **2008**, *99*, 4044–4064. [CrossRef]
111. McKeown, M.R.; Hughes, D.; Collins, G.; Mahony, T.; O'Flaherty, V. Low-temperature anaerobic digestion for wastewater treatment. *Curr. Opin. Biotechnol.* **2012**, *23*, 444–451. [CrossRef]
112. Xia, A.; Cheng, J.; Murphy, D.J. Innovation in biological production and upgrading of methane and hydrogen for use as gaseous transport biofuel. *Biotechnol. Adv.* **2016**, *34*, 451–472. [CrossRef]
113. Monnet, F. An Introduction to Anaerobic Digestion of Organic Wastes. 2003. Available online: http://www.remade.org.uk/media/9102/an-introduction-to-anaerobic-digestion-nov2003.pdf (accessed on 1 September 2021).
114. Khanal, S.K. *Anaerobic Biotechnology for Bioenergy Production: Principles and Applications*; Wiley-Blackwell: Ames, IA, USA, 2008.
115. Colón, J.; Cadena, E.; Pognani, M.; Barrena, R.; Sánchez, A.; Font, X.; Artola, A. Determination of the energy and environmental burdens associated with the biological treatment of source-separated municipal solid wastes. *Energy Environ. Sci.* **2012**, *5*, 5731–5741. [CrossRef]
116. Atelge, M.R.; Krisa, D.; Kumar, G.; Eskicioglu, C.; Nguyen, D.D.; Chang, S.W.; Atabani, A.E.; Al-Muhtaseb, A.H.; Unalan, S. Biogas Production from Organic Waste: Recent Progress and Perspectives. *Waste Biomass Valorization* **2018**, *11*, 1019–1040. [CrossRef]
117. Turan, N.G.; Akdemir, A.; Ergun, O.N. Emission of Volatile Organic Compounds during Composting of Poultry Litter. *Water Air Soil Pollut.* **2007**, *184*, 177–182. [CrossRef]
118. Sironi, S.; Capelli, L.; Céntola, P.; Del Rosso, R. Development of a system for the continuous monitoring of odors from a composting plant: Focus on training, data processing and results validation, methods. *Sens. Actuators B Chem.* **2007**, *124*, 336–346. [CrossRef]
119. Font, X.; Artola, A.; Sánchez, A. Detection, composition and treatment of volatile organic compounds from waste treatment plants. *Sensors* **2011**, *11*, 4043–4059. [CrossRef]
120. Wiśniewska, M.; Kulig, A.; Lelicińska-Serafin, K. The importance of the microclimatic conditions inside and outside of plant buildings in odorants emission at municipal waste biogas installations. *Energies* **2020**, *13*, 6463. [CrossRef]
121. Rincón, C.A.; De Guardia, A.; Couvert, A.; Le Roux, S.; Soutrel, I.; Daumoin, M.; Benoist, J.C. Chemical and odor characterisation of gas emissions released during composting of solid wastes and digestates. *J. Environ. Manag.* **2019**, *233*, 39–53. [CrossRef]
122. Zheng, G.; Liu, J.; Shao, Z.; Chen, T. Emission characteristics and health risk assessment of VOCs from a food waste anaerobic digestion plant: A case study of Suzhou, China. *Environ. Pollut.* **2020**, *257*, 113546. [CrossRef]
123. Wiśniewska, M.; Kulig, A.; Lelicińska-Serafin, K. The Impact of Technological Processes on Odorant Emissions at Municipal Waste Biogas Plants. *Sustainability* **2020**, *12*, 5457. [CrossRef]
124. Wiśniewska, M.; Kulig, A.; Lelicińska-Serafin, K. Odour Emissions of Municipal Waste Biogas Plants—Impact of Technological Factors, Air Temperature and Humidity. *Appl. Sci.* **2020**, *10*, 1093. [CrossRef]
125. Fang, J.; Zhang, H.; Yang, N.; Shao, L.; He, P. Gaseous pollutants emitted from a mechanical biological treatment plant for municipal solid waste: Odor assessment and photochemical reactivity. *J. Air Waste Manag. Assoc.* **2013**, *63*, 1287–1297. [CrossRef]
126. Tsai, C.-J.; Chen, M.-L.; Ye, A.-D.; Chou, M.-S.; Shen, S.-H.; Mao, I.-F. The relationship of odor concentration and the critical components emitted from food waste composting plants. *Atmos. Environ.* **2008**, *42*, 8246–8251. [CrossRef]
127. Komilis, D.P.; Ham, R.K.; Park, J.K. Emission of volatile organic compounds during composting of municipal solid wastes. *Water Res.* **2004**, *38*, 1707–1714. [CrossRef]
128. Scaglia, B.; Orzi, V.; Artola, A.; Font, X.; Davoli, E.; Sanchez, A.; Adani, F. Odours and volatile organic compounds emitted from municipal solid waste at different stage of decomposition and relationship with biological stability. *Bioresour. Technol.* **2011**, *102*, 4638–4645. [CrossRef] [PubMed]

129. IARC. IARC Monographs: On the Evaluation of Carcinogenic Risks to Humans. *Volume 120: Benzene.* 2018. Available online: http://publications.iarc.fr/Book-And-Report-Series/Iarc-Monographs-On-The-Identification-Of-Carcinogenic-Hazards-To-Humans/Benzene-2018 (accessed on 1 September 2021).
130. ATSDR. Toxicological Profile for Xylene. Agency for Toxic Substances and Disease Registry, Department of Health and Human Services, Public Health Service, USA. 2007. Available online: https://www.atsdr.cdc.gov/toxprofiles/tp71.pdf (accessed on 1 September 2021).
131. ATSDR. Toxicological Profile for Ethylbenzene. Agency for Toxic Substances and Disease Registry, Department of Health and Human Services, Public Health Service, USA. 2010. Available online: https://www.atsdr.cdc.gov/toxprofiles/tp110.pdf (accessed on 1 September 2021).
132. ATSDR. Toxicological Profile for Toluene. Agency for Toxic Substances and Disease Registry, Department of Health and Human Services, Public Health Service, USA. 2017. Available online: https://www.atsdr.cdc.gov/toxprofiles/tp56.pdf (accessed on 1 September 2021).
133. Bolden, A.L.; Kwiatkowski, C.F.; Colborn, T. New loot at BTEX: Are ambient levels a problem? *Environ. Sci. Technol.* **2015**, *49*, 5261–5276. [CrossRef] [PubMed]
134. Uchida, Y.; Nakatsuka, H.; Ukai, H. Symptoms and signs in workers exposed predominantly to xylenes. *Int. Arch. Occup. Environ. Health.* **1993**, *64*, 597–605. [CrossRef]
135. Byliński, H.; Barczak, R.J.; Gębicki, J.; Namieśnik, J. Monitoring of odors emitted from stabilised dewatered sludge subjected to aging using proton transfer reaction–mass spectrometry. *Environ. Sci. Pollut. Res.* **2019**, *26*, 5500–5513. [CrossRef]
136. Costa, J.A.V.; de Morais, M.G. The role of biochemical engineering in the production of biofuels from microalgae. *Bioresour. Technol.* **2011**, *102*, 2–9. [CrossRef]
137. World Health Organization. *Guidelines for Air Quality*; WHO: Geneva, Switzerland, 2000.
138. Wiśniewska, M.; Kulig, A.; Lelicińska-Serafin, K. Olfactometric testing as a method for assessing odour nuisance of biogas plants processing municipal waste. *Arch. Environ. Prot.* **2020**, *46*, 60–68. [CrossRef]
139. EUR—Lex. Commission Implementing Decision (EU) 2018/1147 of 10 August 2018 Establishing Best Available Techniques (BAT) Conclusions for Waste Treatment, under Directive 2010/75/EU of the European Parliament and of the Council. Available online: https://eur-lex.europa.eu/legal-content/EN/TXT/?toc=OJ:L:2018:208:TOC&uri=uriserv:OJ.L_.2018.208.01.0038.01.ENG (accessed on 17 August 2018).
140. Umberto, A. Process and technological aspects of municipal solid waste gasification. *Rev. Waste Manag.* **2012**, *32*, 625–639. [CrossRef]
141. Kamuk, B.; Haukohl, J. ISWA Guidelines: Waste to Energy in Low and Middle Income Countries. 2013. Available online: https://www.iswa.org/index.php?eID\protect$\relax\protect{\begingroup1\endgroup\@@over4}$tx_iswaknowledgebase_download&documentUid\protect$\relax\protect{\begingroup1\endgroup\@@over4}$3252 (accessed on 1 September 2021).
142. WEC. *World Energy Resources, Waste to Energy*; World Energy Council: London, UK, 2013. [CrossRef]
143. Santoyo-Castelazo, E.; Azapagic, A. Sustainability assessment of energy systems: Integrating environmental, economic and social aspects. *J. Clean. Prod.* **2014**, *80*, 119–138. [CrossRef]
144. Astrup, T.F.; Tonini, D.; Turconi, R.; Boldrin, A. Life cycle assessment of thermal Waste-to-Energy technologies: Review and recommendations. *Waste Manag.* **2015**, *37*, 104–115. [CrossRef]
145. Jeswani, H.K.; Azapagic, A. Assessing the environmental sustainability of energy recovery from municipal solid waste in the UK. *Waste Manag.* **2016**, *50*, 346–363. [CrossRef]
146. Kumar, A.; Samadder, S.R. A review on technological options of waste to energy for effective management of municipal solid waste. *Waste Manag.* **2017**, *69*, 407–422. [CrossRef] [PubMed]
147. Dong, J.; Tang, Y.; Nzihou, A.; Chi, Y.; Weiss-Hortala, E.; Ni, M. Life cycle assessment of pyrolysis, gasification and incineration waste-to-energy technologies: Theoretical analysis and case study of commercial plants. *Sci. Total Environ.* **2018**, *626*, 744–753. [CrossRef]
148. Zhou, Z.; Tang, Y.; Chi, Y.; Ni, M.; Buekens, A. Waste-to-energy: A review of life cycle assessment and its extension methods. *Waste Manag. Res.* **2018**, *36*, 3–16. [CrossRef] [PubMed]
149. Ng, H.S.; Kee, P.E.; Yim, H.S.; Chen, P.T.; Wei, Y.H.; Lan, J.C.W. Recent advances on the sustainable approaches for conversion and reutilization of food wastes to valuable bioproducts. *Bioresour. Technol.* **2020**, *302*, 122889. [CrossRef] [PubMed]
150. Cottafava, D.; Ritzen, M. Circularity indicator for residential buildings: Addressing the gap between embodied impacts and design aspects. *Resour. Conserv. Recycl.* **2021**, *164*, 105120. [CrossRef]
151. Mazur-Wierzbicka, E. Towards Circular Economy—A Comparative Analysis of the Countries of the European Union. *Resources* **2021**, *10*, 49. [CrossRef]
152. Padilla-Rivera, A.; Carmo, B.B.T.; Arcese, G.; Merveille, N. Social circular economy indicators: Selection through fuzzy delphi method. *Sustain. Prod. Consum.* **2021**, *26*, 101–110. [CrossRef]
153. Haupt, M.; Hellweg, S. Measuring the environmental sustainability of a circular economy. *Environ. Sustain. Indic.* **2019**, *1–2*, 100005. [CrossRef]
154. Hoornweg, D.; Bhada-Tata, P. *What a Waste: A Global Review of Solid Waste Management*; World Bank: Washington, DC, USA, 2012.

155. Rajaeifara, M.A.; Ghanavati, H.; Dashti, B.B.; Heijungs, R.; Aghbashlo, M.; Tabatabaei, M. Electricity generation and GHG emission reduction potentials through different municipal solid waste management technologies: A comparative review. *Renew. Sustain. Energy Rev.* **2017**, *79*, 414–439. [CrossRef]
156. Shirzad, M.; Panahi, H.K.S.; Dasht, B.B.; Rajaeifar, M.A.; Aghbashlo, M.; Tabatabaei, M. A comprehensive review on electricity generation and GHG emission reduction potentials through anaerobic digestion of agricultural and livestock/slaughterhouse wastes in Iran. *Renew. Sustain. Energy Rev.* **2019**, *111*, 571–594. [CrossRef]
157. Schlegelmilch, M.; Streese, J.; Biedermann, W.; Herold, T.; Stegmann, R. Odour control at biowaste composting facilities. *Waste Manag.* **2005**, *25*, 917–927. [CrossRef]
158. Wojnarowska, M.; Sołtysik, M.; Turek, P.; Szakiel, J. Odour Nuisance as a Consequence of Preparation for Circular Economy. *Eur. Res. Stud. J.* **2020**, *23*, 128–142. [CrossRef]
159. Rodríguez-Navas, C.; Forteza, R.; Cerdà, V. Use of thermal desorption-gas chromatography-mass spectrometry (TD-GC-MS) on identification of odorant emission focus by volatile organic compounds characterisation. *Chemosphere* **2012**, *89*, 1426–1436. [CrossRef]
160. World Health Organization. Circular Economy and Health: Opportunities and Risks. 2018. Available online: https://www.euro.who.int/__data/assets/pdf_file/0004/374917/Circular-Economy_EN_WHO_web_august-2018.pdf (accessed on 1 September 2021).
161. Nikonorova, M.; Imoniana, J.O.; Stankeviciene, J. Analysis of social dimension and well-being in the context of circular economy. *Int. J. Glob. Warm.* **2020**, *21*, 3. [CrossRef]
162. Adani, F.; Tambone, F.; Gotti, A. Biostabilization of municipal solid waste. *Waste Manag.* **2004**, *24*, 775–783. [CrossRef]
163. Titta, G.; Viviani, G.; Sabella, D. Biostabilization and Biodrying of Municipal Solid Waste. In Proceedings of the Eleventh International Waste Management and Landfill Symposium, Cagliari, Sardinia, Italy, 1–5 October 2007; pp. 1085–1086.
164. Velis, C.A.; Longhurst, P.J.; Drew, G.H.; Smith, R.; Pollard, S.J.T. Biodrying for mechanical–biological treatment of wastes: A review of process science and engineering. *Bioresour. Technol.* **2009**, *100*, 2747–2761. [CrossRef] [PubMed]
165. Jędrczak, A.; Suchowska-Kisielewicz, M. A comparison of waste stability indices for mechanical-biological waste treatment and composting plant. *Int. J. Environ. Res. Public Health* **2018**, *15*, 2585. [CrossRef] [PubMed]
166. Gliniak, M.; Grabowski, Ł.; Wołosiewicz-Głąb, M.; Polek, D. Influence of ozone aeration on toxic metal content and oxygen activity in green waste compost. *J. Ecol. Eng.* **2017**, *18*, 90–94. [CrossRef]
167. Voberková, S.; Vaverková, M.D.; Burešová, A.; Adamcová, D.; Vršanská, M.; Kynický, J.; Brtnický, M.; Adam, V. Effect of inoculation with white-rot fungi and fungal consortium on the composting efficiency of municipal solid waste. *Waste Manag.* **2017**, *61*, 157–164. [CrossRef] [PubMed]
168. Żygadło, M.; Dębicka, M. The mechanical-biological treatment (MBT) of waste under polish law. *Struct. Environ.* **2014**, *6*, 37–42.
169. Vaverková, M.D.; Elbl, J.; Voběrková, S.; Koda, E.; Adamcová, D.; Gusiatin, Z.M.; Rahman, A.A.; Radziemska, M.; Mazur, Z. Composting versus mechanical–biological treatment: Does it really make a difference in the final product parameters and maturity. *Waste Manag.* **2020**, *106*, 173–183. [CrossRef]
170. Kasiński, S.; Slota, M.; Markowski, M.; Kamińska, A. Municipal solid waste stabilisation in a reactor with an integrated active and passive aeration system. *Waste Manag.* **2016**, *50*, 31–38. [CrossRef]
171. Fourti, O. The maturity tests during the composting of municipal solid wastes. Resour. *Conserv. Recycl.* **2013**, *72*, 43–49. [CrossRef]
172. Cesaro, A.; Belgiorno, V.; Guida, M. Compost from organic solid waste: Quality assessment and European regulations for its sustainable use. Resour. *Conserv. Recycl.* **2015**, *94*, 72–79. [CrossRef]
173. Pavlas, M.; Dvořáček, J.; Pitschke, T.; Peche, R. Biowaste treatment and waste-to-energy-environmental benefits. *Energies* **2020**, *13*, 1994. [CrossRef]
174. Garfi, M.; Gelman, P.; Comas, J.; Carrasco, W.; Ferrer, I. Agricultural reuse of the digestate from low-cost tubular digesters in rural Andean communities. *Waste Manag.* **2011**, *31*, 2584–2589. [CrossRef]
175. Koszel, M.; Lorencowicz, E. Agricultural use of biogas digestate as a replacement fertilisers. *Agric. Agric. Sci. Procedia* **2015**, *7*, 119–124. [CrossRef]
176. Czekała, W.; Lewicki, A.; Pochwatka, P.; Czekała, A.; Wojcieszak, D.; Jóźwiakowski, K.; Waliszewska, H. Digestate management in polish farms as an element of the nutrient cycle. *J. Clean. Prod.* **2020**, *242*, 1184545. [CrossRef]
177. O-Thong, S.; Boe, K.; Angelidaki, I. Thermophilic anaerobic codigestion of oil palm empty fruit bunches with palm oil mill effluent for efficient biogas production. *Appl. Energy* **2012**, *93*, 648–654. [CrossRef]
178. Borowski, S.; Domański, J.; Weatherley, L. Anaerobic co-digestion of swine and poultry manure with municipal sewage sludge. *Waste Manag.* **2014**, *34*, 513–521. [CrossRef]
179. Zhang, C.; Xiao, G.; Peng, L.; Su, H.; Tan, T. The anaerobic co-digestion of food waste and cattle manure. *Bioresour. Technol.* **2013**, *129*, 170–176. [CrossRef]
180. Wei, Y.; Li, X.; Yu, L.; Zou, D.; Yuan, H. Mesophilic anaerobic codigestion of cattle manure and corn stover with biological and chemical pre-treatment. *Bioresour. Technol.* **2015**, *198*, 431–436. [CrossRef]
181. Kougias, P.G.; Kotsopoulos, T.A.; Martzopoulos, G.G. Effect of feedstock composition and organic loading rate during the mesophilic co-digestion of olive mill wastewater and swine manure. *Renew. Energy* **2014**, *69*, 202–207. [CrossRef]
182. Hartmann, H.; Angelidaki, I.; Ahring, B.K. Co-digestion of the organic fraction of municipal waste with other waste types. In *Biomethanization of the Organic Fraction of Municipal Solid Wastes*; IWA Publishing: London, UK, 2002.

183. Heo, N.H.; Park, S.C.; Lee, J.S.; Kang, H.; Park, D.H. Single-Stage Anaerobic Codigestion for Mixture Wastes of Simulated Korean Food Waste and Waste Activated Sludge. In *Biotechnology for Fuels and Chemicals*; Davison, B.H., Lee, J.W., Finkelstein, M., McMillan, J.D., Eds.; Humana Press: Totowa, NJ, USA, 2003; pp. 567–579, ISBN 978-1-4612-6592-4.
184. Sosnowski, P.; Wieczorek, A.; Ledakowicz, S. Anaerobic co-digestion of sewage sludge and organic fraction of municipal solid wastes. *Adv. Environ. Res.* **2003**, *7*, 609–616. [CrossRef]
185. Heo, N.H.; Park, S.C.; Kang, H. Effects of Mixture Ratio and Hydraulic Retention Time on Single-Stage Anaerobic Co-digestion of Food Waste and Waste Activated Sludge. *J. Environ. Sci. Health Part A* **2004**, *39*, 1739–1756. [CrossRef] [PubMed]
186. Sosnowski, P.; Klepacz-Smolka, A.; Kaczorek, K.; Ledakowicz, S. Kinetic investigations of methane cofermentation of sewage sludge and organic fraction of municipal solid wastes. *Bioresour. Technol.* **2008**, *99*, 5731–5737. [CrossRef] [PubMed]
187. Tsapekos, P.; Kougias, P.G.; Treu, L.; Campanaro, S.; Angelidaki, I. Process performance and comparative metagenomic analysis during co-digestion of manure and lignocellulosic biomass for biogas production. *Appl. Energy* **2017**, *185*, 126–135. [CrossRef]
188. Liu, C.; Li, H.; Zhang, Y.; Liu, C. Improve biogas production from loworganic-content sludge through high-solids anaerobic co-digestion with food waste. *Bioresour. Technol.* **2016**, *219*, 252–260. [CrossRef] [PubMed]
189. Mata-Alvarez, J.; Dosta, J.; Macé, S.; Astals, S. Codigestion of solid wastes: A review of its uses and perspectives including modeling. *Crit. Rev. Biotechnol.* **2011**, *31*, 99–111. [CrossRef] [PubMed]
190. Tsapekos, P.; Kougias, P.G.; Angelidaki, I. Anaerobic mono- and codigestion of mechanically pretreated meadow grass for biogas production. *Energy Fuels* **2015**, *29*, 4005–4010. [CrossRef]
191. Søndergaard, M.M.; Fotidis, I.A.; Kovalovszki, A.; Angelidaki, I. Anaerobic co-digestion of agricultural byproducts with manure for enhanced biogas production. *Energy Fuels* **2015**, *29*, 8088–8094. [CrossRef]
192. Dennehy, C.; Lawlor, P.G.; Gardiner, G.E.; Jiang, Y.; Cormican, P.; McCabe, M.S.; Zhan, X. Process stability and microbial community composition in pig manure and food waste anaerobic co-digesters operated at low HRTs. *Front. Environ. Sci. Eng.* **2017**, *11*, 4. [CrossRef]
193. De Baere, L.; Mattheeuws, B.; Anaerobic Digestion of the Organic Fraction of Municipal Solid Waste in Europe. Anaerobic Digestion of the Organic Fraction of Municipal Solid Waste in Europe—Status, Experience and Prospects. Available online: https://www.ows.be/wp-content/uploads/2013/02/Anaerobic-digestion-of-the-organic-fraction-of-MSW-in-Europe.pdf (accessed on 1 September 2021).
194. Stucki, M.; Jungbluth, N.; Leuenberger, M. *Life Cycle Assessment of Biogas Production from Different Substrates*; Final Report; Federal Department of Environment, Transport, Energy and Communications, Federal Office of Energy: Bern, Switzerland, 2011.
195. Sustainable Energy Authority of Ireland. *Gas Yields Table*; Sustainable Energy Authority of Ireland: Dublin, Ireland, 2002.
196. Conti, C.; Guarino, M.; Bacenetti, J. Measurements techniques and models to assess odor annoyance: A review. *Environ. Int.* **2020**, *134*, 105261. [CrossRef]
197. Toledo, M.; Gutiérrez, M.; Siles, J.; Martín, M. Full-scale composting of sewage sludge and market waste: Stability monitoring and odor dispersion modeling. *Environ. Res.* **2018**, *167*, 739–750. [CrossRef]
198. Toledo, M.; Gutiérrez, M.; Siles, J.; Martín, M. Odor mapping of an urban waste management plant: Chemometric approach and correlation between physico-chemical, respirometric and olfactometric variables. *J. Clean. Prod.* **2019**, *210*, 1098–1108. [CrossRef]
199. Kamaruddin, M.A.; Yusoff, M.S.; Aziz, H.A.; Hung, Y.-T. Sustainable treatment of landfill leachate. *Appl. Water Sci.* **2015**, *5*, 113–126. [CrossRef]
200. Gao, J.; Oloibiri, V.; Chys, M.; Audenaert, W.; Decostere, B.; He, Y.; Van Langenhove, H.; Demeestere, K.; Van Hulle, S.W.H. The present status of landfill leachate treatment and its development trend from a technological point of view. *Rev. Environ. Sci. Biotechnol.* **2015**, *14*, 93–122. [CrossRef]
201. Yuan, M.; Chen, Y.; Tsai, J.; Chang, C. Ammonia removal from ammonia-rich wastewater. *Process. Saf. Environ. Prot.* **2016**, *102*, 777–785. [CrossRef]
202. Wang, D.; Liu, D.; Tao, L.; Li, Z. The impact on the effects of leachate concentrates recirculation for different fill age waste. *J. Mater. Cycles Waste Manag.* **2016**, *19*, 1211–1219. [CrossRef]
203. Hashisho, J.; El-Fadel, M. Membrane bioreactor technology for leachate treatment at solid waste landfills. *Rev. Environ. Sci. Bio/Technol.* **2016**, *15*, 441–463. [CrossRef]
204. Wiśniewska, M.; Szyłak-Szydłowski, M. The Air and Sewage Pollutants from Biological Waste Treatment. *Processes* **2021**, *9*, 250. [CrossRef]
205. Stretton-Maycock, D.; Merrington, G. *The Use and Application to Land of MBT Compost-Like Output—Review of Current European Practice in Relation to Environmental Protection*; Environment Agency: Bristol, UK, 2009.
206. Jamal, M.; Szefler, A.; Kelly, C.; Bond, N. Commercial and Household Food Waste Separation Behaviour and the Role of Local Authority: A Case Study. *Int. J. Recycl. Org. Waste Agric.* **2019**, *8*, 281–290. [CrossRef]

Article

Modern Use of Water Produced by Purification of Municipal Wastewater: A Case Study

Giorgia Tomassi [1,*], Pietro Romano [2] and Gabriele Di Giacomo [2,*]

1 Department of Industrial Engineering, University of Padova, Via Gradenigo 6/a, 35131 Padova, Italy
2 Department of Industrial and Information Engineering and of Economics, Campus of Roio, University of L'Aquila, 67100 L'Aquila, Italy; pietro.romano@graduate.univaq.it
* Correspondence: giorgia.tomassi@studenti.unipd.it (G.T.); gabriele.digiacomo@univaq.it (G.D.G.); Tel.: +39-3381595634 (G.D.G.)

Abstract: All the urban areas of developed countries have hydric distribution grids and sewage systems for collecting municipal wastewater to treatment plants. In this way, the municipal wastewater is purified from human excreta and other minor contaminants while producing excess sludges and purified water. In arid and semi-arid areas of the world, the purified water can be used, before discharging, to enhance the energy efficiency of seawater desalination and solve the problems of marine pollution created by desalination plants. Over the past half-century, seawater desalination has gradually met demand in urbanized, oil-rich, arid areas. At the same time, technological evolution has made it possible to significantly increase the energy efficiency of the plants and reduce the unit cost of the produced water. However, for some years, these trends have flattened out. The purified water passes through the hybridized desalination plant and produces renewable osmotic energy before the final discharge in the sea to restart the descent behaviour. Current technological development of reverse osmosis (RO), pressure retarded osmosis (PRO) and very efficient energy recovery devices (ERDs) allows this. Furthermore, it is reasonable to predict that, in the short-medium term, a new generation of membranes specifically designed for improving the performance of the pressure retarded osmosis will be available. In such circumstances, the presently estimated 13-20% decrease of the specific energy consumption will improve up to more than 30%. With the hybrid plant, the salinity of the final discharged brine is like that of seawater, while the adverse effect of GHG emission will be significantly mitigated.

Keywords: freshwater from the sea; purification of municipal wastewater; osmotic energy; reverse osmosis; pressure retarded osmosis; process integration; energy efficiency; environmental sustainability

Citation: Tomassi, G.; Romano, P.; Di Giacomo, G. Modern Use of Water Produced by Purification of Municipal Wastewater: A Case Study. *Energies* **2021**, *14*, 7610. https:///doi.org10.3390/en14227610

Academic Editor: José Carlos Magalhães Pires

Received: 12 October 2021
Accepted: 11 November 2021
Published: 14 November 2021

1. Introduction

The total volume of water on Earth is estimated at 1.386 billion km^3, with 97.5% being saltwater, having an average concentration of 35g/L. The remaining water is freshwater, of which only 0.3% is in liquid form on the surface [1].

Freshwater availability is naturally produced by evaporation from the oceans and subsequent recondensation and precipitation, as was already observed by Aristotle [2,3]. Unfortunately, it is unevenly distributed on the planet and rarely meets the requirements for many uses by humans and animals.

In developed countries, the per-capita human consumption of freshwater is about 200 L/d, sometimes even double. Both agricultural and industrial activities also require a large amount of freshwater [4]. To overcome freshwater scarcity, oil-rich countries started to build and operate evaporative desalination plants like multi-stage flash (MSF), multiple-effect distillation (MED), and multi-effect thermal vapor compression desalination (TVC). Shortly, the production of freshwater from the sea became popular in other developed countries. Furthermore, the use of semipermeable membranes to reject dissolved salts was demonstrated to be a better way for seawater desalination [5]. In 2018, the global share of the RO for seawater desalination was about 65%, with an increasing trend [6].

The history of industrial desalination dates to the beginning of 1950, with the birth of the Office of Saline Water (OSW) in the United States of America, which gave the first strong impulse to develop efficient technologies for industrial production of freshwater from the sea. In 1963, all the installed capacity of desalination plants was about a thousand m^3/d of freshwater [2]. In 1975, the first automatically controlled MSF desalination plant, with a production of two thousand m^3/h of freshwater (the largest in the world at that time), was built and put into operation in Italy (with Italian technology), serving the water needs of the Porto Torres industrial site in Sardinia [7,8]. Subsequently, numerous other plants were built worldwide, particularly in the emirates bordering the Persian Gulf. According to the latest inventory (30th Desal Data) published in October 2017 by the International Desalination Association (IDA) together with Global Water Intelligence (GWI), the global production of freshwater from the sea was 92.5 billion L/d with 19,372 plants located in 150 countries, against 88.6 billion L/d e and 18,983 plants of the previous year. It is worth noting that ten years earlier, the production of desalted water was about half. This strong growth, associated with the introduction of modern and cheaper technologies, resulted in a significant decrease in the unit cost of freshwater and an equally considerable increase in the energy efficiency of desalination plants. In recent years, the produced water unit cost (0.53 to 1.2 US$/$m^3$) and energy efficiency (2 to 3 kWh/m^3 of freshwater) have reached stable values [2,9,10].

On the other hand, environmental problems related to the discharge of very concentrated brines and global warming of the planet have arisen [11,12].

It remains that water security is one of the principal global risks, particularly in developing and underdeveloped countries that have a tremendous and urgent need for large quantities of freshwater produced at sustainable cost and low environmental impact. About a third of the world's population already lives in countries considered to be in water emergencies. If this trend continues, two-thirds of the earth's population will be thirsty in twenty years [2,13]. For this, the international exposition held in Milan in 2015, "EXPO Milano 2015," organized and hosted the "Water Forum," aiming to make public and private entities aware of this serious issue and stimulate the adaption of the legislation and new investments capable of tackling the problem adequately [14]. The availability and sustainable management of water were also featured prominently in the 6th Sustainable Developments Goals (SDGs) of the 2030 Agenda for Sustainable Development, agreed to by 193 nations in 2015 [15]. Undoubtedly, desalination is essential to bridge the water supply gap expected in the increasingly water-demanding world. The research and development activity to re-establish the growth of seawater desalination at a sustained pace is very intense. To this purpose, it is necessary to change the paradigm that brine and water contaminated by organic compounds are not wastes but sources for additional freshwater, energy, and useful materials. This may be realized, in the short-medium term, by accomplishing seawater desalination using processes obtained through the integration of two or more technologies already consolidated, and through the integration of two or more already existing systems. An example is the one described in an old article concerning the recovery of Magnesium from the brine of an MSF plant [16]. As highlighted in the article, with the hybrid process, Magnesium is produced from a solution that has a concentration approximately double that of seawater. Furthermore, this solution does not require any pre-treatment as it is the product of pre-treated seawater. The advantages of sustainability and environmental impact are equally evident as the ocean represents an inexhaustible and renewable source of Magnesium while mineral resources are limited. The same goes for other metals of increasing commercial importance, such as lithium [17]. The status and prospects of hybrid processes using brine, thermal waste, and water from wastewater purification are described in detail in several recent publications [18–22]. These processes aim to increase the productivity and energy efficiency of desalination, reducing the environmental impact. To this purpose, it is worth citing the ongoing research and technological development activities for improving the performance of the membranes through the homogenization of the density of the filtering surface, as well as the reduction of the bilateral polarization effects by modification of their hydrophobicity at the nanoscale. It seems reasonable to predict that such membranes could be available for full-scale application in the short-medium term [23]. When applied to seawater desalination, a different but not ready applicable way is to use new separation technologies characterized by better performance and economic feasibility [20,23–31].

Everyone agrees that process integration is the primary way to increase productivity, diversify the use of raw materials and products, and reduce the cost of desalination in the context of general sustainability. At the same time, the relentless substantial increase of RO desalination plants' market share is a fact. We did not find any evidence on the construction of new plants of the classic evaporative (MSF), multiple effects distillation (MED), or thermocompression (VTC) types. Of these, the RO-PRO is the one that appears most interesting both in terms of feasibility on an industrial scale and for future prospects [22,30,32–36]. This integration can be achieved with different schemes and with different types of impaired, low salinity water. These options have been analysed using theoretical models to find the best way of combining RO with PRO [32]. However, the conclusion led to conflicting results as a consequence of the many assumptions required by the models. As supported by others [37], we believe that combining the impaired, low salted water with the most concentrated brine is the best configuration for the RO-PRO hybridization.

In arid, urban places, the freshwater produced by seawater desalination is first distributed and used, collected through the sewage grid, purified in the wastewater treatment plants (WWTPs), and finally discharged into the sea at the net of losses. Therefore, there are two comparable streams, i.e., one of desalted water (PW) and one of very concentrated brine. These two streams can meet in the PRO section of the hybrid RO-PRO plant to directly transfer the resulting osmotic power to a portion of the incoming pre-treated seawater (PSW). RO-PRO integration can be achieved in several ways, just as the renewable energy produced within the hybrid plant can be transferred to PSW, with different equipment [18,38,39]. The results that can be obtained are different and depend on the specific conditions in which the desalination plant operates. Regarding the location, there is a vast choice, since, in the world, there are many urban areas where a new seawater desalination plant exists or is necessary. The most critical and, at the same time, most favourable situation is that of the urban areas of the Persian Gulf where there is the highest concentration of desalination plants, both of the evaporative and of the RO type. The criticality depends on the fact that seawater salinity is significantly higher than the average. The advantage derives from a wider choice, from the simultaneous presence of plants based on different technologies and old plants. The results of the research in those conditions of salinity and aridity are undoubtedly conservative. Therefore, once a well-defined configuration for the hybrid RO-PRO desalination processes was selected, this research aimed to describe and evaluate its performance in an arid urban area of the Persian Gulf. More specifically, we aimed to quantify the amount of renewable osmotic energy produced inside the hybrid desalination plant, integrating already available services and technologies. Then, we wanted to quantify the difference between the hybrid process's specific energy consumption (SEC) by comparison with a modern RO stand-alone desalination process, along with other beneficial effects. Furthermore, we aimed to quantitatively show how the energy efficiency of desalination improves, increasing the operating pressure of the PRO section.

2. Materials, Methods, Processes and Products

The salinity (TDS) of the seawater was assumed to be 43 g/L, following the assumption that the desalination plant is in an arid and urban area of the United Arab Emirates (UAE) in the Persian Gulf [39]. This choice is because there is the highest density of desalination plants (equal to 48% of the world total [40,41]), and it is one of the world's driest places exposed to a chronic freshwater shortage [42].

It is further assumed that the desalting plant is of the RO stand-alone type, having a recovery factor of 40% and an operating pressure of 70 bar. The plant is equipped with Pressure Exchangers (PXs) as energy recovery devices with 95% efficiency. Inside the PX, a balanced flow is maintained (i.e., high and low-pressure flow rates are equal) to maximize pressure exchange [43,44]. The TDS of the brine is about 7.2%, while its pressure is 70 bar. Freshwater is produced at atmospheric pressure with a TDS lower than 0.02%. The operating temperature is 30 °C.

Apart from freshwater and very diluted salted solution, the osmotic pressure (π) of the seawater and brine cannot be quantitatively calculated by the classical van't Hoff equation due to the significant deviation from the ideal behaviour resulting from the high concentration of the salt. As shown in previous work [45], the following non-simplified van't Hoff equation can satisfactorily account for this phenomenon up to very high values of TDS, as well as for the operating temperature (T).

$$\pi = \frac{-RT \cdot ln(a_w)}{\overline{V}_w} \tag{1}$$

Here a_w represents the activity of the water and $\overline{V_w}$ represents the partial molar volume of the water in the seawater solution (SW), which could be approximated with the molar volume of the pure water, V_w (L/mol) at the current T (K). When the value of the universal gas constant R is equal to 0.0831446261815324, π results in bar [46].

Alternatively, Figure 1, obtained by fitting experimental data available in the literature [47,48] can be confidently used at 30 °C.

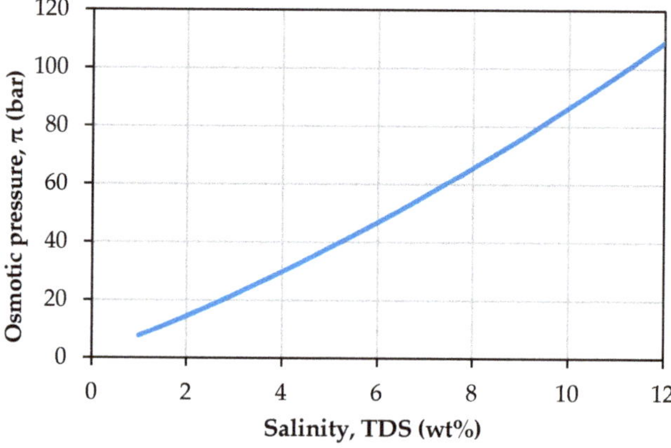

Figure 1. Osmotic pressure of seawater and its diluted and concentrated solutions as a function of the TDS at 30 °C.

Considering the flow rate PW entering the PRO section, it is assumed to be equal to 85% of the freshwater (FW) produced by the RO of seawater, accounting for evaporation and possible leaking in the hydric distribution grid, and in the sewage system. It comes from the activated sludge or similar (WWTPs) serving the urban area. To justify this assumption, it is worth considering that this technology is by far the most diffused and consolidated, with an efficiency higher than 99%. It is also worth emphasizing that when evaporative desalting plants co-exist in the same area with RO based plants, the unsalted stream can be even much higher. Furthermore, it is assumed that the TDS of PW is equal to zero, the operating pressure of PRO is in the range 22–31 bar, and both purified water and brine at atmospheric pressure do not need any pre-treatment before entering the PRO plant. In addition, high-pressure and booster pumps efficiencies were considered equal to 85% and 75%, respectively [44].

Finally, the (SEC) as kWh/m^3 of freshwater produced by the hybrid RO-PRO desalination process was calculated with the following equation [38].

$$SEC = \frac{P_{pumps}^{TOT}}{Q_{FW}} \quad (2)$$

where Q_{FW} (m^3/h) is the constant flow rate of the freshwater produced by the RO section and P_{pumps}^{TOT} (kW) is the sum of the power required by each pump.

Figure 2 schematically shows a typical activated sludge WWTP widely used to purify municipal wastewater mainly contaminated by human excreta. As can be seen, there are two effluent streams: very wet sewage sludge (SS) and purified water (PW), which is usually discharged into some superficial water body or directly into the sea. Several books, scientific papers, and reviews were published and are ongoing on the final disposal of SS [37,48–56], but much less attention was addressed to better use of PW.

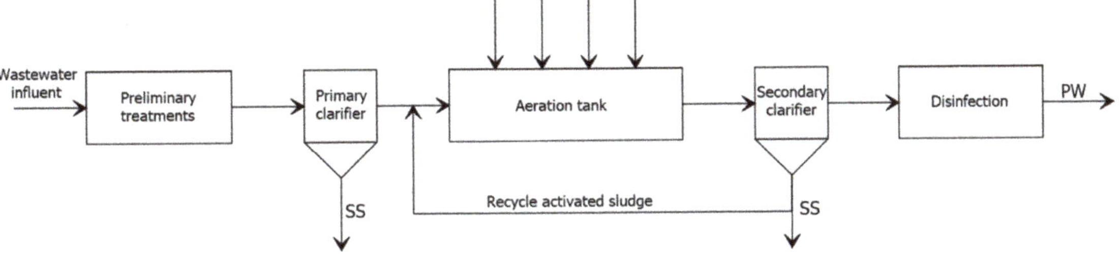

Figure 2. Schematic diagram of a typical activated sludge WWTP.

Figure 3 schematically shows a Traditional Seawater Reverse Osmosis (TSWRO) process.

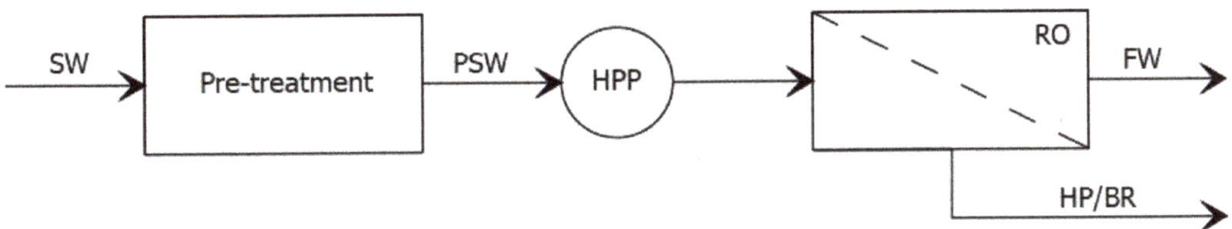

Figure 3. Schematic diagram of a traditional RO seawater desalination process TSWRO.

The pre-treatment unit removes particles and suspended solids in the feed stream that may cause fouling and scaling on the RO membrane surface. The high-pressure pump (HPP) raises the feed pressure from atmospheric to a value adequately higher than the osmotic pressure of the final brine. Then, PSW enters the RO system, where FW flows through the semipermeable membranes while the salts are almost entirely rejected. The Permeate FW is taken as a product after minor treatment at atmospheric pressure, while the pressurized retentate or brine, HP/BR, is a waste characterized by high salt concentration ($TDS_{brine} >> TDS_{seawater}$). TSWRO initially operated at the lowest possible pressure to avoid membrane modules' mechanical stress and minimize the energy for feed pressurization. The recovery, operating in this condition, was low, and the TDS of brine was not significantly different from the TDS of PSW. With better performing membranes and new sophisticated Energy Recovery Devices (ERDs), the modern RO desalination plants adopted more convenient operating conditions to reduce the value of the SEC [57].

In fact, the value of the SEC, and related operating cost, for RO desalination is mainly determined by the energy required for PSW pressurization [13]. Investment and operating costs, as well as other energy-consuming operations (e.g., pre and post-treatments, construction, maintenance), are not susceptible to further improvements, as they are definitively optimized together with the potential of the plants. For this, the new RO based SW desalination plants are equipped with a section for recovering most of the energy of the pressurized brine.

Figure 4 schematically shows an improved seawater desalination process by Reverse Osmosis (ISWRO). As can be seen, an isobaric Pressure Exchanger (PX) pressurizes a portion of PSW by using the HP/BR stream. The rate of PSW_2 is the same as that of the brine to maximize the performance of PX. Then, the low-pressure brine (LP/BR) is discharged, whereas the pressurized PSW_2 is added by a booster pump (BP) to PSW_1 and enters the RO modules. Pressure exchangers were introduced in 2001 [57], and their efficiency is over 96% [58]. The value of the SEC for an industrial desalination plant built and operated according to Figure 4 is still significantly higher than the calculated theoretical minimum for seawater desalination (1.06 kWh/m^3) [57]. There is, therefore, ample room for improvement concerning energy efficiency. Furthermore, a serious environmental problem is associated with the discharge of highly concentrated brines into the sea [11].

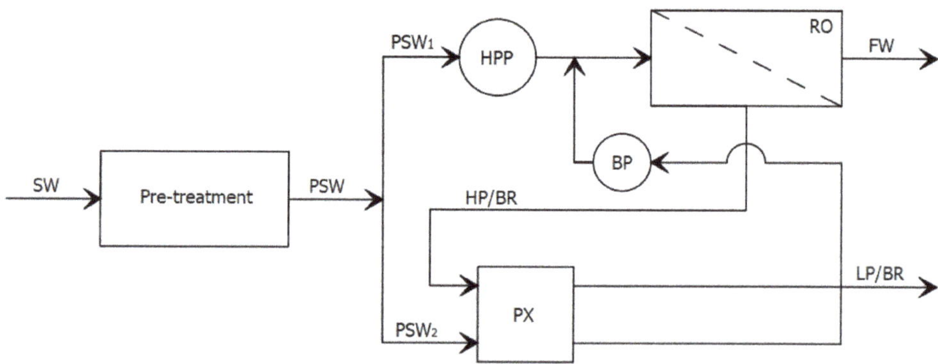

Figure 4. Schematic diagram of a typical modern RO seawater desalination process ISWRO with energy recovery from the brine.

For these reasons, it may be advantageous to consider a natural osmosis process, known as PRO, that can be carried out with the same membranes and filter modules used in RO based desalination plants. Figure 5 schematically shows the working principle of the PRO.

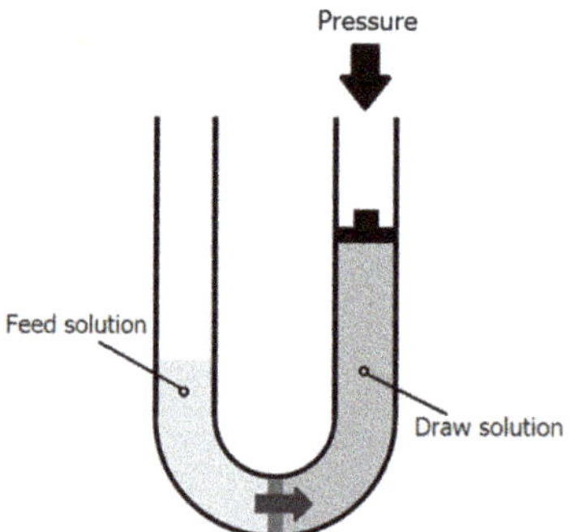

Figure 5. PRO working principle.

As can be seen, the pressurization of a stream of salty water (draw solution) up to a value lower than its osmotic pressure [26] allows freshwater (feed solution) to flow through the semipermeable osmotic membrane to mix with the draw solution [59]. The result of this simple osmotic process is a pressurized stream of brackish water that can be used for producing renewable energy, as schematically shown in Figure 6 [11,60].

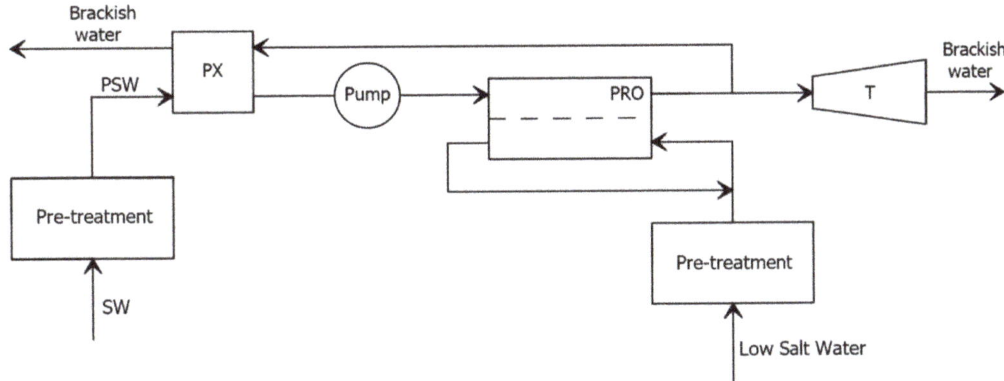

Figure 6. Schematic diagram of typical PRO process for the production of renewable osmotic energy.

The necessary condition for the operation of this technology is the presence of two currents with different salinity, such as river and seawater [61]. Although the PRO stand-alone process has been conceived as a technology for making renewable energy, currently, it is not economically competitive compared to other forms of renewable energy available, as stated by the Norwegian renewable energy company Statkraft [11,62].

It may be viable, using the osmotic energy produced by PRO, to directly pressurize, by a Dual Work Exchanger Energy device (DWEER), a liquid stream portion of PSW to be desalted by RO, from the atmospheric pressure up to a pressure close to the pressurized draw solution in the PRO [11,63]. It is worth underling that, contrary to PX, the DWEER completely avoids any mixing between the pressurized and the pressuring streams, while maintaining an efficiency above 95%. In such a way, the quality of the FW is the same as that produced by the ISWRO stand-alone. Figure 7 schematically shows the hybrid desalination process obtained by properly interconnecting RO and PRO along with a PX and a DWEER, and all the required pumps.

As can be seen, in this way, the PSW_2 stream is partially pressurized by renewable energy before entering the RO section. In addition, the TDS of the discharged low pressure brackish (LP/BW) is considerably lower than the LP/BR discharged when using the process schematically depicted in Figure 4. The draw solution feeding the PRO section is the medium-pressure brine, MP/BR, leaving PX. Considering the purified water stream, PW, feeding the same section, it is worthy of highlighting that, in an arid urban area, it usually comes from the sewage water treatment plant, which receives all the polluted water from the urban sewage system. At least in principle, this process should be more energetically efficient than the ISWRO desalination one, and simultaneously, it should represent the solution to the environmental concerns related to the discharge of very concentrated brine. Both these effects will be quantified and critically analysed in the following section.

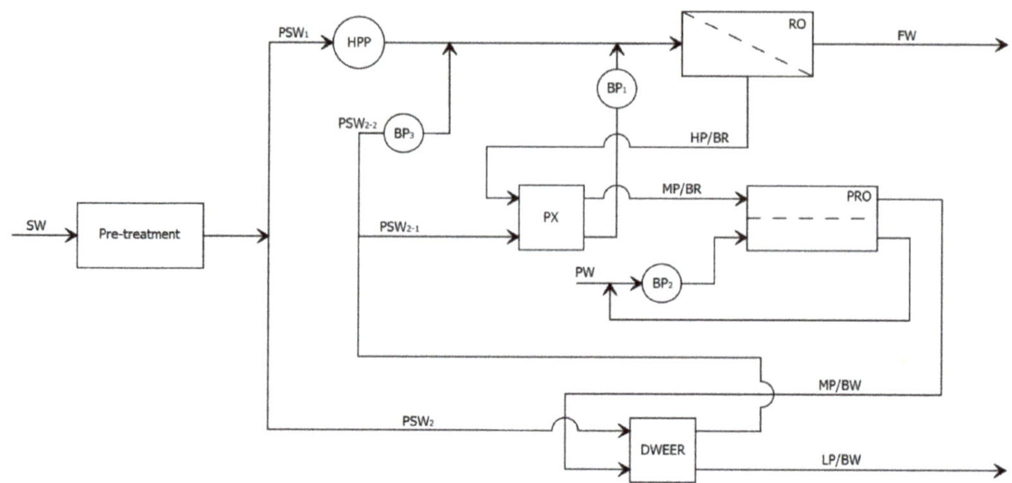

Figure 7. Schematic diagram of the selected RO-PRO hybrid desalination process for an arid urban area.

3. Results and Discussion

The volumetric flow rate, temperature, pressure, salinity, and osmotic pressure are reported in Tables 1 and 2; the first refers to an ISWRO process (Figure 4), while the second refers to the RO-PRO hybrid process (Figure 7). As can be seen, starting with an arbitrarily fixed flow rate of a PSW equal to 1000 m^3/h, and according to the assumption made in the previous section, FW is always equal to 400 m^3/h, since the recovery factor was equal to 40%, in both cases. Considering the hybrid process, PW is 85% of FW, or 340 m^3/h. It is worth emphasizing that the arbitrary assumption made for the flow rate of FW does not represent any limitation for the generality of our approach to quantify the value of the SEC and the concentration of the discharged saline solution in both cases.

Table 1. Material balances and relevant properties of the incoming and the outcoming liquid streams for the ISWRO desalination process.

Stream	Volumetric Flow Rate, Q (m^3/h)	T (°C)	P (bar)	π (bar)	TDS (wt%)
PSW	1000	30	1	32	4.3
FW	400	30	1	-	0.02
LP/BR	600	30	1	58	7.2

Table 2. Material balances and relevant properties of the incoming and the outcoming liquid streams for the selected RO-PRO hybrid desalination process.

Stream	Volumetric Flow Rate, Q (m^3/h)	T (°C)	P (bar)	π (bar)	TDS (wt%)
PSW	1000	30	1	32	4.3
PW	340	30	1	-	0
FW	400	30	1	-	0.02
LP/BW	940	30	1	35	4.6

In addition to what was underlined earlier, it is worth emphasizing that the flow rate of the discharged stream from the hybrid RO-PRO process is higher than that of the ISWRO process. However, the salinity of this stream is quite different: the ISWRO process is about 1.7 times higher, while for the RO-PRO hybrid process it is only 1.07 times higher, thus solving marine pollution.

Considering the energy efficiency, the blue line in Figure 8 shows the behaviour of the SEC for the selected RO-PRO process as a function of the PRO pressure (P_{PRO}), from zero up to an ideal value equal to the osmotic pressure of the LP/BR solution. As can be seen, the SEC decreases from the value associated with an ISWRO process to that of the hybrid RO-PRO process having an ideal PRO section. Quantitatively, this trend is given by the following equation:

$$SEC = 9E - 0.6\ P_{PRO}^3 - 0.0009\ P_{PRO}^2 + 0.0004\ P_{PRO} + 2.45 \quad (3)$$

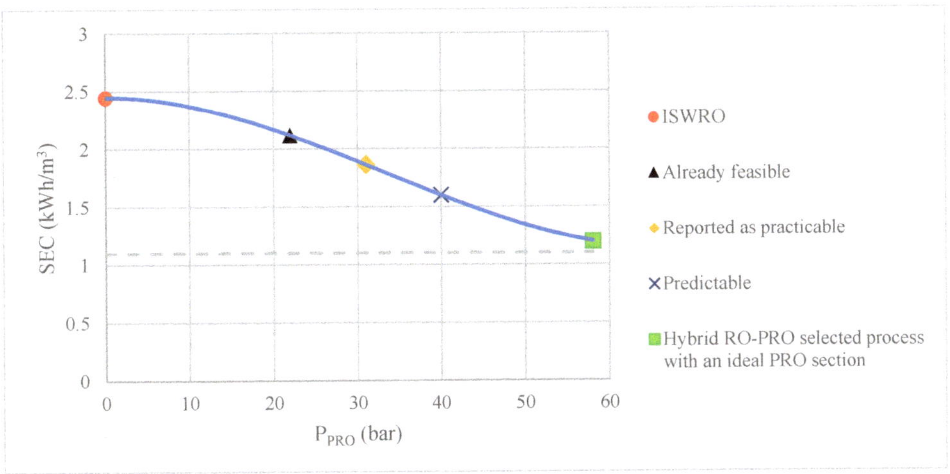

Figure 8. SEC behaviour of the selected hybrid RO-PRO desalination process as a function of the operating pressure of PRO. The dashed line represents the theoretical energy required for desalting seawater with an average TDS of 3.5 wt%.

The SEC was calculated by Equation (2) according to the assumption made earlier on about the overall efficiency of each pump, of PX, and of DWEER. Overall:

- when $P_{PRO} = 0$, there is no improvement in the energetic efficiency of the desalination. The value of the SEC would be the same as that of the ISWRO since the hybrid process would generate no indigenous osmotic power. In this condition, PRO acts as a forward osmosis section.
- If PRO could operate in the other ideal limiting condition (P_{PRO} equal to the osmotic pressure of LP/BR), the energy efficiency of desalination would increase drastically. In fact, in this case, much renewable energy would be produced inside the desalination plant in the form of osmotic pressure and be directly used for the partial pumping of seawater to the reverse osmosis plant.
- With PRO working between 22 and 31 bar (already applicable in conditions [37,38]), the value of the SEC would be in the range from 12.7% to 24% lower than that corresponding to that of ISWRO.
- With PRO working at 40 bar (foreseeably applicable at short-medium term), the value of the SEC would be about 30% lower than that corresponding to that of ISWRO.

As a consequence of the redaction of the SEC, the GHG pollution is also mitigated since the present reduction of CO_2 produced by the hybrid process is in the range of 0.2–0.45 kg/m^3 of FW. At the short-medium time, it is reasonably predictable as 0.6 kg/m^3 of FW.

4. Conclusions

In arid urban areas, such as those of the Persian Gulf, seawater desalination is the only system of supplying the drinking water needed. Currently, this inevitable activity is associated with high consumption of non-renewable energy and, in many cases, with non-marginal pollution problems linked to the discharge of excessively concentrated brines into the sea.

The results of this research highlighted the possibility of mitigating or solving these problems by the rational use of purified municipal wastewater. Instead of discharging it directly into the sea, this aqueous current, practically unsalted, can be employed to produce renewable osmotic energy inside the desalination plant. This energy is used directly, by a very efficient DWEER, to partially pressurize the seawater fed to the Reverse Osmosis for producing freshwater. The amount of renewable energy produced by PRO depends on the water flow from the WWTPs, and from the operating pressure of this section of the desalination hybrid process. Presently, the reduction of the SEC is between 12.7% and 24% compared to that of ISWRO. At the same time, the problem of discharging excessively concentrated brines into the sea is practically solved.

The ongoing research and technological development activities aim to improve the performance of the membranes through the homogenization of the density of the filtering surface and the reduction of polarization effects by modifying their hydrophobicity.

In the short-medium term, when better-performing membranes will likely be available for full-scale application, PRO could be operated with a pressure significantly higher than $\Delta\pi/2$. In such a case, the reduction of the SEC would be close to 30%, while the atmospheric pollution related to GHG emission would be further reduced as the CO_2 produced would be lesser than now of about 0.6 kg/m^3 of FW.

The proposed strategy can be applied in all arid urban areas of the world to alleviate or eliminate the adverse environmental effects of seawater desalination as it is currently carried out. Nevertheless, what appears more important is the possibility of restarting a downward trend for the SEC and, presumably, for the unit cost of desalinated water, after years of stagnation. The latter can only be reliably estimated on a case-by-case basis through a rigorous economic analysis that considers the investments necessary for the interconnection of services and the RO-PRO hybridization.

Author Contributions: Conceptualization, G.D.G.; methodology, G.D.G. and G.T.; investigation, G.D.G., G.T. and P.R.; resources, G.D.G., G.T. and P.R.; writing—original draft preparation, G.T.; writing—review and editing, G.D.G., G.T. and P.R.; supervision, G.D.G.; visualization, P.R. All authors have read and agreed to the published version of the manuscript.

Funding: This research received no external funding.

Institutional Review Board Statement: Not applicable.

Informed Consent Statement: Not applicable.

Data Availability Statement: Not applicable.

Acknowledgments: The authors thank the administrative and technical staff of the department of Industrial and Information Engineering and of Economics of the University of L'Aquila for helpful support.

Conflicts of Interest: The authors declare no conflict of interest.

References

1. Eakins, B.W.; Sharman, G.F. *Volumes of the World's Oceans from ETOPO1*; NOAA National Geophysical Data Center: Boulder, CO, USA, 2010.
2. Bennett, A. 50th Anniversary: Desalination: 50 Years of Progress. *Filtr. Sep.* **2013**, *50*, 32–39. [CrossRef]
3. Yfantis, D.; Yfantis, A. Aristotle and seawater desalination: A new explanation of an experiment described in meteorologica and historia animalium. In *The Capital of Knowledge*; Society for the Propagation of Useful Books: Athens, Greece, 2020; pp. 167–173, ISBN 978-960-8351-83-7.

4. Micale, G.; Cipollina, A.; Rizzuti, L. Seawater Desalination for Freshwater Production. In *Seawater Desalination: Conventional and Renewable Energy Processes*; Micale, G., Rizzuti, L., Cipollina, A., Eds.; Green Energy and Technology; Springer: Berlin/Heidelberg, Germany, 2009; pp. 1–15, ISBN 978-3-642-01150-4.
5. Altaee, A.; Millar, G.J.; Zaragoza, G. Integration and Optimization of Pressure Retarded Osmosis with Reverse Osmosis for Power Generation and High Efficiency Desalination. *Energy* **2016**, *103*, 110–118. [CrossRef]
6. Aende, A.; Gardy, J.; Hassanpour, A. Seawater desalination: A review of forward osmosis technique, its challenges, and future prospects. *Processes* **2020**, *8*, 901. [CrossRef]
7. Barba, D.; Liuzzo, G.; Tagliaferri, G. Multi Stage Flash-Evaporator. U.S. Patent No. 3,763,014, 2 October 1973.
8. Barba, D.; D'Agostino, C.; Liuzzo, G. Desalination of Sea or Brackish Water by Multi-Stage Flash Evaporation. U.S.Patent No. 3,684,661, 15 August 1972.
9. Bennett, A. Desalination Trends: What's the Future for Desalination? *Filtr. Sep.* **2012**, *49*, 12–15. [CrossRef]
10. Tomassi, G. Integration of Pressure Retarded Osmosis with Reverse Osmosis for Energy Recovery and High Efficiency Desalination of Sea Water. Bachelor's Thesis, University of L'aquila, L'Aquila AQ, Italy, 2021.
11. Bargiacchi, E.; Orciuolo, F.; Ferrari, L.; Desideri, U. Use of Pressure-Retarded-Osmosis to Reduce Reverse Osmosis Energy Consumption by Exploiting Hypersaline Flows. *Energy* **2020**, *211*, 118969. [CrossRef]
12. Elsaid, K.; Kamil, M.; Sayed, E.T.; Abdelkareem, M.A.; Wilberforce, T.; Olabi, A. Environmental Impact of Desalination Technologies: A Review. *Sci. Total. Environ.* **2020**, *748*, 141528. [CrossRef]
13. World Bank. *The Role of Desalination in an Increasingly Water-Scarce World*; Water Papers; World Bank: Washington, DC, USA, 2019.
14. Quanta Acqua c'è Nel Mondo. Available online: http://www.expo2015.org/magazine/it/sostenibilita/quanta-acqua-c-e-nel-mondo.html (accessed on 20 June 2021).
15. Transforming Our World: The 2030 Agenda for Sustainable Development | Department of Economic and Social Affairs. Available online: https://sdgs.un.org/2030agenda (accessed on 22 June 2021).
16. Barba, D.; Brandani, V.; Di Giacomo, G.; Foscolo, P.U. Magnesium oxide production from concentrated brines. *Desalination* **1980**, *33*, 241–250. [CrossRef]
17. Kim, S.; Joo, H.; Moon, T.; Kim, S.H.; Yoon, J. Rapid and selective lithium recovery from desalination brine using an electrochemical system. *Environ. Sci. Process. Impacts* **2019**, *21*, 667–676. [CrossRef] [PubMed]
18. Lee, S.; Choi, J.; Park, Y.G.; Shon, H.; Ahn, C.H.; Kim, S.H. Hybrid desalination processes for beneficial use of reverse osmosis brine: Current status and future prospects. *Desalination* **2019**, *454*, 104–111. [CrossRef]
19. El-Hady, B.; Kashyout, A.; Hassan, A.; Hassan, G.; El-Banna Fath, H.; El-Wahab Kassem, A.; Elshimy, H.; Shaheed, M.H. Hybrid renewable energy/hybrid desalination potentials for remote areas: Selected cases studied in Egypt. *RSC Adv.* **2021**, *11*, 13201–13219. [CrossRef]
20. Esmaeilion, F. Hybrid renewable energy systems for desalination. *Appl. Water Sci.* **2020**, *10*, 1–47. [CrossRef]
21. Vanoppen, M.; Blandin, G.; Derese, S.; Le Clech, P.; Post, J.; Verliefde, A.R.D. Salinity gradient power and desalination. In *Sustainable Energy from Salinity Gradients*; Elsevier: Amsterdam, The Netherlands, 2016.
22. Makabe, R.; Ueyama, T.; Sakai, H.; Tanioka, A. Commercial pressure retarded osmosis systems for seawater desalination plants. *Membranes* **2021**, *11*, 69. [CrossRef] [PubMed]
23. Yip, N.Y.; Tiraferri, A.; Phillip, W.A.; Schiffman, J.D.; Elimelech, M. High performance thin-film composite forward osmosis membrane. *Environ. Sci. Technol.* **2010**, *44*, 3812–3818. [CrossRef]
24. Li, L.; Shi, W.; Yu, S. Research on forward osmosis membrane technology still needs improvement in water recovery and wastewater treatment. *Water* **2020**, *12*, 107. [CrossRef]
25. Morillo, J.; Usero, J.; Rosado, D.; El Bakouri, H.; Riaza, A.; Bernaola, F.J. Comparative study of brine management technologies for desalination plants. *Desalination* **2014**, *336*, 32–49. [CrossRef]
26. Ghaffour, N.; Missimer, T.M.; Amy, G.L. Technical Review and Evaluation of the Economics of Water Desalination: Current and Future Challenges for Better Water Supply Sustainability. *Desalination* **2013**, *309*, 197–207. [CrossRef]
27. Barba, D.; Di Giacomo, G.; Evangelista, F.; Tagliaferri, G. High temperature distillation process with sea water feed decalcification pretreatment. *Desalination* **1982**, *40*, 347–355. [CrossRef]
28. Wang, J.; Tanuwidjaja, D.; Bhattacharjee, S.; Edalat, A.; Jassby, D.; Hoek, E.M.V. Produced water desalination via pervaporative distillation. *Water* **2020**, *12*, 3560. [CrossRef]
29. Wang, P.; Chung, T.S. A conceptual demonstration of freeze desalination-membrane distillation (FD-MD) hybrid desalination process utilizing liquefied natural gas (LNG) cold energy. *Water Res.* **2012**, *46*, 4037–4052. [CrossRef]
30. Feria-Díaz, J.J.; Correa-Mahecha, F.; López-Méndez, M.C.; Rodríguez-Miranda, J.P.; Barrera-Rojas, J. Recent desalination technologies by hybridization and integration with reverse osmosis: A review. *Water* **2021**, *13*, 1369. [CrossRef]
31. Bland, E. *Study: Bacteria Can Make Salt Water Drinkable*; NBC, 2009. Available online: https://www.nbcnews.com/id/wbna32558231. (accessed on 11 June 2021).
32. Wang, Q.; Zhou, Z.; Li, J.; Tang, Q.; Hu, Y. Investigation of the Reduced Specific Energy Consumption of the RO-PRO Hybrid System Based on Temperature-Enhanced Pressure Retarded Osmosis. *J. Membr. Sci.* **2019**, *581*, 439–452. [CrossRef]
33. Tamburini, A.; Giacalone, F.; Cipollina, A.; Grisafi, F.; Vella, G.; Micale, G. Pressure retarded osmosis: A membrane process for environmental sustainability. *Chem. Eng. Trans.* **2016**, *47*, 355–360. [CrossRef]

34. Qasim, M.; Badrelzaman, M.; Darwish, N.N.; Darwish, N.A.; Hilal, N. Reverse osmosis desalination: A state-of-the-art review. *Desalination* **2019**, *459*, 59–104. [CrossRef]
35. Darwish, M.A.; Abdel-Jawad, M.; Hauge, L.J. A new dual-function device for optimal energy recovery and pumping for all capacities of RO systems. *Desalination* **1989**, *75*, 25–39. [CrossRef]
36. Andrews, W.T.; Laker, D.S. A twelve-year history of large scale application of work-exchanger energy recovery technology. *Desalination* **2001**, *138*, 201–206. [CrossRef]
37. Chung, H.W.; Banchik, L.D.; Swaminathan, J.; Lienhard, V.J.H. On the Present and Future Economic Viability of Stand-Alone Pressure-Retarded Osmosis. *Desalination* **2017**, *408*, 133–144. [CrossRef]
38. Aumesquet-Carreto, M.; Ortega-Delgado, B.; Garcia-Rodriguez, L. *Improving the Performance of Reverse Osmosis Desalination Process with Pressure Retarded Osmosis*. EERES4WATER PROJECT (EAPA 1058/2018), European Regional Development Fund. Seville, ES. 2019. Available online: https://www.eeres4water.eu/wp-content/uploads/2020/05/Report-Action-6.2-PRO-SWRO.pdf (accessed on 11 June 2021).
39. Evans, Graham. "Persian Gulf". Encyclopedia Britannica. Available online: https://www.britannica.com/place/Persian-Gulf (accessed on 11 June 2021).
40. Escobar, I.C.; Schäfer, A. *Sustainable Water for the Future: Water Recycling Versus Desalination*; Elsevier: Amsterdam, The Netherlands, 2009; ISBN 978-0-08-093217-0.
41. Barau, A.S.; Al Hosani, N. Prospects of Environmental Governance in Addressing Sustainability Challenges of Seawater Desalination Industry in the Arabian Gulf. *Environ. Sci. Policy* **2015**, *50*, 145–154. [CrossRef]
42. Lattemann, S. Protecting the Marine Environment. In *Seawater Desalination: Conventional and Renewable Energy Processes*; Micale, G., Rizzuti, L., Cipollina, A., Eds.; Green Energy and Technology; Springer: Berlin/Heidelberg, Germany, 2009; pp. 273–299, ISBN 978-3-642-01150-4.
43. Guirguis, M. *Energy Recovery Devices in Seawater Reverse Osmosis Desalination Plants with Emphasis on Efficiency and Economical Analysis of Isobaric versus Centrifugal Devices*; University of South Florida: Tampa, FL, USA, 2011.
44. Sawaki, N.; Chen, C.-L. Cost Evaluation for a Two-Staged Reverse Osmosis and Pressure Retarded Osmosis Desalination Process. *Desalination* **2021**, *497*, 114767. [CrossRef]
45. Di Giacomo, G.; Scimia, F.; Taglieri, L. Solvent Activity and Osmotic Pressure of Binary Aqueous and Alcoholic Solutions of Calcium Chloride up to 368 K and High Salt Concentration. *Indian J. Chemistry-Sect. A (IJCA)* **2020**, *56*, 297–304.
46. Gekas, V.; Gonzalez, C.; Sereno, A.; Chiralt, A.; Fito, P. Mass Transfer Properties of Osmotic Solutions. I. Water Activity and Osmotic Pressure. *Int. J. Food Prop.* **1998**, *1*, 95–112. [CrossRef]
47. Nayar, K.G.; Sharqawy, M.H.; Banchik, L.D.; Lienhard, V.J.H. Thermophysical Properties of Seawater: A Review and New Correlations That Include Pressure Dependence. *Desalination* **2016**, *390*, 1–24. [CrossRef]
48. Sastry, S. Study of Parameters before and after Treatment of Municipal Waste Water from an Urban Town. *Int. J. Appl. Environ. Sci.* **2013**, *3*, 41–48.
49. Christodoulou, A.; Stamatelatou, K. Overview of Legislation on Sewage Sludge Management in Developed Countries Worldwide. *Water Sci. Technol.* **2015**, *73*, 453–462. [CrossRef] [PubMed]
50. Aragón-Briceño, C.I.; Grasham, O.; Ross, A.B.; Dupont, V.; Camargo-Valero, M.A. Hydrothermal Carbonization of Sewage Digestate at Wastewater Treatment Works: Influence of Solid Loading on Characteristics of Hydrochar, Process Water and Plant Energetics. *Renew. Energy* **2020**, *157*, 959–973. [CrossRef]
51. Hara, K.; Kuroda, M.; Yabar, H.; Kimura, M.; Uwasu, M. Historical Development of Wastewater and Sewage Sludge Treatment Technologies in Japan—An Analysis of Patent Data from the Past 50 Years. *Environ. Dev.* **2016**, *19*, 59–69. [CrossRef]
52. Djandja, O.S.; Wang, Z.-C.; Wang, F.; Xu, Y.-P.; Duan, P.-G. Pyrolysis of Municipal Sewage Sludge for Biofuel Production: A Review. *Ind. Eng. Chem. Res.* **2020**, *59*, 16939–16956. [CrossRef]
53. Hong, J.; Hong, J.; Otaki, M.; Jolliet, O. Environmental and Economic Life Cycle Assessment for Sewage Sludge Treatment Processes in Japan. *Waste Manag.* **2009**, *29*, 696–703. [CrossRef]
54. Jatav, H.S.; Singh, S.K.; Singh, Y.; Kumar, O. Biochar and Sewage Sludge Application Increases Yield and Micronutrient Uptake in Rice (*Oryza Sativa* L.). *Commun. Soil Sci. Plant Anal.* **2018**, *49*, 1617–1628. [CrossRef]
55. Onaka, T. Sewage Can Make Portland Cement: A New Technology for Ultimate Reuse of Sewage Sludge. *Water Sci. Technol.* **2000**, *41*, 93–98. [CrossRef]
56. Werther, J.; Ogada, T. Sewage Sludge Combustion. *Prog. Energy Combust. Sci.* **1999**, *25*, 55–116. [CrossRef]
57. Kim, J.; Park, K.; Yang, D.R.; Hong, S. A Comprehensive Review of Energy Consumption of Seawater Reverse Osmosis Desalination Plants. *Appl. Energy* **2019**, *254*, 113652. [CrossRef]
58. Al-Hazmi, A.A. IDA Global Connections-Fall 2019. Available online: https://issuu.com/idadesal/docs/ida_fall19_digital (accessed on 22 June 2021).
59. Tawalbeh, M.; Al-Othman, A.; Abdelwahab, N.; Alami, A.H.; Olabi, A.G. Recent Developments in Pressure Retarded Osmosis for Desalination and Power Generation. *Renew. Sustain. Energy Rev.* **2021**, *138*, 110492. [CrossRef]
60. Touati, K.; Tadeo, F.; Kim, J.H.; Silva, O.A.A.; Chae, S.H. *Pressure Retarded Osmosis: Renewable Energy Generation and Recovery*; Academic Press: Cambridge, MA, USA, 2017; ISBN 978-0-12-812315-7.
61. Pattle, R.E. Production of Electric Power by Mixing Fresh and Salt Water in the Hydroelectric Pile. *Nature* **1954**, *174*, 660. [CrossRef]

62. Sarp, S.; Li, Z.; Saththasivam, J. Pressure Retarded Osmosis (PRO): Past Experiences, Current Developments, and Future Prospects. *Desalination* **2016**, *389*, 2–14. [CrossRef]
63. Prante, J.L.; Ruskowitz, J.A.; Childress, A.E.; Achilli, A. RO-PRO Desalination: An Integrated Low-Energy Approach to Seawater Desalination. *Appl. Energy* **2014**, *120*, 104–114. [CrossRef]